Experimental Methods in the Physical Sciences

VOLUME 36

MAGNETIC IMAGING AND ITS APPLICATIONS TO MATERIALS

EXPERIMENTAL METHODS IN THE PHYSICAL SCIENCES

Volume 36

Magnetic Imaging and Its Applications to Materials

Edited by

Marc De Graef
Department of Materials Science and Engineering
Carnegie Mellon University
Pittsburgh, Pennsylvania

Yimei Zhu
Department of Applied Science
Brookhaven National Laboratory
Upton, New York

ACADEMIC PRESS

A Harcourt Science and Technology Company

San Diego San Francisco New York Boston London Sydney Tokyo

This book is printed on acid-free paper. ♾

Academic Press
A Harcourt Science and Technology Company
525 B Street, Suite 1900, San Diego, CA 92101-4495, USA
http://www.academicpress.com

Academic Press
Harcourt Place, 32 Jamestown Road, London NW1 7BY, UK

International Standard Serial Number: 1079-4042/01

International Standard Book Number: 0-12-475983-1

PRINTED IN THE UNITED STATES OF AMERICA
00 01 02 03 QW 9 8 7 6 5 4 3 2 1

CONTENTS

CONTRIBUTORS ix

VOLUMES IN SERIES xi

PREFACE xv

1. Micromagnetic Modeling of Domain Structures in Magnetic Thin Films
by JIAN-GANG ZHU

1.1. Introduction 1

1.2. Model and Computation 2

1.3. Comparisons of Simulated Domain Structures with Experimental Observations 8

1.4. Domain Configurations in Patterned Magnetic Thin Film Elements 15

1.5. Magnetization Processes in Thin Film Recording Media 18

1.6. Summary and Remarks 25

2. Lorentz Microscopy: Theoretical Basis and Image Simulations
by MARC DE GRAEF

2.1. Introduction 27

2.2. Basic Lorentz Microscopy (Classical Approach) 28

2.3. Quantum Mechanical Formulation 39

2.4. Magnetic Induction Mapping Methods 58

2.5. The Transport-of-Intensity Equation 61

3. Recent Advances in Magnetic Force Microscopy: Quantification and *in Situ* Field Measurements
by Romel D. Gomez

3.1. Introduction 69
3.2. Review of Magnetic Force Microscopy 70
3.3. Quantification Techniques: Models and Experiments 86
3.4. Advanced Techniques: Imaging in the Presence of an External Field 91
3.5. Applications 96

4. Electron Holography and Its Application to Magnetic Materials
by M. R. McCartney, R. E. Dunin-Borkowski, and David J. Smith

4.1. Introduction 111
4.2. Development of Electron Holography 112
4.3. The Technique of Off-Axis Electron Holography 113
4.4. Applications to Magnetic Materials 118
4.5. Prospects and Concluding Remarks 134

5. Phase Retrieval in Lorentz Microscopy
by Anton Barty, David Paganin, and Keith Nugent

5.1. Introduction 137
5.2. Phase Retrieval Theory 138
5.3. Practical Considerations in Electron Phase Microscopy 144
5.4. Application to Lorentz Microscopy 152
5.5. Critical Evaluation of the Technique 160
5.6. Summary and Conclusions 164

6. Scanning Electron Microscopy with Polarization Analysis (SEMPA) and Its Applications
by JOHN UNGURIS

6.1. Introduction 167
6.2. Spin Polarized Secondary Electron Magnetic Contrast 168
6.3. Instrumentation 171
6.4. Examples of SEMPA Applications 179
6.5. Conclusion 193

7. High Resolution Lorentz Scanning Transmission Electron Microscopy and Its Applications
by YUSUKE YAJIMA

7.1. Introduction 195
7.2. Principle 196
7.3. Instrument 213
7.4. Applications 214
7.5. Conclusions 225

8. Magnetic Structure and Magnetic Imaging of $Re_2Fe_{14}B$ (RE = Nd, Pr) Permanent Magnets
by YIMEI ZHU AND VYACHESLAV VOLKOV

8.1. Introduction 227
8.2. Microstructure 229
8.3. Grain Boundaries in Die-Upset Magnets 231
8.4. Domain Structure 236
8.5. Effects of Stray Fields on Magnetic Imaging 249
8.6. *In Situ* Experiments 255
8.7. Structure and Properties Correlation 264

REFERENCES 271

INDEX 287

CONTRIBUTORS

Numbers in parentheses indicate the pages on which the authors' contributions begin.

ANTON BARTY (137), *School of Physics, The University of Melbourne, Victoria 3010, Australia*

MARC DE GRAEF (27), *Department of Materials Science and Engineering, Carnegie Mellon University, Pittsburgh, Pennsylvania 15213*

R. E. DUNIN-BORKOWSKI (111), *Center for Solid State Science, Arizona State University, Tempe, Arizona 85287; and Department of Materials, University of Oxford, Oxford OX1 3PH, United Kingdom*

ROMEL D. GOMEZ (69), *Department of Electrical and Computer Engineering, University of Maryland, College Park, Maryland 20742; and Laboratory for Physical Sciences, College Park, Maryland 20740*

M. R. McCARTNEY (111) *Center for Solid State Science, Arizona State University, Tempe, Arizona* 85287

KEITH NUGENT (137), *School of Physics, The University of Melbourne, Victoria 3010, Australia*

DAVID PAGANIN (137), *School of Physics, The University of Melbourne, Victoria 3010, Australia*

DAVID J. SMITH (111), *Center for Solid State Science, and Department of Physics and Astronomy, Arizona State University, Tempe, Arizona 85287*

JOHN UNGURIS (167), *Electron Physics Group, National Institute of Standards and Technology, Gaithersburg, Maryland 20872*

VYACHESLAV VOLKOV (227) *Department of Applied Science, Brookhaven National Laboratory, Upton, New York 11973*

YUSUKE YAJIMA (195), *Hitachi Research Laboratory, Hitachi, Ltd., Kokubunji, Tokyo 185-8601, Japan*

JIAN-GANG ZHU (1), *Department of Electrical and Computer Engineering, Carnegie Mellon University, Pittsburgh, Pennsylvania 15213*

YIMEI ZHU (227), *Department of Applied Science, Brookhaven National Laboratory, Upton, New York 11973*

VOLUMES IN SERIES

EXPERIMENTAL METHODS IN THE PHYSICAL SCIENCES

(formerly Methods of Experimental Physics)

Editors-in-Chief
Robert Celotta and Thomas Lucatorto

Volume 1. Classical Methods
Edited by Immanuel Estermann

Volume 2. Electronic Methods, Second Edition (in two parts)
Edited by E. Bleuler and R. O. Haxby

Volume 3. Molecular Physics, Second Edition (in two parts)
Edited by Dudley Williams

Volume 4. Atomic and Electron Physics—Part A: Atomic Sources and Detectors; Part B: Free Atoms
Edited by Vernon W. Hughes and Howard L. Schultz

Volume 5. Nuclear Physics (in two parts)
Edited by Luke C. L. Yuan and Chien-Shiung Wu

Volume 6. Solid State Physics—Part A: Preparation, Structure, Mechanical and Thermal Properties; Part B: Electrical, Magnetic and Optical Properties
Edited by K. Lark-Horovitz and Vivian A. Johnson

Volume 7. Atomic and Electron Physics—Atomic Interactions (in two parts)
Edited by Benjamin Bederson and Wade L. Fite

Volume 8. Problems and Solutions for Students
Edited by L. Marton and W. F. Hornyak

Volume 9. Plasma Physics (in two parts)
Edited by Hans R. Griem and Ralph H. Lovberg

Volume 10. Physical Principles of Far-Infrared Radiation
By L. C. Robinson

Volume 11. Solid State Physics
Edited by R. V. Coleman

Volume 12. Astrophysics—Part A: Optical and Infrared Astronomy
Edited by N. Carleton

Part B: Radio Telescopes; Part C: Radio Observations
Edited by M. L. Meeks

Volume 13. Spectroscopy (in two parts)
Edited by Dudley Williams

Volume 14. Vacuum Physics and Technology
Edited by G. L. Weissler and R. W. Carlson

Volume 15. Quantum Electronics (in two parts)
Edited by C. L. Tang

Volume 16. Polymers—Part A: Molecular Structure and Dynamics; Part B: Crystal Structure and Morphology; Part C: Physical Properties
Edited by R. A. Fava

Volume 17. Accelerators in Atomic Physics
Edited by P. Richard

Volume 18. Fluid Dynamics (in two parts)
Edited by R. J. Emrich

Volume 19. Ultrasonics
Edited by Peter D. Edmonds

Volume 20. Biophysics
Edited by Gerald Ehrenstein and Harold Lecar

Volume 21. Solid State: Nuclear Methods
Edited by J. N. Mundy, S. J. Rothman, M. J. Fluss, and L. C. Smedskjaer

Volume 22. Solid State Physics: Surfaces
Edited by Robert L. Park and Max G. Lagally

Volume 23. Neutron Scattering (in three parts)
Edited by K. Sköld and D. L. Price

Volume 24. Geophysics—Part A: Laboratory Measurements; Part B: Field Measurements
Edited by C. G. Sammis and T. L. Henyey

Volume 25. Geometrical and Instrumental Optics
Edited by Daniel Malacara

Volume 26. Physical Optics and Light Measurements
Edited by Daniel Malacara

Volume 27. Scanning Tunneling Microscopy
Edited by Joseph Stroscio and William Kaiser

Volume 28. Statistical Methods for Physical Science
Edited by John L. Stanford and Stephen B. Vardaman

Volume 29. Atomic, Molecular, and Optical Physics—Part A: Charged Particles; Part B: Atoms and Molecules; Part C: Electromagnetic Radiation
Edited by F. B. Dunning and Randall G. Hulet

Volume 30. Laser Ablation and Desorption
Edited by John C. Miller and Richard F. Haglund, Jr.

Volume 31. Vacuum Ultraviolet Spectroscopy I
Edited by J. A. R. Samson and D. L. Ederer

Volume 32. Vacuum Ultraviolet Spectroscopy II
Edited by J. A. R. Samson and D. L. Ederer

Volume 33. Cumulative Author Index and Tables of Contents, Volumes 1–32

Volume 34. Cumulative Subject Index

Volume 35. Methods in the Physics of Porous Media
Edited by Po-zen Wong

Volume 36. Magnetic Imaging and Its Applications to Materials
Edited by Marc De Graef and Yimei Zhu

PREFACE

Magnetic materials are among the most exotic classes of materials, and are especially important in today's rapidly advancing technology and telecommunication era. The fascinating properties of these materials and their applications are exclusively determined by their underlying magnetic structures. Despite their importance, characterization of the magnetic structure, especially quantitative characterization, is not a trivial undertaking, compared with characterizing the more traditional microstructure of materials. The recent rise in interest in magnetic imaging of recording-media devices and magnetic sensors suggests there is an urgent need to document the availability, advantages, and drawbacks of various magnetic imaging techniques and to exchange ideas to meet future challenges in analyzing magnetic structure.

Thus, in 1999, we organized a symposium on "Magnetic Imaging and Its Applications to Materials" at the 57th Annual Conference of the Microscopy Society of America held in Portland, Oregon. The symposium was the first of its kind for the Society and attracted much attention from physicists, chemists, and materials scientists as well as from electron microscopists. The extremely positive response we received indicated that a reference book covering today's advanced magnetic microscopy techniques would be beneficial to researchers in the field and to the material science community at large. Hence, we assembled experts to review the status of forefront research in magnetic imaging and its applications to a variety of magnetic materials.

The aim of this volume is to cover a wide range of magnetic microscopies, ranging from classical Lorentz microscopy including *in situ* Fresnel and Foucault methods, to state-of-the-art magnetic force microscopy and electron holography. This book also gives an extensive introduction to magnetic structure and properties in various soft- and hard-magnetic materials and emphasizes the application of experimental methods and theoretical modeling to the study of technologically important complex systems. The effects of structural defects, including interfaces and precipitates, crystal orientations, external magnetic fields, ambient temperatures, and sample history on the behavior of magnetic domain, both in thin film and bulk magnetic materials, are discussed.

Chapter 1 of this volume provides a fundamental basis of micromagnetic simulation of domain structures in patterned thin film elements. Chapter 2 is a comprehensive guide to the theory and practice of Lorentz microscopy, both qualitative and quantitative. Chapter 3 and Chapter 4 describe recent advances in magnetic imaging, with Chapter 3 focusing on magnetic force microscopy and Chapter 4 focusing on off-axis electron holography. Extensive applications of these techniques are given in both chapters. In Chapter 5, elementary concepts of a newly developed phase reconstruction technique are presented. High-resolution Lorentz microscopy and related methods using a specially designed scanning electron microscope are described in Chapter 6. Chapter 7 is devoted to high resolution STEM-based observation modes, and Chapter 8 describes recent progress in understanding magnetic structure and properties relationship in $Nd(Pr)_2Fe_{14}B$ permanent magnets using various magnetic imaging techniques. Emphasis is placed on the magnetic domain structure and dynamic behavior including interactions with crystal defects for various crystal orientations, ambient temperature, and magnetic fields.

With such wide coverage of topics, researchers who may be interested in only a single magnetic imaging technique, or concerned with only one type of magnetic structure problem will be encouraged to explore more than one chapter and discover a variety of approaches. The one best suited to their particular needs and compatible with their resources can be selected. On the other hand, we have to bear in mind that, very often, several microscopy techniques have to be used to successfully tackle one single magnetic material problem.

This volume is intended to serve as a reference source for all those who are interested in magnetic imaging and want to understand the magnetic structure of materials. The experimental imaging techniques covered in this volume are mainly focused on, but not limited to, the use of transmission electron microscopes. The book reviews extensively the principle and theoretical background of various magnetic imaging techniques on different length scales. In addition to scientific researchers, upper-level undergraduate students and entry-level graduate students majoring in physics or materials science should also find the book useful as it consistently summarizes the recent progress in the field of magnetic imaging and imaging simulations.

It is our hope that this book will draw new inspiration and enthusiasm to the area of magnetic imaging and magnetic materials science and technology in which fruitful research is being conducted and many challenging problems still remain.

We would like to thank the authors for their splendid contributions. The assistance of Lori Asbury and Greg Franklin of Academic Press in preparing the manuscript is gratefully acknowledged.

MARC DE GRAEF AND YIMEI ZHU

1. MICROMAGNETIC MODELING OF DOMAIN STRUCTURES IN MAGNETIC THIN FILMS

Jian-Gang Zhu

Department of Electrical and Computer Engineering
Carnegie Mellon University
Pittsburgh, Pennsylvania

1.1 Introduction

With the wave of new generation magnetic technology, that is, magneto-electronic devices such as magnetoresistive random access memory (MRAM), the need to understand the magnetization process on a length scale much below a micron is becoming increasingly important [1, 2]. Micromagnetic modeling has become a powerful tool in understanding the complicated microscopic magnetization processes in magnetic thin films and thin film devices. It has helped to guide material microstructure engineering for applications such as thin film recording media for disk drives in data storage, and it is aiding the designs and development of MRAM devices. Previously, micromagnetic theory focused on static and transient magnetization configurations in fine magnetic particles and domain wall structures in soft magnetic films and provided fundamental understanding of particle magnetization reversal processes and the inner structure of magnetic domain walls [3–6]. Today, with the ability to perform large scale computing, complicated magnetization configurations and domain structures during magnetization reversal processes in magnetic thin films and patterned thin film devices can be simulated in a reasonable time using desktop workstations and even personal computers. Successful applications of the micromagnetic modeling studies not only have greatly enriched our knowledge of the complexity of microscopic magnetization processes [7–11], but they also have led to the realization that such capability can be crucial in aiding material microstructure engineering and device design for desired magnetic properties and performance.

The complicated magnetization configurations and magnetic domains in both bulk magnetic materials and magnetic thin films arise from the cooperative behavior of the electron spins deriving from an exchange interaction against local and global constraints to the spin orientation, such as anisotropy and magnetostatic fields. Domain patterns in soft magnetic films can be of the order of the film sample size, which could be over

EXPERIMENTAL METHODS IN THE PHYSICAL SCIENCES
Vol. 36
ISBN 0-12-475983-1

ISSN 1079-4042/01 $35.00

hundreds of microns. However, correct modeling of the detailed structure of the domain walls, which is of the order of tens of nanometers, is absolutely needed for obtaining correct results. Such scaling creates tremendous difficulty for large scale micromagnetic simulations.

In this chapter, the theory of micromagnetic modeling is reviewed in Section 1.2 along with computation methods. In Section 1.3, comparisons of various simulation results with experimental observations are given. Although later chapters in this book will detail the various observation techniques, this chapter emphasizes the relation between observations and modeling results. In Section 1.4, examples of simulated domain structures in various patterned thin film elements are discussed. In Section 1.5, an example of utilizing micromagnetic modeling to aid the engineering of thin film microstructure is provided. A brief summary concludes this chapter.

1.2 Model and Computation

In micromagnetic theory [2, 3], a macroscopic view is adopted where the atomic spin configuration has been averaged to a continuous magnetization **M(r)**. The magnitude of **M(r)** is taken to be the spontaneous magnetization at the temperature of interest; only its orientation at each location **r** needs to be determined. For a numerical simulation, the material is spatially discretized into a mesh of fine cells. Within each cell of the spatial discretization mesh, the magnetization is assumed to be uniform. The size of each cell has to be small enough so that the spatial gradient of a magnetization orientation in a simulated magnetization configuration is described with sufficient accuracy. The effective magnetic field, which exerts a torque on the magnetization, can be calculated as follows [12, 13]:

$$\mathbf{H}(\mathbf{r}) = -\frac{\partial E(\mathbf{r})}{\partial \mathbf{M}} = -\frac{1}{M}\frac{\partial E(\mathbf{r})}{\partial \mathbf{m}}, \tag{1}$$

where $E(\mathbf{r})$ is the total energy density and $\mathbf{m} = \mathbf{M}/M$ is the unit vector of the magnetization.

In a typical ferromagnetic system, the total energy density E is usually composed of the following terms:

$$E = E_{\text{ani}} + E_{\text{ex}} + E_{\text{mag}} + E_{\text{Zeeman}}, \tag{2}$$

where E_{ani} is the magnetic anisotropy energy density, E_{ex} is the ferromagnetic exchange energy density, E_{mag} is the magnetostatic energy density, and E_{Zeeman} is the magnetic potential energy due to any external field. For a multilayer structured thin film, possible interfacial exchange coupling needs to be included. For a strained system with significant magnetostriction, magnetoelastic energy density also needs to be included.

1.2.1 Magnetic Anisotropy Energy

The magnetic anisotropy energy either arises from the interaction of electron spin moments with the lattice via spin–orbit coupling (referred to as magnetocrystalline anisotropy) or is induced due to local atomic ordering. It acts as a local constraint on the magnetization orientation. The simplest form of magnetic anisotropy is uniaxial; for example in materials such as hexagonal close-packed (hcp) Co, it is given as follows:

$$E_{\text{ani}}(\mathbf{r}) = K_1 \sin^2 \theta(\mathbf{r}) + K_2 \sin^4 \theta(\mathbf{r}), \tag{3}$$

where $\theta(\mathbf{r})$ is the angle between the magnetization orientation and the local easy axis of the magnetic anisotropy and K_1 and K_2 are energy density constants. The magnetocrystalline anisotropy in cubic magnetic materials, such as face-centered cubic (fcc) Co and body centered cubic (bcc) Fe, is described as

$$E_{\text{ani}}(\mathbf{r}) = K_1(\alpha_1^2\alpha_2^2 + \alpha_2^2\alpha_3^2 + \alpha_3^2\alpha_1^2) + K_2\alpha_1^2\alpha_2^2\alpha_3^2, \tag{4}$$

where α_1, α_2, and α_3 are the magnetization components along the three principal axes of the lattice. For modeling of a polycrystalline magnetic film, a single-crystal grain may be modeled with a cluster of mesh cells with the same crystalline axes orientation [14–16].

1.2.2 Exchange Energy

In the case of a continuous magnetization distribution, the exchange energy density can be written as

$$E_{\text{ex}} = A[(\nabla\alpha)^2 + (\nabla\beta)^2 + (\nabla\gamma)^2], \tag{5}$$

where A is the exchange constant (e.g. $A = 1.6 \times 10^{-11}$ J/m for Co) and α, β, and γ are the direction cosines of the continuous magnetization components:

$$\alpha = m_x = \frac{M_x}{M}, \qquad \beta = m_y = \frac{M_y}{M}, \qquad \gamma = m_x = \frac{M_x}{M}. \tag{6}$$

In a cubic mesh, the exchange energy density can be reduced to the following expression:

$$E_{ex}(\mathbf{r}_i) = -\frac{2A}{\Delta^2}\mathbf{m}(\mathbf{r}_i)\cdot \sum_{j=NN} \mathbf{m}(\mathbf{r}_i), \tag{7}$$

where Δ is the length of the mesh cell and the sum is taken over all the nearest neighbor cells (NN) of the ith cell. In the case of fine grain polycrystalline thin films, such as a thin film recording medium, where grain boundaries are weak magnetic materials, the above expression is also used, but the exchange constant A is replaced with an effective exchange constant A^* [6]:

$$E_{ex}(\mathbf{r}_i) = -\frac{2A^*}{\Delta^2}\mathbf{m}(\mathbf{r}_i)\cdot \sum_{j=NN} \mathbf{m}(\mathbf{r}_i), \tag{8}$$

where Δ becomes the grain diameter.

1.2.3 Magnetostatic Energy

Magnetostatic energy arises from the surface magnetic poles at the surface of the material and the volume poles due to the gradient of the magnetization orientation. The energy for a mesh cell at location $\mathbf{r}_i$ can be written as

$$E_{mag}(\mathbf{r}_i) = -\frac{M}{2V_i}\int_{V_i} d^2\mathbf{r}'\mathbf{m}(\mathbf{r}_i)\cdot \mathbf{H}_{mag}(\mathbf{r}'), \tag{9}$$

where the integration is over the volume of the ith cell, V_i, and the magnetostatic field is given by

$$\mathbf{H}_{mag}(\mathbf{r}') = M \sum_{j=\text{all cells}} \int_{S_j} d^2\mathbf{r}''\hat{\mathbf{n}}\cdot \mathbf{m}(\mathbf{r}_i)\frac{(\mathbf{r}'-\mathbf{r}'')}{|\mathbf{r}'-\mathbf{r}''|^3}. \tag{10}$$

In the above expression, $\hat{\mathbf{n}}$ is the unit surface normal of the jth cell, and the summation is over the entire material modeled, because the magnetostatic interaction is long range. Uniform magnetization within each cell is assumed so the energy density can be rewritten as:

$$E_{mag}(\mathbf{r}_i) = -M^2\mathbf{m}(\mathbf{r}_i)\cdot\left[\sum_{j\neq i} \tilde{D}_{ij}\mathbf{m}(\mathbf{r}_j) + \frac{1}{2}\tilde{D}_{ii}\mathbf{m}(\mathbf{r}_i)\right], \tag{11}$$

where

$$\tilde{D}_{ij} = \frac{1}{V_i}\int_{V_i} \mathrm{d}^3\mathbf{r}' \int_{S_i} \mathrm{d}^2\mathbf{r}'' \cdot \frac{\hat{\mathbf{n}}\cdot(\mathbf{r}'-\mathbf{r}'')}{|\mathbf{r}'-\mathbf{r}''|^3} \tag{12}$$

is often referred to as the magnetostatic field tensor. These tensor coefficients can be calculated at the beginning of a simulation program to simplify the calculation of the magnetostatic field during the simulation of magnetization processes. It is to be noted here that the most computation extensive procedure in a micromagnetic simulation is the calculation of magnetostatic fields, because it involves $(N_x N_y N_z)^2$ calculation steps for each time iteration, where $N_x N_y N_z$ is the total number of mesh cells. If a regular mesh with spatial displacement symmetry can be used, the fast Fourier transform (FFT) can be used for calculating Eq. (11) to reduce the calculation steps to $8(N_x \log 2N_x)(N_y \log 2N_y)(N_x \log 2N_x)$ calculation steps. For large size meshes, the FFT method is indeed necessary for obtaining solutions within a reasonable time frame even with present computer power.

1.2.4 Zeeman Energy

The Zeeman energy is the magnetic potential energy due to an externally applied magnetic field, which may be either uniform or spatially varying. The general expression of the energy is

$$E_{\text{Zeeman}}(\mathbf{r}_i) = -\frac{M}{V_i}\int_{V_i} \mathrm{d}^3\mathbf{r}'\mathbf{m}(\mathbf{r}_i)\cdot\mathbf{H}_{\text{ext}}(\mathbf{r}'), \tag{13}$$

where the integration is taken over the volume of the ith mesh cell.

1.2.5 Modeling of Magnetization Processes

To correctly describe dynamic or transient magnetization processes, the Landau–Lifshitz equations are used to solve magnetization configurations at various times as well as reached-static states. For the magnetization $\mathbf{M}$ of each mesh cell with a total effective magnetic field $\mathbf{H}$ we have

$$\frac{\mathrm{d}\mathbf{M}_i}{\mathrm{d}t} = -\gamma\mathbf{M}_i \times \mathbf{H}_i - \frac{\lambda}{M}\mathbf{M}_i \times \mathbf{M}_i \times \mathbf{H}_i \qquad i = 1, \ldots, N, \tag{14}$$

where N is the total number of mesh cells, γ is the electron gyromagnetic ratio, and λ is the energy damping constant. The first term on the right-hand side of Eq. (14) describes the gyro motion of the magnetization vector due to the fact that the magnetization arises from electron spins. The second term describes the damping of the magnetization toward the effective magnetic field direction. The rate of energy damping is linearly proportional to λ. Note in this equation that $d\mathbf{M}$ is always perpendicular to $\mathbf{M}$; thus, local magnetization magnitude is indeed an invariant.

Now, the time evolution of the magnetization vector configuration can be obtained by integrating the above set of Landau–Lifshitz equations that are coupled through the effective magnetic field. The equation set is highly nonlinear, and coupling between the equations is nonlocal.

Figure 1 shows a two-dimensional array of square-shaped mesh cells for modeling a patterned thin film element. A single layer of mesh cells can be used for a single-layer magnetic film if the film is sufficiently thin. For example, for Permalloy film, the critical thickness is approximately 20–30 nm [17]. For film thicknesses greater than the critical thickness, discretization in the film thickness direction is needed, as magnetization variation within the film depth will be sufficiently significant not to be neglected. Such regular mesh can be used for modeling polycrystalline magnetic films provided that either the magnetic anisotropy energy is sufficiently low in comparison to other energy terms or the magnetic grain size is significantly smaller than the exchange length of the material. Otherwise, realistic geometry, including both shapes and sizes, of the crystallites must be considered.

Figure 2 shows a two-dimensional array of hexagonal mesh with adjacent hexagonal mesh cells separated by voided boundaries. This mesh has been used very successfully to study the micromagnetics of thin film recording media, in which case each hexagon represents an actual magnetic grain with uniaxial anisotropy [6]. One of the key aspects to its success is the fact that the number of the nearest neighboring grains in the hexagonal array is very similar to that in practical thin film recording media. It is also believed that the individual crystalline grains in the real films actually do behave as a single-domain particle, with magnetization always kept uniform within each grain.

Figure 3 shows a mesh for modeling a polycrystalline soft magnetic thin film with relative large crystallites. In this case, a crystalline grain consists

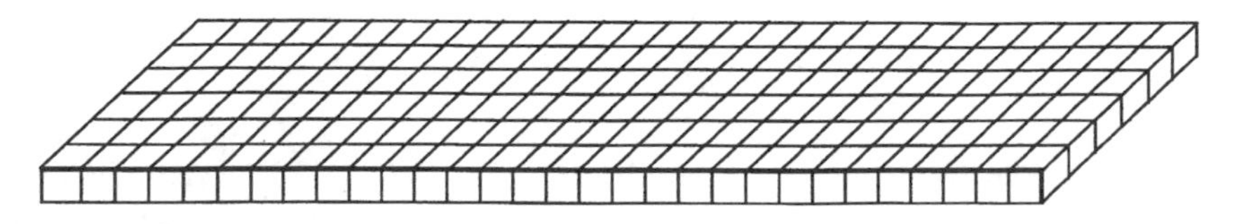

FIG. 1. A single-layer square discretization mesh for a rectangular soft thin film element.

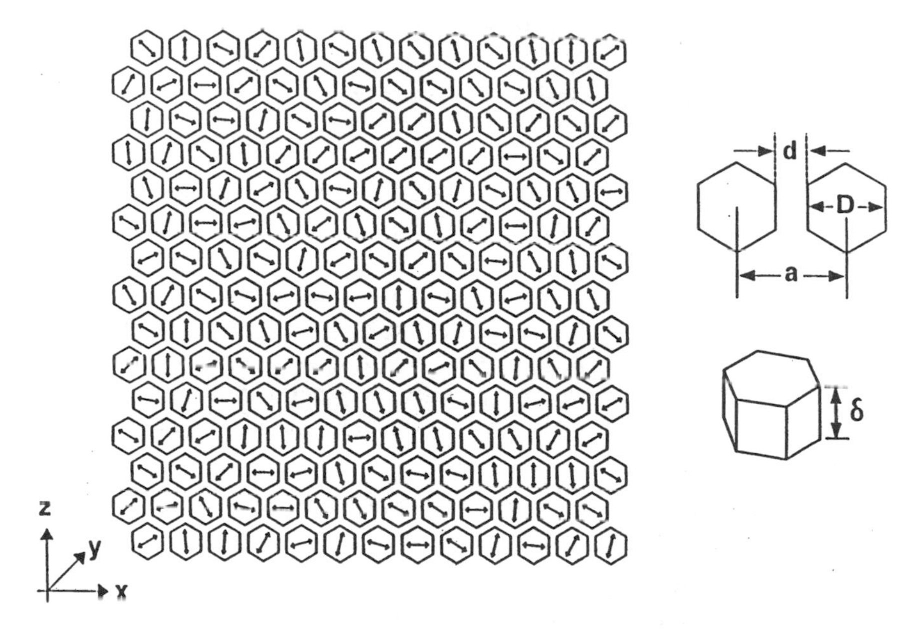

FIG. 2. Hexagonal mesh used in modeling of thin film recording media. Each hexagon represents a single crystal grain in the film, and the double-headed arrows indicate the magnetic easy axes of the grains.

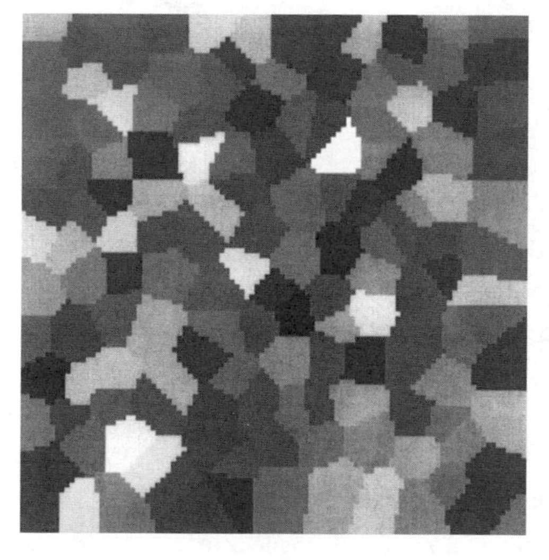

FIG. 3. Square mesh used for modeling polycrystalline films with nonnegligible magnetocrystalline anisotropy. The different gray scales represent different crystalline orientations. Each grain is modeled by multiple number of mesh cells.

of a group of mesh cells with the same crystalline axis orientation. Such a mesh has been used to study polycrystalline fcc Co films because of their relatively large magnetocrystalline anisotropy value [15, 16].

1.3 Comparisons of Simulated Domain Structures with Experimental Observations

To be able to rely on the simulations provided by micromagnetic modeling, it is important to validate the modeling results with careful experimental measurements. In this section, a few examples of these comparison studies are presented. We present the experimental results without discussing the techniques used to obtain them; that is left for later chapters.

1.3.1 Comparison with Magnetic Force Microscopy Images for Patterned Permalloy Film Elements

Figure 4 shows a comparison between a magnetic force microscopy (MFM) image and the corresponding micromagnetic simulation results

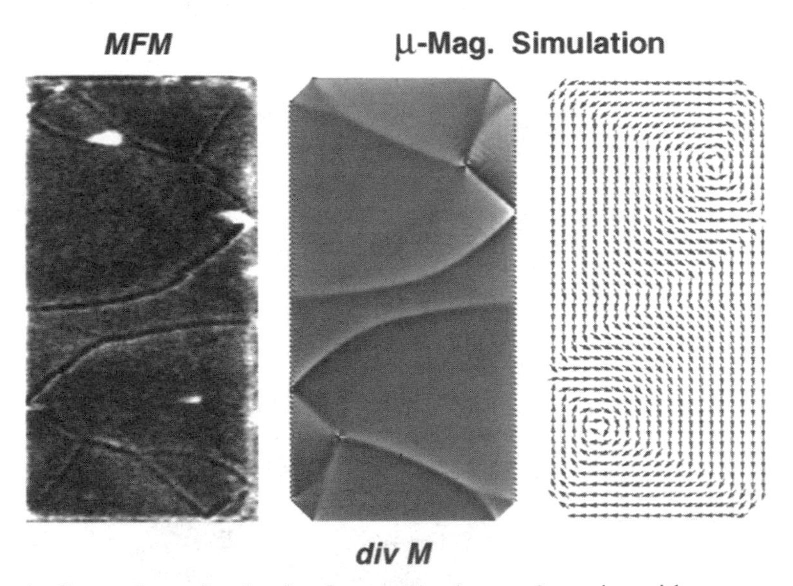

FIG. 4. Comparison of a simulated magnetization configuration with a corresponding MFM image. The gray scale in the middle picture represents the divergence of the simulated magnetization configuration. The film element is a 30-nm-thick $Ni_{81}Fe_{19}$ (Permalloy) film with dimensions of $10 \times 5\ \mu m^2$.

presented as magnetization vector fields and as a gray scale showing magnetic pole density distribution. The latter is calculated as $p = -\nabla \cdot \mathbf{M}$, since the magnetic pole density distribution is what the MFM image measures [18, 19]. As shown in Fig. 4, the simulated magnetic pole density distribution closely resembles the MFM image. With the corresponding magnetization vector field configuration, the understanding of the MFM image is directly provided: the vertex at which the edge domain walls join corresponds to a magnetization vortex. The curved low-angle domain walls that extend to the middle of the element correspond to the transverse orientation of the magnetization.

Figure 4 is selected out of a series of MFM images and the corresponding simulation results that are shown in Figs. 5 and 6. The figures show one-to-one comparisons between a set of MFM images and the corresponding micromagnetic simulation results during the magnetization reversal. The MFM images were taken on an $10 \times 5\,\mu\text{m}^2$ patterned Permalloy thin film element 30 nm thick with an external field applied *in situ*. The Permalloy film was deposited on a 30-nm-thick NiO film with a thin Ta underlayer. The exchange bias field arising from the NiO/Permalloy interface on the Permalloy film is approximately 56 Oe. The initial applied field is in the opposite direction of the exchange bias field. The gray scale in the simulated picture represents the magnetic pole density for comparison with the MFM images.

As shown in Figs. 5 and 6, the simulated domain configurations of the entire magnetization reversal process are almost exact replicas of the corresponding MFM images. It should be noted here that the simulations had no adjustable parameters. The excellent agreement between the MFM images and the micromagnetic simulation is indeed remarkable. The validity of the modeling calculations is demonstrated with significant confidence.

Because of the dynamic approach, the micromagnetic simulations also provide transient magnetization configurations that are very challenging to obtain with present imaging techniques. Figure 7 shows a sequence of simulated transient domain configurations during an irreversible magnetization switching between the two static domain configurations, shown as frames 3 and 4 in Fig. 5. No static states were observed between these two states in MFM imaging. With micromagnetic simulation, transient states can be obtained for analysis and understanding. The curved domain walls resulted from partially formed transverse domains (with magnetization oriented along the horizontal direction). As the external field exceeds a critical value, the two transverse domains expand and connect (frames 2 and 3 in Fig. 7), forming the middle diamond-shaped transverse domain for a complete magnetization flux-closure configuration and reaching the demagnetized state (frame 4 in Fig. 7). The transient wall ripples shown in

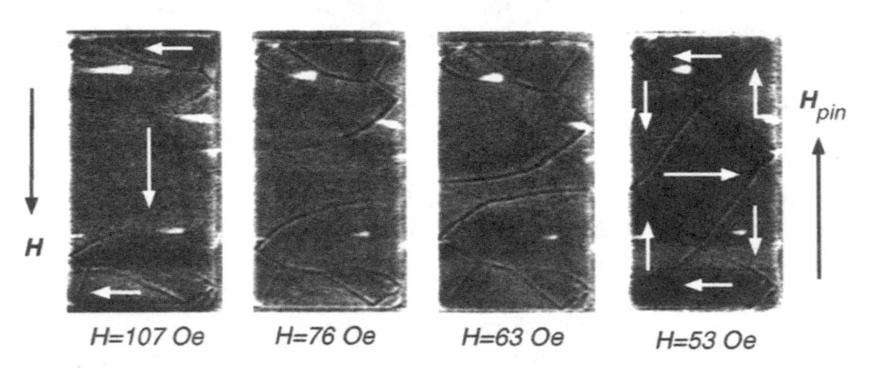

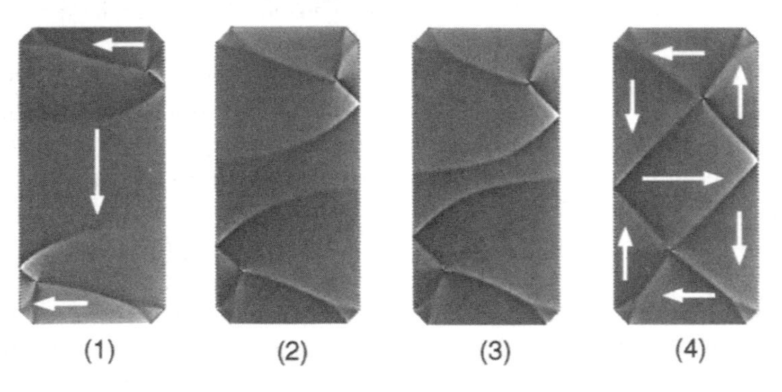

FIG. 5. One-to-one comparison of a series of MFM images of a 30-nm-thick $Ni_{81}Fe_{19}$ (Permalloy) film with dimensions of $10 \times 5\ \mu m^2$, taken at different external fields during a magnetization reversal process, and the corresponding micromagnetic simulations. The gray scale in the simulated pictures represents the magnetization divergence.

frame 3 are characteristic for dynamic domain wall motion processes in these films.

Figure 8 shows a series of MFM images and the magnetic pole density images of the corresponding micromagnetic simulation results. Note that the MFM imaged Permalloy element has its right-bottom corner missing. For the calculation, the same corner is purposely cut to mimic reality. The cut corner yields a corner domain with a split domain wall, as evident in both the measured and simulated images.

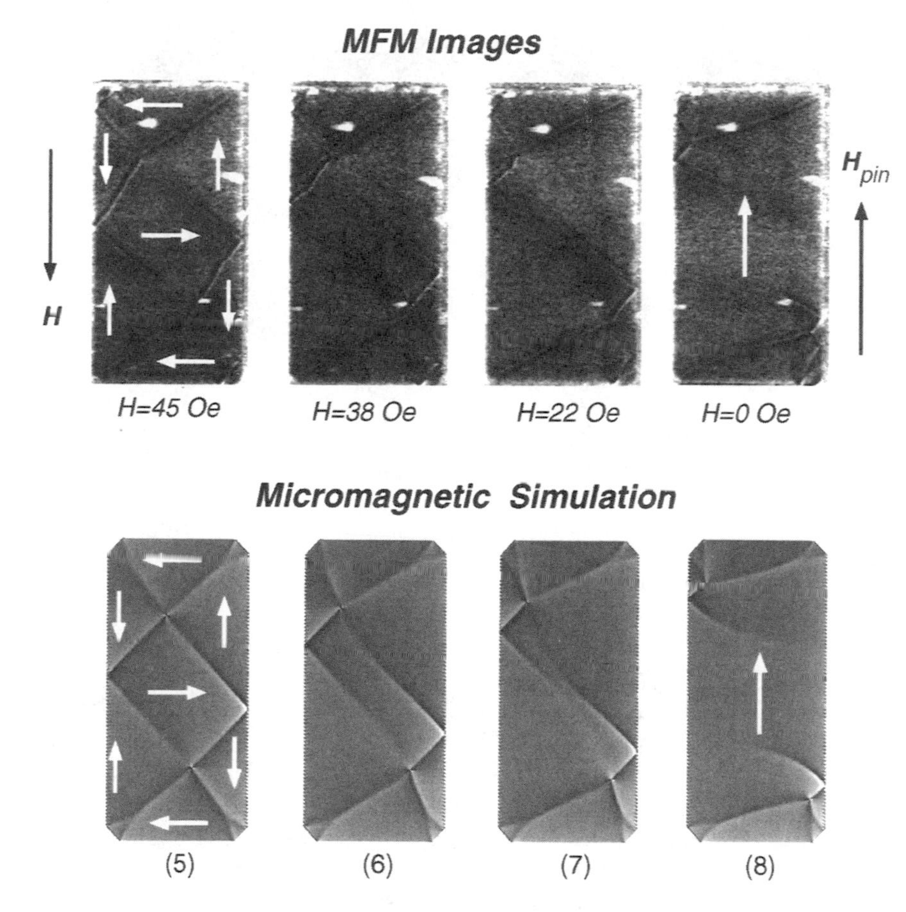

FIG. 6. One-to-one comparison of a series of MFM images of a 30-nm-thick $Ni_{81}Fe_{19}$ (Permalloy) film with dimensions of $10 \times 5\ \mu m^2$, taken at different external fields during a magnetization reversal process, and the corresponding micromagnetic simulations. The gray scale in the simulated pictures represents the magnetization divergence.

A magnetization vector field can be expressed in terms of its curl-free component and its divergence-free component [20]. Although only the curl-free component can be obtained with an MFM image, the divergence-free component cannot be recovered with just an MFM image. As a complementary tool, transmission electron microscopy recovers the divergence-free component of the imaged magnetization vector field, although, in principle, the curl-free component usually is difficult to recover. Comparison

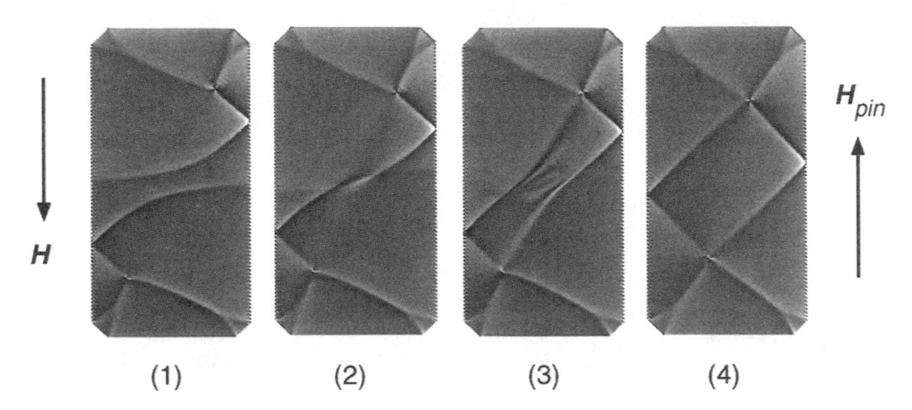

FIG. 7. Simulated transient domain configurations during the formation of flux-closure domain configurations.

with both imaging techniques gives more complete verification of the modeling results.

1.3.2 Comparison with Differential Phase-Contrast Microscopy Images

Differential phase-contrast (DPC) electron microscopy essentially provides a direct mapping of the magnetization vector field in a thin film element [21, 22] (except where the magnetization curl-free component is dominating) through the recovery of the divergence-free component of the magnetization vector field. Figure 9 shows a series of DPC images and the corresponding micromagnetic simulation results for a 30-nm-thick Permalloy film element $4 \times 0.75\,\mu m$, in size. The images were taken at various stages during a magnetization reversal with a magnetic field applied along the element length direction. The gray scale in both the measurement and the simulation represents the magnetization component along the element width direction in the film plane. In this case, excellent agreement is also achieved between the corresponding micromagnetic calculations and the DPC images. No adjustable parameter was used in the simulation. Prior to the reversal, edge domains at the ends of the element have formed. The reversal starts when magnetization vortices form as the transverse domain at the ends of the element evolves. The vortices travel toward the middle of the elements, resulting in the magnetization reversal of the entire element.

To illustrate the magnetization reversal process in a submicrometer size film element with large aspect ratio, Figure 10 shows a transient sequence

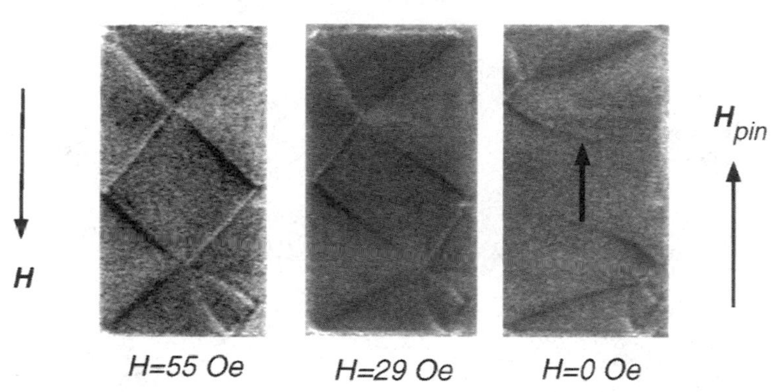

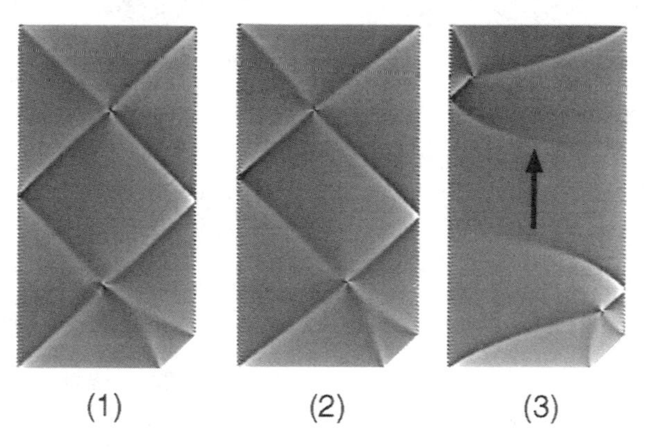

FIG. 8. Comparison of MFM images and simulated domain configurations for a 30-nm-thick $Ni_{81}Fe_{19}$ (Permalloy) and $10 \times 5\,\mu m^2$ size film element with a missing corner.

of simulated magnetization vector configurations during a magnetization reversal of a bar of Ni film 1 μm long and 0.1 μm wide with a thickness of 35 nm [23]. Only the magnetization vector of half of the bar is plotted, with the right edges in the figure corresponding to the middle of the bar element. As shown, at the early stage of the reversal process, a magnetization vortex is formed at the end of the element along the lower edge. The polarity of

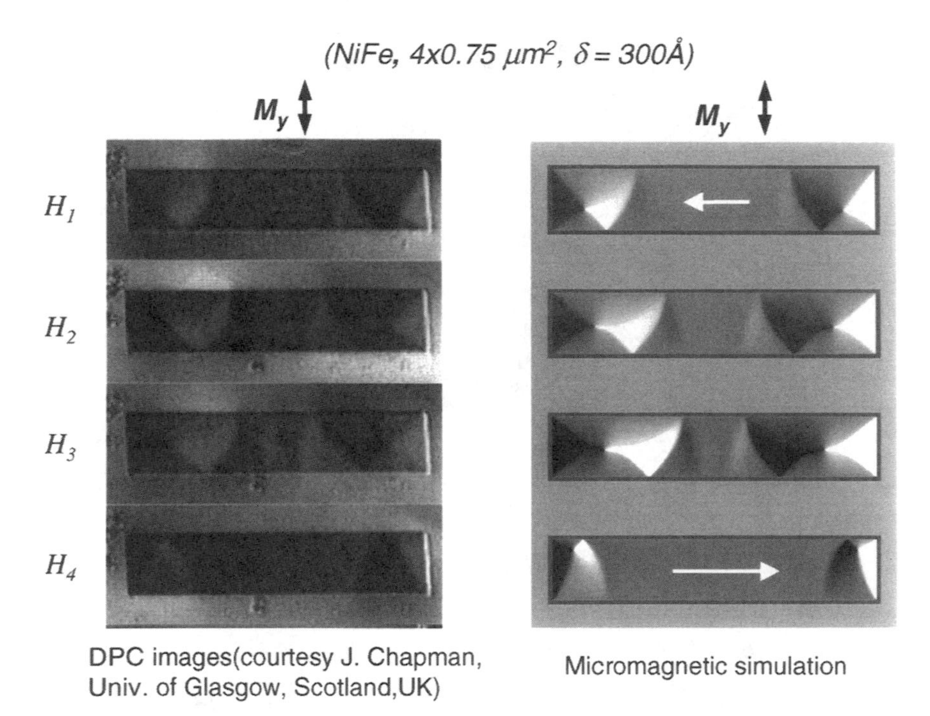

FIG. 9. Comparison of a sequence of DPC images, taken at different external reversing field amplitudes, and the corresponding micromagnetic simulation results. For all pictures, the gray scale represents the magnetization component along the film element width direction. DPC images courtesy of Prof. J. Chapman, University of Glasgow, Scotland.

the magnetization circulation is such that the bottom part of the magnetization vortex is in the same direction as the applied reversing field. As time evolves, the vortex moves to the upper edge of the bar element, yielding the reversal of the end region. As the vortex approaches the upper edge, a new vortex with opposite magnetization circulation forms long the upper edge and travels down toward the lower edge of the bar element, resulting in expansion of the end reversed region into the interior of the bar element. As the vortex approaches the lower edge of the element, a new vortex forms along the lower edge of the bar element, further into the middle of the bar element. This vortex would travel up toward upper edge of the element. The subsequent formation and motion of the magnetization vortices yield the complete magnetization reversal of the element.

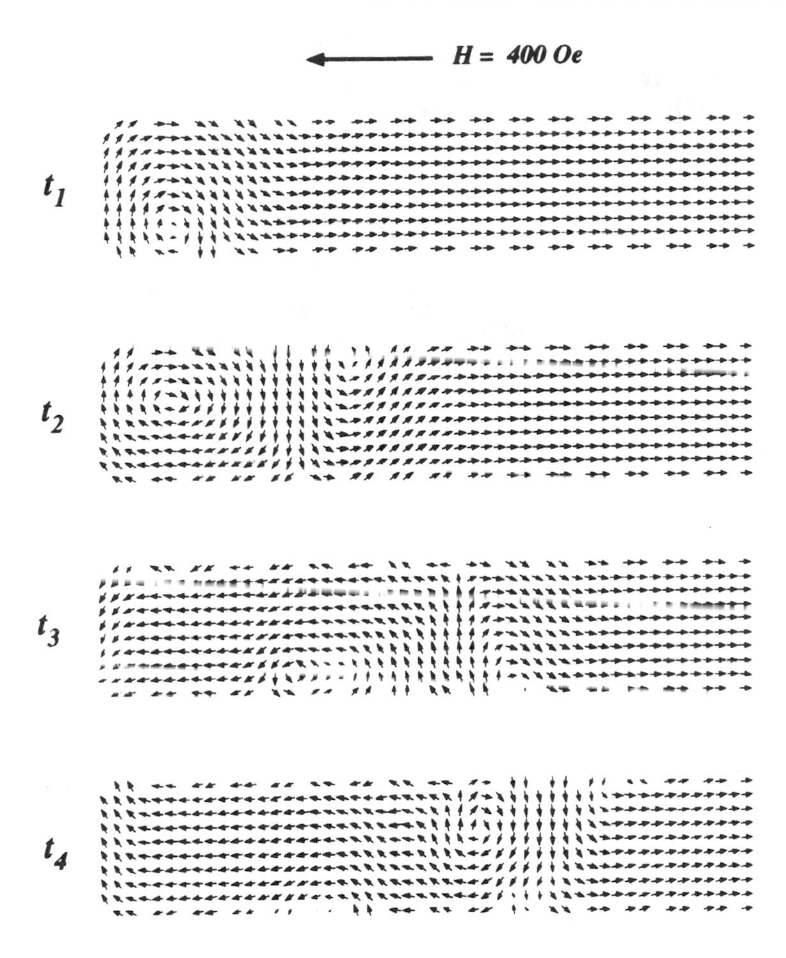

FIG. 10. Simulated transient magnetization configurations during a magnetization reversal of a 35-nm-thick Ni thin film element $1.0 \times 0.1\ \mu m^2$ in size.

1.4 Domain Configurations in Patterned Magnetic Thin Film Elements

Figure 11 shows a series of simulated static domain configurations in 20-nm-thick Permalloy thin film elements with various geometric shapes [24]. The magnetization configurations are presented by mapping their in-plane components with a gray scale, and the corresponding divergence of the magnetization vector configuration is also shown. For a 1-μm-diameter

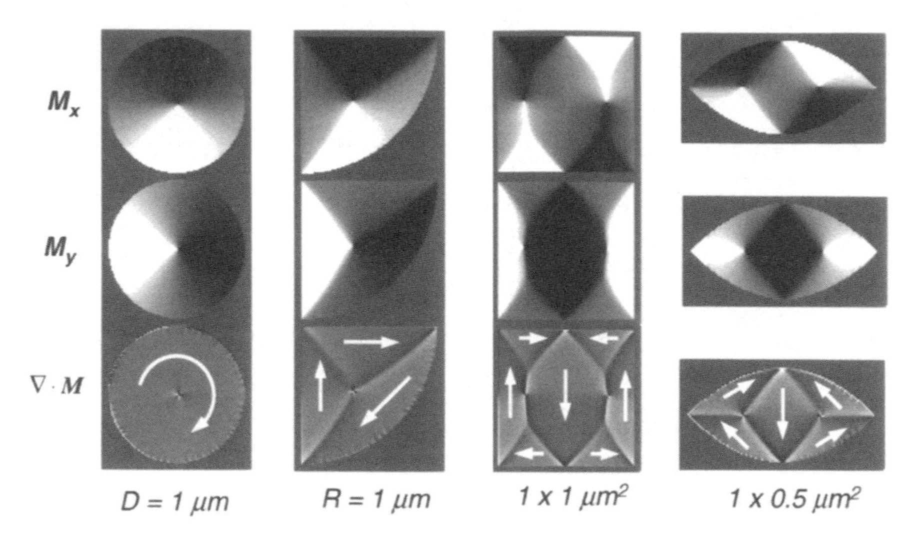

FIG. 11. Simulated remanent magnetization domain configurations for a 20-nm-thick Permalloy film patterned into various shapes with 1 μm as the bigger dimension.

disk, the circular magnetization configuration is extremely stable. Roughness on the edge of the disk yields excessive magnetic poles near the rim of the disk. For the quarter-circle element, three domains form a magnetization flux closure, and the magnetization direction within each domain is mainly determined by the corresponding element edge. The vertex where the three domain walls join is the center of the magnetization vortex, which signifies the flux closure of the three-domain structure. For Permalloy films at 1 μm dimension and 20 nm thickness, magnetostatic energy in the film is still the dominant term, as we can see closure domains within the square and eye-shaped elements. Such domain configurations have been confirmed by MFM images on the same film elements [25].

Saturating the 1-μm-diameter disk with a sufficiently large magnetic field by driving the center magnetization singularity shown in Fig. 1 out of the disk edge, results in the remanent state shown in the first column of Fig. 12. A three-domain configuration is formed with a football-shaped domain in the middle. Two magnetization vortices form in the middle of the two domain walls, yielding local magnetization flux closure and thereby a reduction of the wall energy. It is interesting to note that in the absence of intrinsic magnetic anisotropy, the orientation of the middle football-shaped domain can be rather arbitrary.

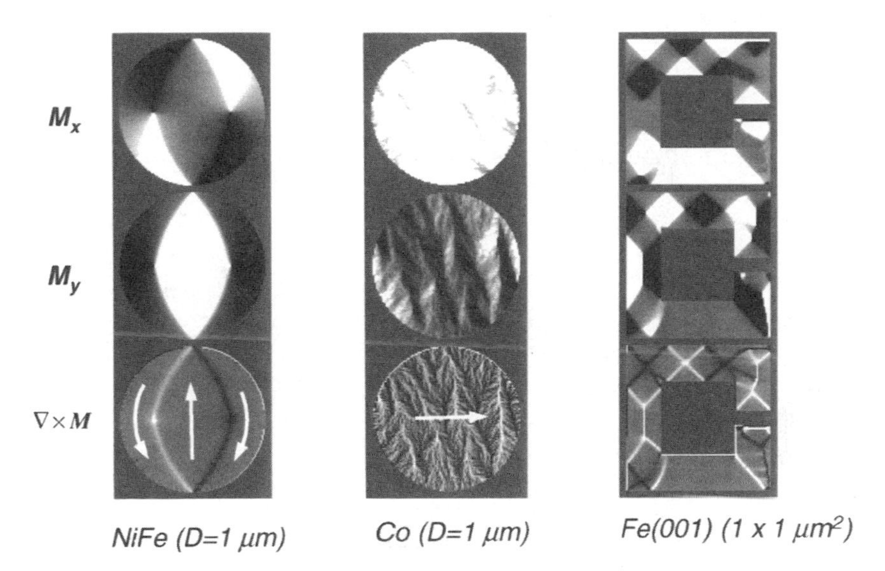

FIG. 12. Simulated remanent magnetization domain configurations for various thin film elements of different materials. The film thickness is 20 nm for all three cases: first column, Permalloy disk, 1 μm in diameter; second column, polycrystalline *hcp* Co disk, 1 μm in diameter; third column, single-crystal Fe C-shaped structure with outer dimension 1 *μm*.

The second column of Fig. 12 shows the simulated magnetization configuration of a remanent magnetization state in a polycrystalline *hcp* Co film patterned into a 1-μm-diameter disk. In this case, each Co crystalline grain has an uniaxial anisotropy with $K_1 = 4 \times 10^5$ J/m^3. However, the easy axis orientations of the grains are completely random because of the polycrystalline nature of the film, which assumes no particular film texture. The grain size here is assumed to be the same as the mesh cell size, namely 10 nm. Relative to that of the Permalloy film, the magnetocrystalline anisotropy of the Co grains is two orders of magnitude larger. Local magnetic anisotropy energy becomes comparable to the ferromagnetic exchange energy and the magnetostatic energy. The net effect is the ripple structure formed at the remanent states: grain clusters form in terms of their magnetization orientations. The magnetization of the grains within a cluster is rather uniform, whereas the magnetization orientation between different clusters varies over relatively large angles. Such ripple structure in *hcp* Co or Co alloy films has been observed using transmission electron microscopy [26].

The third column of Fig. 12 shows a 20-nm-thick single-crystal Fe film with (001) texture. Although each leg in the C-shaped structure is only

0.25 μm wide, closure domains still form because the high magnetic moment of Fe yields high magnetostatic energy.

1.5 Magnetization Processes in Thin Film Recording Media

Magnetic thin film has been used as recording media for disk drive applications. The thin film recording media are polycrystalline films, consisting of a magnetic layer deposited a nonmagnetic underlayer for facilitating proper texture of the magnetic layer. The magnetic layer is usually a Co alloy with *hcp* crystalline structure, and the Co alloy crystallites have relatively large magneto-crystalline anisotropy, with the *c*-axis being the magnetic easy axis. To model films in this category, hexagonal cells on a hexagonal mesh, as shown in Fig. 2, have been used with great success [6, 27].

Figure 13 shows a simulated recording track with gray scale representing the normal component of the magnetization curl, which mimics the Fresnel image of the Lorentz electron microscopy image. A recording head with

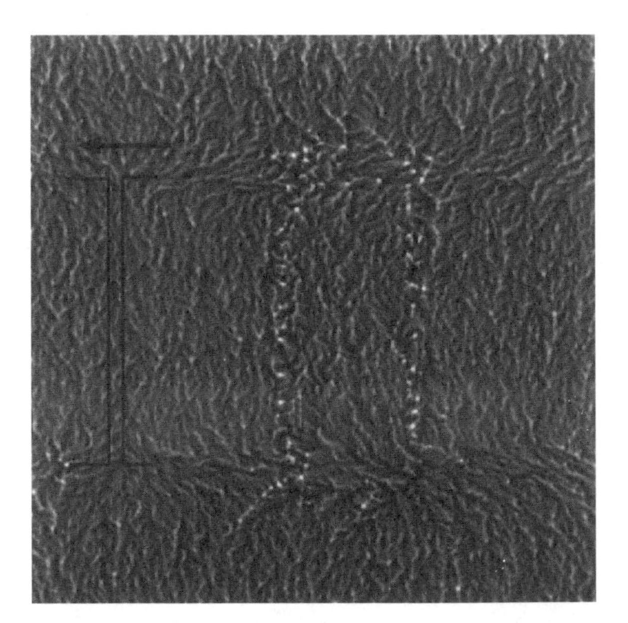

FIG. 13. Micromagnetic simulation of recording on a conventional thin film recording medium. The gray scale here represents the normal component of the magnetization curl, for the purpose of mimicking the Lorentz transmission electron microscopy image.

sufficient field magnitude traveled upward, and two transitions were recorded along the track. The bright starlike structures are the magnetization vortices formed at the transition centers. A track edge field from the recording head resulted in magnetization structures along the edge of the recording track. The simulation bears a remarkable resemblance to the corresponding experimental observations utilizing Lorentz transmission electron microscopy [26].

The magnetization vortices at transition centers are the result of local demagnetization. Figure 14 shows a simulated magnetization vector configuration at the center of a transition. In this case, the ferromagnetic exchange coupling strength between adjacent magnetic grains is assumed to be 50% of that if a perfect magnetic grain boundary is assumed. Magnetization vortices at the center of the transition are evident, and the sizes of the vortices are relatively large, meaning a vortex involves a large number of magnetic grains. The large vortices yield magnetization fluctuations over large length scales and cause severe magnetic flux noise in the readback signal for data retrieving in a disk drive [27].

If the intergranular exchange coupling can be totally eliminated or significantly suppressed by forming nonmagnetic grain boundaries in the

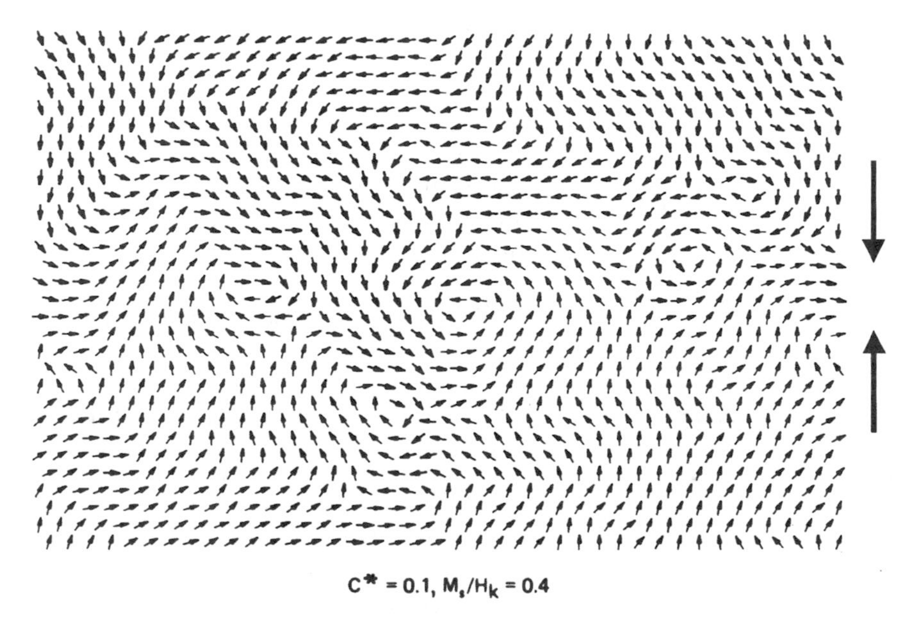

FIG. 14. Simulated magnetization configuration at the center of a recorded transition of a thin film recording medium with intergranular exchange coupling.

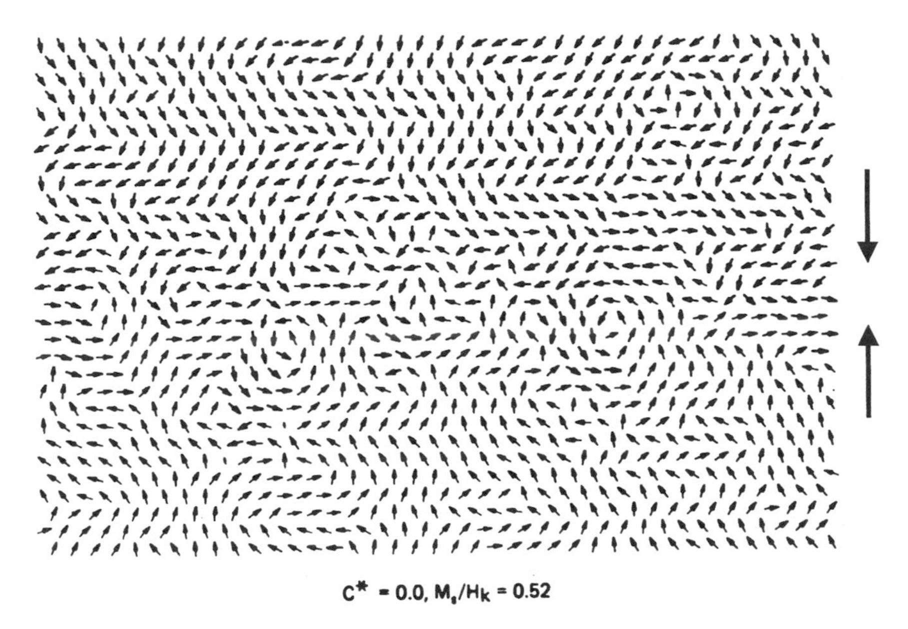

FIG. 15. Simulated magnetization configuration at the center of a recorded transition of a thin film recording medium with zero intergranular exchange coupling.

film making, such noise at the transition center can be greatly reduced. Figure 15 shows the same calculation of Fig. 14 except with zero intergranular exchange coupling. In this case, the sizes of the vortices are significantly smaller, and the associated spatial magnetization fluctuations become significantly smaller, greatly reducing the noise in the readback signal. Today, all fabrication of thin film recording media follows the concept of minimizing intergranular exchange coupling by using various film deposition techniques to generate nonmagnetic or weak magnetic grain boundaries [27–31].

Bicrystal thin films have been studied in the search for more advanced film microstructures that might yield further improved magnetic properties for high density recording applications. In the bicrystal thin film recording medium, the crystalline easy axes of the magnetic crystallites are oriented in either one of the two orthogonal directions in the film plane [32, 33]. Therefore, this structure is referred to as a *bicrystal structure*. In practice, this film microstructure has been realized by epitaxial growth of the magnetic layer with a proper underlayer on cubic single-crystal substrates, such as NaCl [32], LiF [33], MgO [34], and GaAs [35]. By depositing a Cr underlayer with (100) texture on one of the single-crystal substrates

Modeling of Bicrystal Thin Film

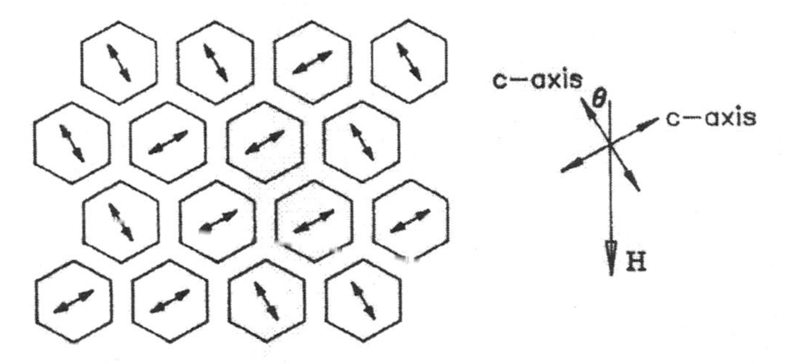

An Example of Practical Bicrystal Film

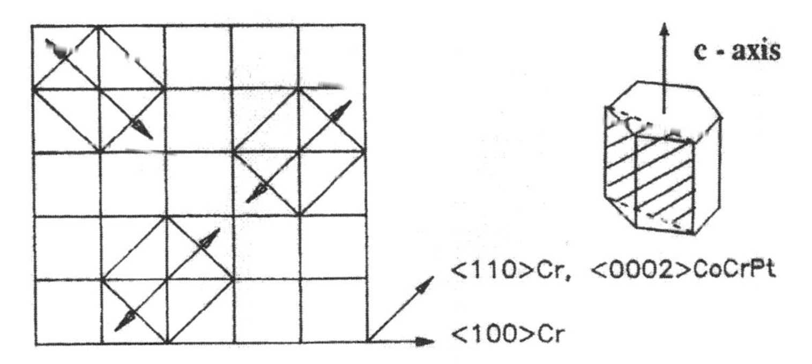

FIG. 16. Sketch of the epitaxial relation between Co and the cubic single-crystal underlayer for a bicrystal thin film recording medium. The hexagonal mesh used for modeling the bicrystal thin film is also shown.

with epitaxial growth, the Cr crystallites have a single orientation. Under appropriate deposition conditions, the magnetic layer can have $(11\bar{2}0)$ texture with the c-axes of the crystallites along either the $\langle 110 \rangle$ or the $\langle 1\bar{1}0 \rangle$ directions of the Cr underlayer grain, usually with equal probability. Figure 16 shows the epitaxial relationship and the modeling mesh of grains with the two magnetic easy axes orientations.

Magnetization reversal processes in the bicrystal films revealed by micromagnetic modeling are fascinating [36]. Figures 17, 18, and 19 show the reversal processes for an external field applied in the film plane at angles of 45°, 25°, and 0°, respectively, with respect to the one of the c-axis directions. The gray scale in the figures represents the normal component of the magnetization curl.

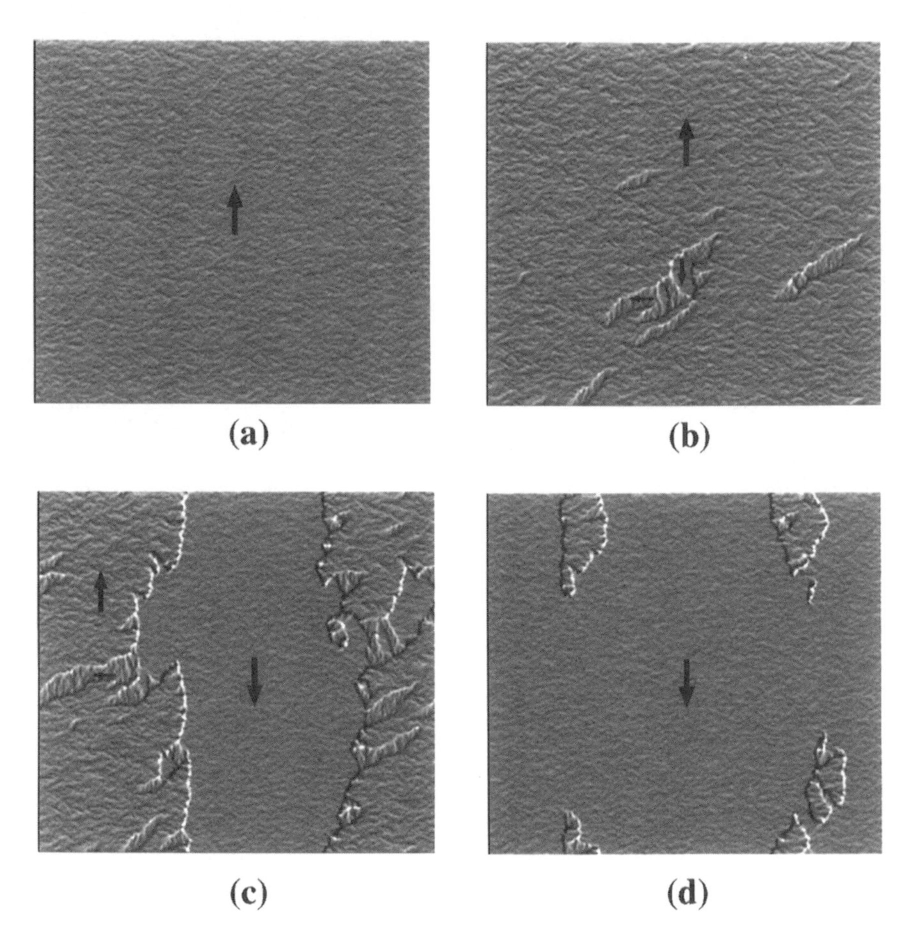

FIG. 17. Simulated magnetization domain configurations during a magnetization reversal for a bicrystal thin film. The gray scale represents the normal component of the magnetization curl. The magnetic field is applied along one of the effective easy axes, a direction in between the two orthogonal *c*-axis orientations.

Figure 17 shows the magnetization reversal with the external applied field along the [100] direction of the Cr layer, at a 45° angle with respect to the two *c*-axis directions of the Co grains. This field direction is along one of the two effective easy axes, and the magnetization remanence is almost the same as the saturation magnetization (i.e., high saturation squareness). At the saturation remanent state (Fig. 17a), reached after removing a large saturation external field, a slight ripple pattern is present, indicating a small local deviation of the magnetization directions along the initial saturation

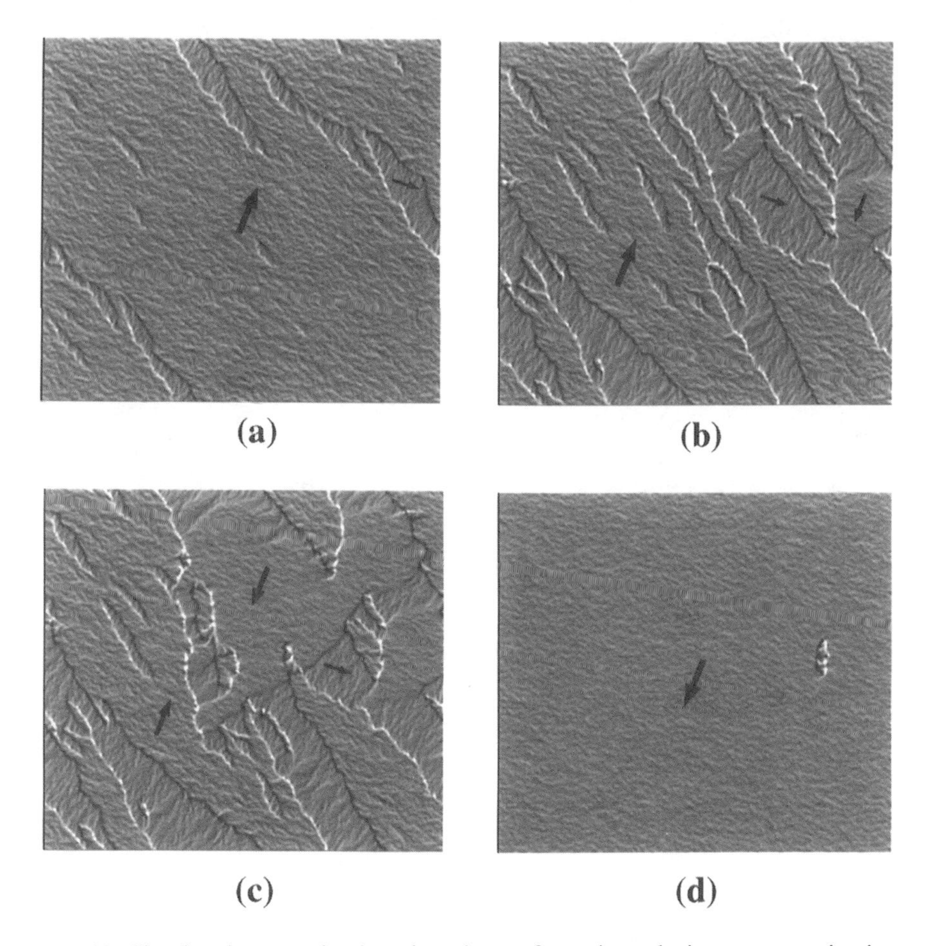

FIG. 18. Simulated magnetization domain configurations during a magnetization reversal for a bicrystal thin film. The gray scale represents the normal component of the magnetization curl. The magnetic field is applied at a 25° angle with respect to one of the *c*-axis orientations.

direction. As the external field starts to increase in the direction opposite to the initial saturation, the magnetization reversal of the film starts as a small domain nucleates with magnetization rotated by 90°. This transverse domain expands as reversal progresses, and then a complete reverse domain nucleates within the transverse domain. Both the reverse domain and transverse domain expand as other transverse domains continue to nucleate in the film. As shown in Fig. 17c, the reverse domain nucleated inside the transverse domain has expanded through nearly half of the entire simulation

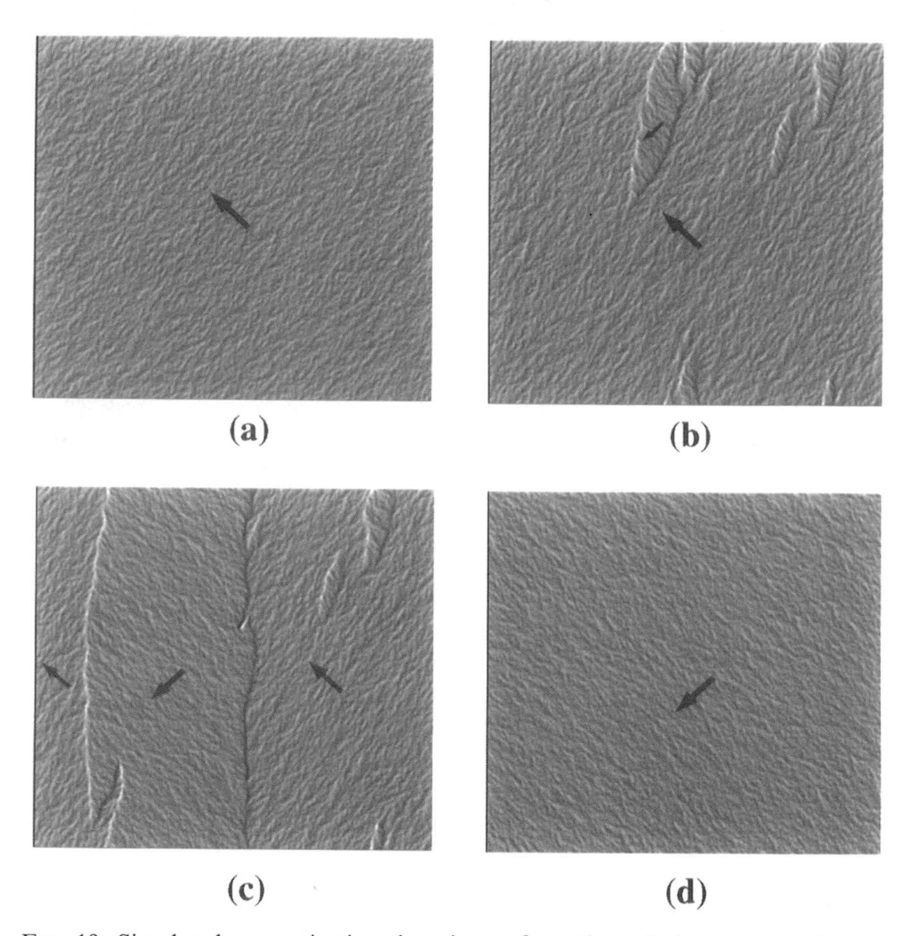

FIG. 19. Simulated magnetization domain configurations during a magnetization reversal for a bicrystal thin film. The gray scale represents the normal component of the magnetization curl. The magnetic field is applied along one of the c-axis orientations.

sample. The magnetization of most of the regions in the film is reversed by expansion of the reverse domains. The expansion and elongation of the transverse domains is always very limited, and the size of these domains is always small.

Figure 18 shows the reversal process for the external field applied at an angle of 25° to one of the crystalline easy axis directions. At the saturation remanent state, the film is essentially magnetized along the nearest effective easy axis, namely, the [100] direction in the film. The reversal starts by the nucleation of transverse domains with the magnetization rotated by 90° in the domain. These nucleated transverse domains elongate along the [110] direction to avoid magnetic poles along domain boundaries. Reversed domains then nucleate within transverse domains and expand within transverse domains as well as beyond through the unreversed regions.

Figure 19 shows the reversal when the field is applied along the effective hard axis, that is, parallel to one of the crystalline easy axes. Prior to the reversal, at the saturation remanent state, the magnetization is virtually along one of the effective easy axes, yielding the squareness $S = \cos 45°$. The reversal starts by nucleating transverse domains. These transverse domains elongate along the reverse field direction and expand through the initial unreversed regions by transverse motion of the domain boundaries. The domain boundary contains a series of half-formed magnetization vortices. After the transverse domains expand through the entire array, further increase of the external field in the reverse direction yields gradual rotation of the magnetization toward the external field direction.

Micromagnetic modeling studies have also shown that recording along any one of the effective easy axes (there are two of them because of the fourfold symmetry) yields sharp transitions and low medium noise [36]. Experimentally, recording disks with bicrystal Co films have been successfully made on a GaAs single-crystal substrate, and actual recording tests have been conducted. The measurements showed that recording along the effective easy axes was significantly superior, with both sharp transitions and low medium noise relative to that of conventional thin film media [35]. However, recordings along the effective hard axes showed more broadened transitions than for the conventional media. Since recording tracks are circumferential, actual application of bicrystal film media in a disk drive requires the ability to orient the effective easy axes along the circumferential direction of the disks.

1.6 Summary and Remarks

Micromagnetic modeling is a powerful tool for understanding the magnetization processes in magnetic films and thin film devices, thereby providing guidelines to material microstructure engineering and device design. However, experimental verifications are often needed to ensure the validity of the modeling results. Magnetic imaging techniques, such as magnetic force microscopy and various transmission electron microscopy techniques, not only can provide direct understanding of the micromagnetic behavior, but also help provide comparisons for either verifying a modeling study or for changing or revising the model [37].

The dynamic approach in micromagnetic modeling is sometimes critical. The inclusion of gyromagnetic motion is often absolutely necessary. For example, the motion of a domain wall in soft magnetic films is facilitated by the gyromotion of the spins. The damping ratio α, defined as

$$\alpha = \frac{\lambda}{\gamma}, \tag{15}$$

where λ is the damping constant in Eq. (14) and γ is the gyromagnetic ratio, is an important parameter. For Permalloy, $\alpha \approx 0.02$–0.08 has been reported via experimental measurements [38, 39].

Acknowledgments

The author would like to thank Dr. Youfeng Zheng for the work on domain structures in soft magnetic thin film elements and Ms. Tzuning Fang and Ms. Kathy Miskinis for help with the preparation of this chapter. The work is supported in part by a Young Investigator Award from the National Science Foundation.

2. LORENTZ MICROSCOPY: THEORETICAL BASIS AND IMAGE SIMULATIONS

Marc De Graef

Department of Materials Science and Engineering
Carnegie Mellon University
Pittsburgh, Pennsylvania

2.1 Introduction

Magnetization configurations in "real" engineering materials can be studied both qualitatively and quantitatively in a variety of ways. In this chapter we review the basic theory of *Lorentz microscopy*, a set of observation modes that are commonly used in either a conventional transmission electron microscope (CTEM) or a scanning transmission electron microscope (STEM). We analyze the behavior of relativistic electrons in a magnetic field, using first a classical and then a quantum mechanical approach. Then we define the various observation modes (Fresnel, Foucault, and derived techniques) and discuss how the transmission electron microscope (TEM) can be used to obtain qualitative pictures of the magnetization configuration in a thin foil.

We then proceed with more quantitative observation modes, which produce direct maps of the sample magnetization configuration. These include differential phase contrast (DPC), which can be used in CTEM and STEM versions. Along the way we also describe how Lorentz images can be simulated for a given magnetization configuration. We conclude with a derivation of the transport-of-intensity equation (TIE) from the standard transfer function formalism of image formation in the electron microscope; this equation will form the basis of the phase reconstruction method described in Chapter 5.

The 1990s saw increased interest in quantitative Lorentz microscopy. The decreasing length scale of modern electronic and magnetic devices must be accompanied by improvements in the observation techniques used to study their microstructures and nanostructures. There is a growing interest in both room temperature and very low temperature magnetostrictive actuator materials. Terfenol-D, a rare earth–transition metal magnet, possesses the largest known room temperature magnetostrictive strain of about 0.2% [40]. Ni_2MnGa exhibits large magnetostrictive strains that arise from a magnetically induced martensitic transformation in a temperature range

EXPERIMENTAL METHODS IN THE PHYSICAL SCIENCES
Vol. 36
ISBN 0-12-475983-1

ISSN 1079-4042/01 $35.00

near room temperature [41]. Polycrystalline rare earth low temperature magnets have been shown to retain a significant fraction of the magnetostriction of the single crystals, and they are considerably less expensive to produce [42]. Determination of the fine scale magnetic structure of these materials is essential for prediction of, among other things, the bulk magnetostrictive strains. In addition, since virtually all engineering parameters of these materials are directly affected by microstructural features, the nature of these materials makes it necessary to closely combine the study of magnetic structure and microstructure. Lorentz microscopy, in particular the Fresnel and Foucault observation modes, has for several decades been the dominant observation technique for qualitative magnetic domain observations.

The magnetic recording industry has a similar need for high spatial resolution magnetic imaging. Along with significant advances in read–write head design and manufacturing, much of the rapid increase in the recording density—the density has increased from 1 Gbit/in^2 (IBM, December 1989) to 35.3 Gbit/in^2 (IBM, October 1999), while the cost per gigabyte has gone from $5,230 in 1991 to less than $10 in 1999—has been made possible by increased understanding of the relation between thin film growth and processing and the resulting microstructure and properties. Advanced materials characterization tools have played an important role in this process. Among the more commonly used tools are thin film X-ray diffractometry (for the study of film texture), conventional and high resolution transmission electron microscopy (for the study of microstructure and defects), energy filtered TEM (to study chemical profiles and segregation), Lorentz transmission electron microscopy (LTEM, this chapter and Chapters 7 and 8), spin-polarized scanning electron microscopy (Chapter 6), electron holography (Chapter 4) and noninterferometric techniques (Chapter 5), magnetic force microscopy (MFM, Chapter 3), and a wide variety of magnetic measurements to characterize basic parameters (Chapter 1) such as coercivity, saturation magnetization, loop squareness, and high frequency response.

2.2 Basic Lorentz Microscopy (Classical Approach)

An electron with charge e moving with velocity $\mathbf{v}$ through a region of space with an electrostatic field $\mathbf{E}$ and a magnetic field $\mathbf{B}$ experiences a velocity-dependent force, commonly known as the *Lorentz force* $\mathbf{F}_L$:

$$\mathbf{F}_L = -e(\mathbf{E} + \mathbf{v} \times \mathbf{B}). \tag{1}$$

We will assume that all fields are static. Because the magnetic component of the Lorentz force acts normal to the travel direction of the electron, a

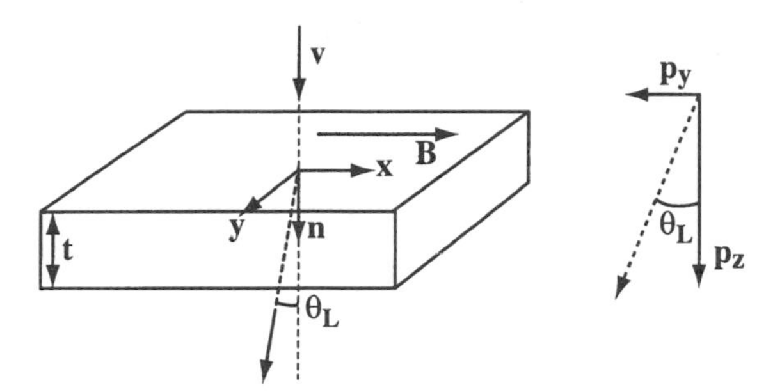

FIG. 1. Schematic of a magnetic thin foil and the resulting deflection of an incident electron beam.

deflection will occur. Only the component of **B** normal to **v** will contribute to the deflection, and we will refer to this component as the *in-plane magnetic induction* $\mathbf{B}_\perp$, that is,

$$\mathbf{B} = \mathbf{B}_\perp + B_z\mathbf{n}, \tag{2}$$

with **n** being a unit vector parallel to the beam direction. As only two of the three components of **B** act on the electron trajectory, it is *a priori* clear that a complete determination of all three components will require the use of at least two independent incident beam directions **n**.

Although it is possible to mount the sample in a field-free region (see Section 2.2.1), it is in general not possible to completely remove the electrostatic field **E**, because the crystal structure itself generates an electrostatic lattice potential. This potential causes a slight acceleration of the electron on entering the sample (a phenomenon commonly known as *refraction*). The mean inner potential (the average of the electrostatic lattice potential) is usually incorporated as a correction factor for the electron wavelength. The magnetic deflection caused by the Lorentz force does not change the electron energy, whereas the (positive) electrostatic component causes the electron to accelerate, which changes its kinetic energy.

Consider a planar thin foil of thickness t, normal to the incident beam. If the in-plane magnetic induction is directed along the x-axis (see Fig. 1) with magnitude $B_\perp$, then the momentum transferred to the electron is given by

$$p_y = \int_0^\tau F_L \,\mathrm{d}\tau, \tag{3}$$

with τ being the time it takes for the electron to traverse the sample. In the absence of electrostatic fields, the magnitude of the Lorentz force along the y direction is given by

$$F_{\mathrm{L}} = ev_z B_{\perp}. \tag{4}$$

The integral can be rewritten in terms of the sample thickness $t = v\tau$ as

$$p_y = e\int_0^t B_{\perp}\,\mathrm{d}z = eB_{\perp}t. \tag{5}$$

The deflection angle θ_L is then to a good approximation (Fig. 1) given by the ratio of the y and z momentum components:

$$\theta_L = \frac{p_y}{p_z} = \frac{eB_{\perp}t}{mv}, \tag{6}$$

where m is the relativistic electron mass. This equation can be rewritten in a more useful form by means of the de Broglie relation between particle momentum and wavelength ($p = h/\lambda$), and we find for the Lorentz deflection angle θ_L:

$$\theta_L = \frac{e\lambda}{h} B_{\perp}t = C_L(E)B_{\perp}t. \tag{7}$$

The constant $C_L(E)$ is determined by the acceleration voltage E of the microscope and is given by

$$C_L(E) = \frac{9.37783}{\sqrt{E + 0.97485 \times 10^{-3}E^2}} \quad \mu\mathrm{rad/T/nm}$$

(to get the Lorentz angle in microradians, $B_{\perp}$ must be stated in teslas, t in nanometers, and E in kilovolts). For the commonly used accelerating voltages we have $C_L(100) = 0.895018$, $C_L(200) = 0.606426$, $C_L(300) = 0.476050$, $C_L(400) = 0.397511$, and $C_L(1000) = 0.210834$ μrad/T/nm. A 100-nm thin foil with an in-plane magnetic induction of $B_{\perp} = 1$ T will give rise to a beam deflection of $\theta_L = 39.75$ μrad at 400 kV. For comparison, typical Bragg angles for electron diffraction are in the range of a few milliradians, that is, two to three orders of magnitude larger than Lorentz deflection angles.

Note that the sample thickness and the in-plane induction component both appear in the expression for the deflection angle. Since the sample thickness is notoriously difficult to quantify experimentally, it should be clear that any measurement based on the Lorentz deflection angle will at best be qualitative and provide only the product $B_{\perp}t$. Localized thickness variations and local out-of-plane excursions of the magnetic induction will produce identical changes in the Lorentz deflection angle θ_L. Only when an independent thickness measurement is available can Lorentz methods provide a direct map of the in-plane induction component.

2.2.1 Experimental Methods

Direct observations of the magnetic substructure of a material in a TEM require that a thin foil be made, using appropriate thinning procedures. As for most TEM experiments, sample preparation is the most crucial and often the most time-consuming step; without a good thin foil, no reliable observations can be made. In addition, one must always be aware that the magnetic structure of the thin foil may be different from the bulk magnetic structure. In particular in soft magnetic materials, the introduction of two free surfaces may completely change the energetics of magnetic domains and, indeed, the very nature of the magnetic domain walls (e.g., Ref [43]).

Although the scanning probe microscopy techniques discussed in Chapter 3 result in images reflecting the magnetization pattern near the surface of the sample, all TEM-based observation modes result in images that represent the *integration* of the magnetic information along the path of the beam electron. A complete three-dimensional reconstruction of the magnetization pattern is therefore similar to a tomography problem, with the added difficulty that a ferromagnetic foil changes the phase of the electron quantum wave function, and the phase is not a directly observable quantity (see Section 2.3.1).

To preserve the magnetic microstructure of the thin foil, the sample must be mounted in a field-free region in the microscope column. Since the objective lens in a standard TEM is an immersion-type lens, the requirement of a field-free region has profound consequences on the electron optical properties of the microscope. The low-field sample environment and consequent increase in the focal length of the objective lens result in a reduced final image magnification as compared to that of conventional transmission electron microscopy. This represents a serious limitation in the quantitative study of nanoscale magnetic structures. Furthermore, inelastic scattering in the sample contributes noise to the images, which further limits the attainable resolution of the standard Lorentz modes.

A number of different sample and lens configurations can be used:

1. Mount the sample above the main objective lens. This generally requires the introduction of a second goniometer stage in between the condenser and objective lenses, and is therefore attempted only in dedicated instruments.

2. Use a dedicated low-field Lorentz pole piece. This is perhaps the most efficient and reliable method, but it may require a pole piece change, which is not always an easy thing to do. Furthermore, if the microscope is used for many different types of materials, the loss of magnification due to the reduced field may be unacceptable to other users. A permanent Lorentz pole piece downstream from the sample holder is therefore preferable, assuming it does not deteriorate the electron optical properties of the column when used in conventional or high resolution observation modes.
3. Turn the objective lens off and use an objective minilens (if present) to obtain a back focal plane at the location of the selected area aperture. If a postcolumn energy filter is available, the loss of image magnification may be partially compensated by the internal magnification of the filter, and zero-loss images may be obtained, as first reported by Dooley and De Graef [44].

Nearly all images shown in this chapter were obtained on a Jeol 4000EX top-entry high resolution TEM, operated at 400 kV in Low Mag mode (with the main objective lens turned off). The objective minilens provides a crossover near the selected area aperture plane, at a maximum magnification of 3000 ×. A Gatan imaging filter (GIF) is then used to remove most inelastically scattered electrons (using an energy selecting slit width of 20 eV) and to magnify the image by an additional factor of about 20 ×. All images were acquired on a 1K × 1K charge-coupled device (CCD) camera, and they are gain normalized and background subtracted. Unless mentioned otherwise, no additional image processing was carried out on the experimental images.

2.2.2 Fresnel Mode

The Fresnel or *out-of-focus* Lorentz mode is perhaps the easiest observation method to carry out experimentally, since it requires only a change in the lens current for the main imaging lens (either a Lorentz lens or an objective minilens). The method is schematically shown in Fig. 2. Consider a sample with three magnetic domains separated from one another by 180° magnetic domain walls. The character of the walls is unimportant at this point. The outer domains have their magnetization pointing into the plane of the drawing, and the central domain has the opposite magnetization direction. The resulting trajectory deflections are toward the positive y direction for the outer domains.

A standard bright-field image, obtained by selecting only the transmitted beam with a sufficiently large aperture in the back focal plane of the imaging lens, does not show any contrast (other than diffraction or absorption

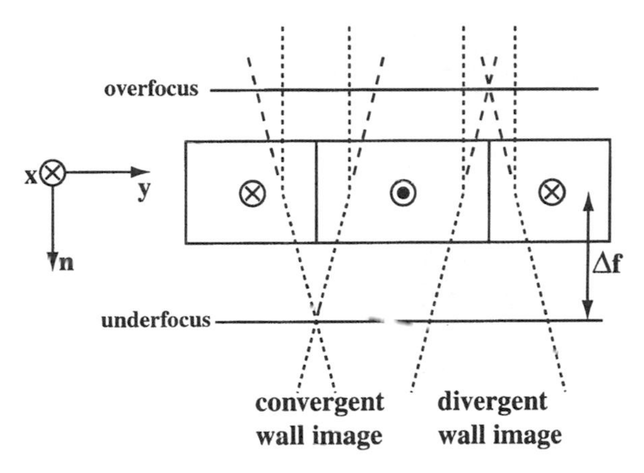

FIG. 2. Schematic illustration of the Fresnel or out-of-focus imaging mode.

contrast) for an *in-focus* condition. In other words, when the object plane of the imaging lens is located at the sample position, no magnetic contrast is observed. If the current through the imaging lens is reduced, the focal length increases (causing a reduction in the lateral image magnification), and the object plane is located at a distance Δf below the sample. Because of the small Lorentz deflection angles, a rather large *defocus* Δf is needed to obtain an overlap between the electrons deflected on either side of the left domain wall in Fig. 2. In the overlap region, an excess of electrons will be detected, and this will show up in the image as a bright line feature at the projected position of the domain wall. The domain wall on the right will show up as a dark feature, since electrons are deflected away from the domain wall position. The bright domain wall image is known as the *convergent wall*, and the dark wall is the *divergent wall.*

If the imaging lens current is increased, so that the focal length is reduced and the image magnification increased, then an overfocus image is obtained. The image contrast of convergent and divergent domain walls reverses, as can be seen from the extended electron trajectories in Fig. 2.

Figure 3 shows a set of experimental Fresnel images for a cross-tie wall arrangement in a 50-nm-thick Ni–20 wt% Fe Permalloy film. The in-focus image is shown in Fig. 3a, and the underfocus and overfocus images are shown in Fig. 3c,d, respectively. A schematic of the magnetization configuration around the cross-tie wall is shown in Fig. 3b. Note that this configuration cannot be deduced from the Fresnel images; the Foucault images shown in the next section (see Fig. 6) were used to derive the magnetization arrangement shown in the schematic (Fig. 3b). The bright points along the main domain wall (Fig. 3c) correspond to the centers of

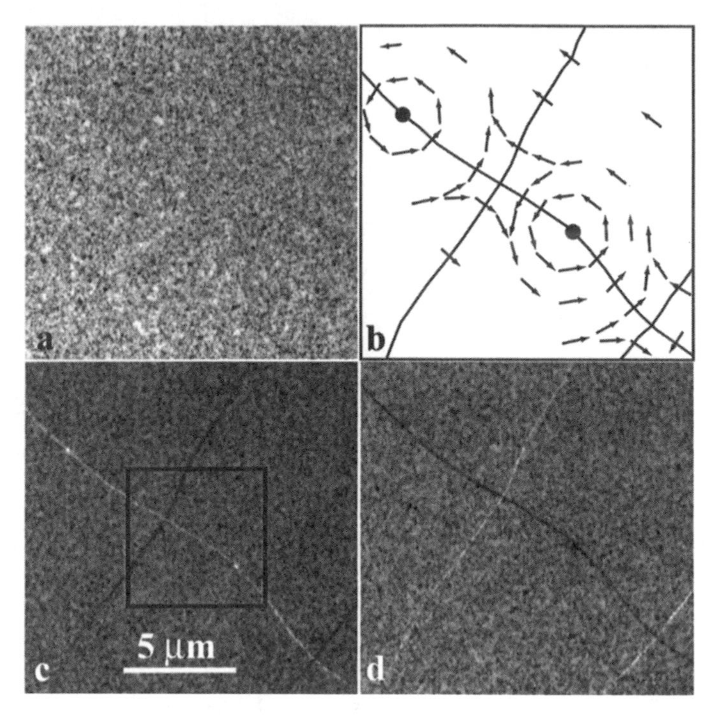

FIG. 3. (a) In-focus, (c) underfocus, and (d) overfocus images (zero-loss filtered with 20 eV slit width) of cross-tie domain walls in a 50-nm-thick Ni–20 wt% Fe Permalloy film. Note that the image magnification changes slightly with defocus. The magnetization pattern in (b) is derived from the Foucault images in Fig. 6. (Sample courtesy of Chang-Min Park, Carnegie Mellon University).

regions around which the magnetization has a circular pattern (vortices); this arrangement acts as a lens that focuses the electrons into a single spot. The magnetization at the vortex is predominantly out of plane (see, e.g., Ref. [43, pp. 402–408]) and opposite to the magnetization at the intersection of the main wall and the divergent walls in Fig. 3c.

Because the Fresnel image mode highlights the domain walls in the thin foil, it should be possible to derive from the images an estimate of the domain wall width δ. This parameter plays a central role in the energetics of magnetic materials, and a direct experimental measurement method would be an extremely valuable tool. One can use the divergent wall image to obtain an estimate of the domain wall width δ; a reasonable estimate of δ is obtained when the measured width of a divergent wall is plotted against the defocus value and extrapolated to zero defocus (e.g., Refs. [45–47]). This

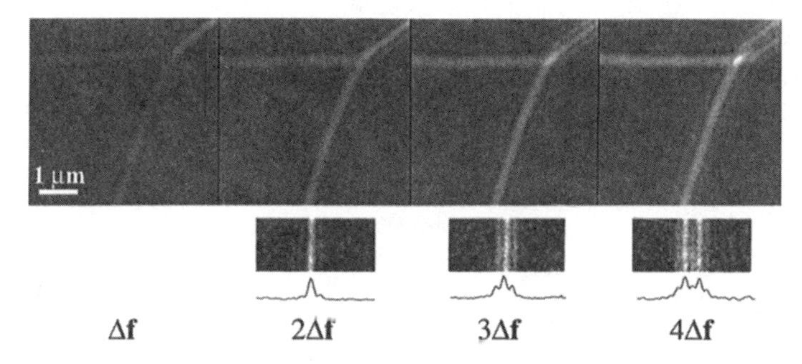

FIG. 4. Underfocus images of a 180° domain wall (top right) that splits into two 90° domain walls in a 50-nm Ni–20 wt% Fe Permalloy thin film. The images were obtained with a strongly underfocused second condenser lens, for increasing underfocus value Δf. The fringe contrast is clearly visible; intensity profiles across the 180° domain wall are shown at the bottom.

extrapolation method works well for large domain wall widths, but it overestimates δ for the narrow walls found in hard magnetic materials. A discussion of the accuracy of the extrapolation method can be found elsewhere [48].

The classical Lorentz deflection theory provides a good first approximation for the understanding of image contrast in the Fresnel imaging mode. However, when a more coherent electron beam is used (for instance, by defocusing the condenser lens), then the convergent wall image reveals the presence of interference fringes that cannot be explained by the classical model. Figure 4 shows a sequence of underfocus images of a 180° domain wall (top right corner) in a Permalloy thin film at increasing defocus Δf; all images were taken with an underfocused second condenser lens or, equivalently, a reasonably coherent electron beam. The traces below the micrographs are averages over 40 pixels along the wall profile. Although the number of fringes increases with increasing defocus, the spacing between the fringes appears to be constant. The classical model for the Fresnel observation mode fails to explain the presence of such fringes. We return to the origin of the fringes in Section 2.3.2; image simulations for coherent Fresnel images are discussed in Section 2.3.5.

Although the Fresnel imaging mode is easy to use and provides a direct, qualitative image of the domain wall configurations in the sample, its quantitative usefulness is rather limited because of the sometimes large magnification changes caused by the defocus. However, advances in noninterferometric observation techniques are bringing the Fresnel mode back to the forefront of Lorentz microscopy, and, as described in detail in Chapter

5 and also in Section 2.5, quantitative Lorentz observations based on the Fresnel method are now possible. The Fresnel mode can also be implemented on a scanning transmission electron microscope (STEM), since STEM and CTEM are related to each other by the reciprocity theorem; for a detailed analysis of implementation and image formation we refer the reader to elsewhere [49].

2.2.3 Foucault Mode

A second commonly used observation mode employs an in-focus image and is somewhat similar to the familiar *dark field* imaging mode of conventional TEM. As illustrated in Fig. 5, an aperture is introduced in the back focal plane of the imaging lens (an objective minilens or a dedicated Lorentz lens), and by translating the aperture normal to the beam a section of the split central beam is cut off. The regions in the sample that give rise to deflected electrons which are passed by the aperture will appear bright in the image. This observation mode produces high contrast in-focus images of magnetic domain configurations, which are known as *Foucault images*. Since a typical aperture stage has two translation controls that are normal to one another, one would acquire typically either two or four images, one or two for each main aperture translation direction. An example of Foucault images for the same sample region as shown in Fig. 3 is shown in Fig. 6; the

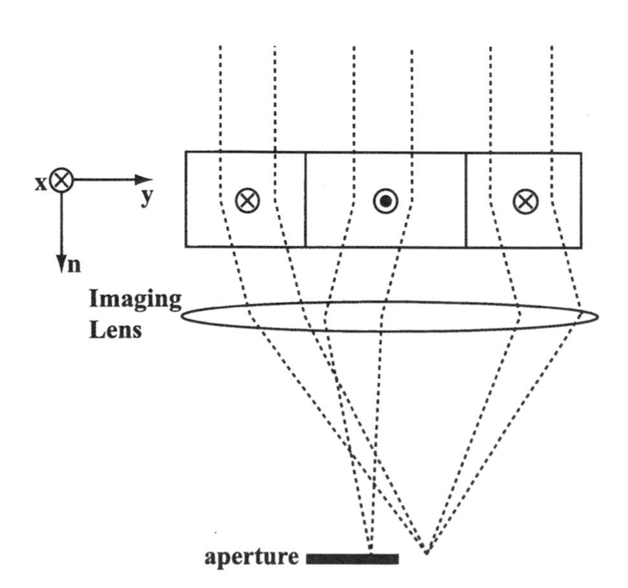

FIG. 5. Schematic illustration of the Foucault imaging mode.

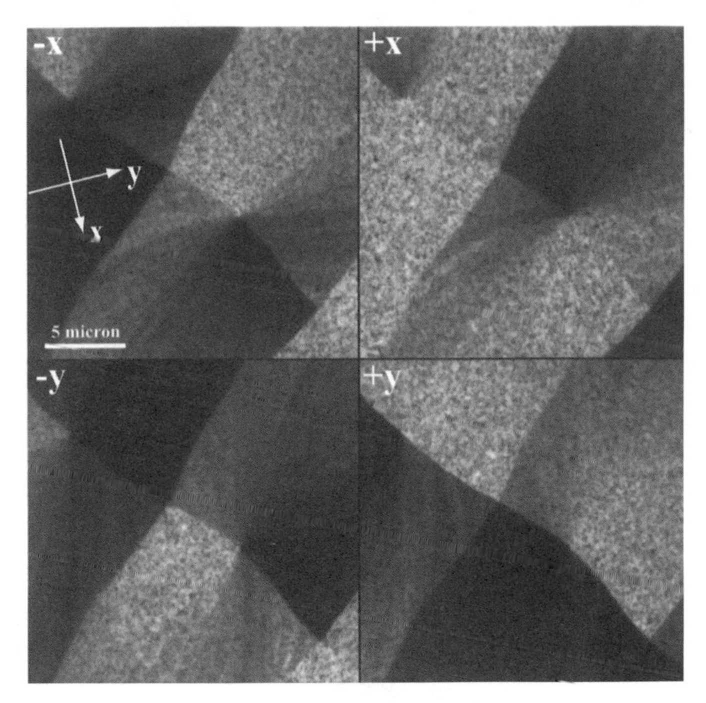

FIG. 6. Foucault images (zero-loss filtered with 20 eV slit width, 400 kV) for four aperture shifts for the same sample region as Fig. 3 ($\pm x$ and $\pm y$ directions, with respect to the aperture directions, indicated by the white arrows). The magnitude of the aperture shift is approximately equal for all four images.

approximate aperture shift directions are indicated by white arrows in the top left image. Since we know that the direction of the Lorentz deflection is normal to the corresponding in-plane induction direction, we can use the four images in Fig. 6 to obtain a schematic (qualitative) drawing of the magnetization configuration, which is shown in Fig. 3b.

When more coherent illumination is used to obtain the Foucault images, fringes are often observed parallel to the domain walls, inside the bright portions of the image (e.g., Ref. [44]). The origin of these fringes is the same as for the Fresnel mode and is addressed in more detail in Section 2.3.2.

The Foucault mode is a highly qualitative observation mode, because it is notoriously difficult to reproducibly position the aperture. Apertures are often not perfectly circular, or they may have debris at the aperture edge. When a set of four Foucault images is acquired, it is difficult if not impossible to manually (for mechanically controlled apertures) position the aperture at equal shifts from the central beam position. This would be a prerequisite for quantitative work. It is also a nontrivial task to determine the location of the optical axis, that is, the position of the beam in the absence of a magnetic sample. This position would be the "origin" for all Foucault images, and one would have to specify the magnitude of the aperture shift (typically in nm^{-1}) with respect to this origin.

2.2.4 Diffraction Mode

Electron diffraction observations are a standard part of TEM experiments, since it is straightforward to create an image of the back focal plane (BFP) of the objective lens on the viewing screen. All electrons that are scattered in a given direction by the thin foil are focused into a single point in the BFP, regardless of where they passed through the sample. This means that the BFP contains a map of the in-plane momentum transfers (transfers normal to the incident beam). This is commonly known as a *diffraction pattern*, and for conventional TEM observations on crystalline materials the pattern is essentially a nearly planar section through the reciprocal space lattice. Bragg scattering angles for high energy electrons are in the range of a few milliradians, which is about two orders of magnitude larger than Lorentz deflection angles. It follows that a direct observation of the Lorentz deflection angles (or in-plane momentum transfer) is possible, provided a sufficiently large camera length can be obtained.

Figure 7a shows an electron diffraction pattern obtained from the divergent part of a cross-tie wall in a 50-nm Permalloy thin film. Since the saturation induction of this film is about 1 T, the Lorentz deflection angle at 400 kV is about 20 μrad. The pattern was acquired directly on the CCD

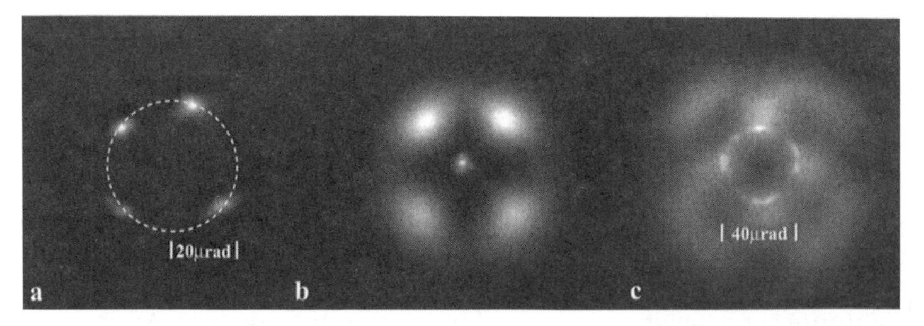

FIG. 7. (a) Electron diffraction pattern for the beam centered on the divergent part of a cross-tie wall; each spot corresponds to the deflection from one of the quadrants of the magnetization pattern. (b) Image of the undersaturated filament tip (LaB_6), for the beam passing through a single magnetic domain. (c) Same as (b), but now the beam passes through the convergent part (vortex) of a cross-tie wall.

camera of the energy filter, using an undersaturated filament to prevent overexposure of the camera. An exposure time of 0.1 s was used, along with an energy selecting slit width of 20 eV. The pattern is obtained by first focusing the second condenser lens so that the filament tip is conjugate to the specimen and then adjusting the objective minilens current to make the viewing screen (CCD plane) conjugate to the specimen. The lateral magnification of the diffraction pattern can be adjusted by using the free-lens control on the microscope, or by using the magnification reduction options on the energy filter.

Figure 7b shows an image of the filament tip (LaB_6) in an undersaturated condition, when the beam passes through a single magnetic domain. The beam is deflected from where it would be in the absence of a magnetic sample, but that precise location is difficult to determine experimentally. In Fig. 7c the incident beam is focused on the vortex part of a cross-tie wall, and a nearly continuous circular intensity distribution is obtained, along with a complex distribution from the outer parts of the filament emission pattern. This pattern contains all the information about the momentum transfer due to the sample magnetization, but it is complicated owing to the fact that all magnetization directions around the vortex contribute simultaneously to the image. If a fine electron probe were scanned across the region containing the cross-tie vortex, then a synchronized measurement of both magnitude and direction of the deflection angle would provide a direct magnetization map of the sample (actually, a map of the product $B_{\perp}t$). This is essentially the so-called *differential phase contrast* (DPC) acquisition mode, which is discussed in more detail in Section 2.4.

2.3 Quantum Mechanical Formulation

The basic theory derived in the preceding sections is adequate to explain some of the more commonly observed contrast features in the Fresnel and Foucault modes. When a partially coherent or a coherent electron beam is used, however, the theory fails to explain the appearance of fringes (Fig. 4) at convergent wall images and intensity oscillations near domain walls in Foucault images. The theoretical explanation of those contrast features requires a wave-mechanical approach, which includes the influence of the electron microscope itself. First we describe which quantum mechanical quantity can be used to replace the classical Lorentz deflection; then we introduce the microscope transfer function and explain how the standard Lorentz observation modes can be described within this formalism. We conclude this section with a general discussion on numerical Lorentz image contrast simulations.

2.3.1 Strong Phase Objects

In 1959 Aharonov and Bohm published a revolutionary paper on the importance of the phase of the electron wave function in the presence of electrostatic and magnetic potentials [50]. They found that even in regions of space where all the fields vanish, the wave function of a charged particle could still experience changes owing to the corresponding electromagnetic potentials. It has taken several decades for the scientific community to accept the notion that the wave function of a particle can be affected by something other than a force, and an extensive review of the quantum effects of electromagnetic fluxes can be found elsewhere [51]. There is now ample evidence that the Aharonov–Bohm (or A–B) effect is indeed real, and a direct experimental proof based on electron holography was provided by Tonomura and co-workers [52] in 1986. The existence of the A–B phase shift is related to the fact that the electric and magnetic fields do not appear directly in the Schrödinger equation, but only in the form of the potentials; although the electromagnetic potentials are not observables in the classical theory, they do become the fundamental quantities in the quantum mechanical framework.

The A–B phase shift imparted on an electron wave with relativistic wavelength λ by the presence of electromagnetic potentials V and $\mathbf{A}$ is given by [50]

$$\phi(\mathbf{r}_\perp) = \frac{\pi}{\lambda E_t}\int_L V(\mathbf{r}_\perp, z)\,\mathrm{d}z - \frac{e}{\hbar}\int_L \mathbf{A}(\mathbf{r}_\perp, z)\cdot \mathrm{d}\mathbf{r}, \tag{8}$$

where $\hbar = h/2\pi$, E_t is the total energy of the beam electron and the integrals are carried out along a straight line L parallel to the incident beam direction [i.e., crossing the plane of the sample at the point $(x, y, 0)$]. The prefactor of the electrostatic component of the phase shift can be written as a constant that depends on the microscope accelerating voltage E (e.g., Ref. [53]):

$$\frac{\pi}{\lambda E_t} = \sigma(E) = \frac{e}{\hbar c}\frac{m_0c^2 + eE}{\sqrt{2m_0c^2eE + e^2E^2}}. \tag{9}$$

The constant σ is known as the *interaction constant*, and it has a limit value of $\lim_{E\to\infty} = e/\hbar c$. At the commonly used accelerating voltages (in kV), the interaction constant takes on the values $\sigma(100) = 0.009244$, $\sigma(200) = 0.007288$, $\sigma(300) = 0.006526$, $\sigma(400) = 0.006121$, and $\sigma(1000) = 0.005385$ $\mathrm{V}^{-1}\,\mathrm{nm}^{-1}$. A 100-nm foil with a mean inner potential V_0 of 30 V will hence give rise to an electrostatic phase shift of 18.36 radians at 400 kV. The prefactor of the magnetic component of the A–B phase shift equals $e/\hbar = 0.00151927\ \mathrm{T}^{-1}\,\mathrm{nm}^{-2}$ and is independent of the electron energy.

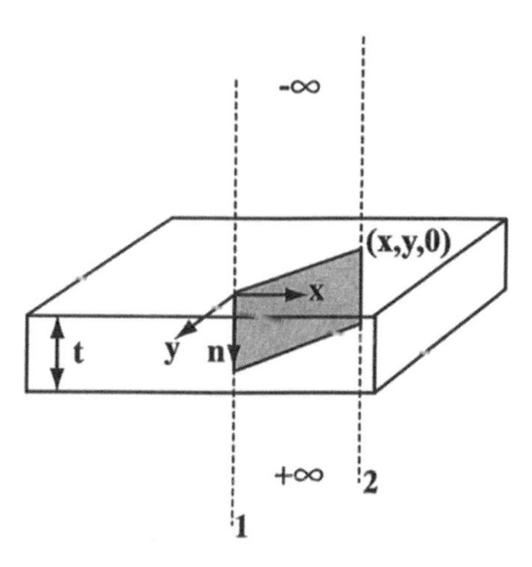

FIG. 8. Schematic of the two trajectories involved in the formulation of the Aharonov–Bohm phase shift.

As it is rather difficult to measure absolute phases, one usually works in terms of the phase difference $\Delta\phi(\mathbf{r}_\perp)$ between a given electron trajectory crossing the sample at the location $\mathbf{r}_\perp$ and a reference trajectory that is conveniently chosen to coincide with the optical axis of the microscope, as schematically indicated in Fig. 8. In the absence of an electrostatic potential, the magnetic phase difference $\phi_2 - \phi_1$ between the trajectories 1 and 2 is then equal to

$$\Delta\phi_m(\mathbf{r}_\perp) = \frac{e}{\hbar}\left(\int_{-\infty}^{+\infty} \mathbf{A}\cdot d\mathbf{r}_2 - \int_{-\infty}^{+\infty} \mathbf{A}\cdot d\mathbf{r}_1\right) = \frac{e}{\hbar}\oint \mathbf{A}\cdot d\mathbf{r}.$$

This last line integral can be converted to a double integral using Stokes' theorem (e.g., Ref. [54]):

$$\Delta\phi_m{}'(\mathbf{r}_\perp) = \frac{e}{\hbar}\iint \nabla \times \mathbf{A}\cdot d\mathbf{S}, \tag{10}$$

with $\mathbf{S}$ being a unit vector normal to the integration surface. This last integral is equal to the *magnetic flux* Φ_m enclosed between the two trajectories:

$$\Delta\phi_m(\mathbf{r}_\perp) = \frac{e}{\hbar}\Phi(m)(\mathbf{r}_\perp) = \pi\frac{\Phi_m(\mathbf{r}_\perp)}{\Phi_0}, \tag{11}$$

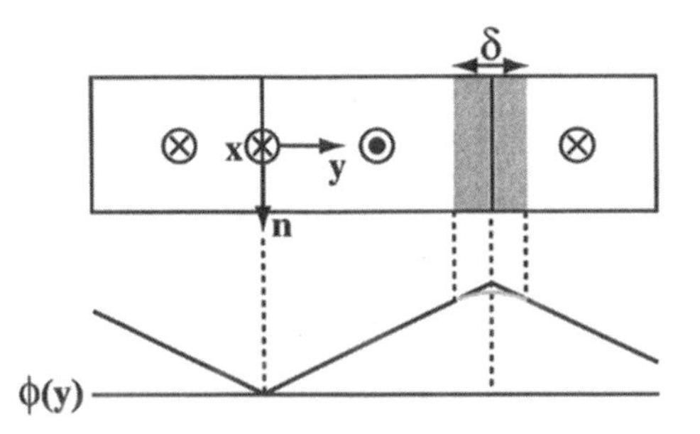

FIG. 9. Schematic illustration of the phase computation for a set of three alternating magnetic domains (magnetization normal to the plane of the drawing). The domain wall profile is taken to be discontinuous (black line) and with a finite domain wall width δ (gray line).

where $\Phi_0 = h/2e$ is the flux quantum. The phase shift is therefore proportional to the enclosed flux, measured in units of the flux quantum. If we take trajectory 1 to be the reference phase (i.e., "zero phase"), then the phase at any other point $\mathbf{r}_\perp$ can be calculated if either the vector potential $\mathbf{A}$ or the magnetic field $\mathbf{B}$ is known. In Sections 2.3.4 and 2.3.5 we discuss two examples of the use of this relation.

Consider a thin foil of thickness t with parallel top and bottom surfaces. The foil contains three magnetic domains, as illustrated in Fig. 9, separated from one another by 180° domain walls. The origin of the reference frame is taken to be in the center of the foil, with a right-handed Cartesian reference frame oriented as shown in Fig. 9. The z-axis coincides with the optical axis of the electron microscope, with positive z in the direction of the beam. The magnetization $B_\perp$ in the domains is uniform and parallel to the x-direction, so that the Lorentz deflection will be in the $\pm y$ direction. In the absence of fringe fields one can calculate the phase shift of the electron wave function along the line $0 - y$ from the A–B equation [Eq. (10)]:

$$\phi(y) = \frac{e}{\hbar}\int_{-t/2}^{+t/2}\int_0^y B_\perp \,\mathrm{d}y\,\mathrm{d}z = \frac{e}{\hbar}B_\perp t y.$$

If the domain wall is infinitely narrow, then the slope of the phase changes discontinuously across the domain wall, as indicated in Fig. 9. For a more realistic domain wall profile, the slope change is continuous (gray line). The range over which the slope changes corresponds to the domain wall width. The phase function is a linear function inside magnetic domains with a uniform magnetization and a constant foil thickness. Small local thickness variations will cause nonlinearities in the phase as well, but they generally do not change the sign of the slope. The same is true for magnetization ripple, which causes fluctuations in the orientation of the gradient vector of the phase function.

It is interesting to note that the magnetic component of the phase shift does not depend on the electron energy. The Lorentz deflection angle [Eq. (7)] can be converted to a phase gradient:

$$\nabla\phi = \frac{2\pi}{\lambda}\theta_L \mathbf{e}, \tag{12}$$

where $\mathbf{e}$ is a suitably oriented unit vector. For a uniformly magnetized foil of thickness t this results in a phase gradient of

$$\nabla\phi = \frac{e}{\hbar}B_\perp t\mathbf{e} = -\frac{e}{\hbar}(\mathbf{B}\times\mathbf{n})t. \tag{13}$$

This expression is indeed independent of the electron energy. Since the electrostatic component of the phase is proportional to the interaction constant σ, and since σ decreases with increasing microscope accelerating voltage E, it follows that magnetic contributions to the phase shift are larger relative to the electrostatic contribution for higher voltage microscopes. Since the Lorentz deflection angle is proportional to the electron wavelength, it is clear that intermediate voltages in the range 200–400 kV provide the best compromise between a strong relative magnetic contribution and reasonably large Lorentz deflection angles.

There is a very simple relation between the Heisenberg uncertainty principle and the measurable magnetic flux difference $\Delta\Phi$ (e.g., Ref. [47]). Consider a film of thickness t, with in-plane magnetic induction $B_\perp$. According to the classical approach in Section 2.2, the momentum transfer p_y is then equal to $etB_\perp$. If the magnetization is not uniform but differs by an amount $\Delta B_\perp$ from a point y_1 to a point y_2, then the uncertainty principle states that $\Delta p_y \Delta y \geqslant h$. Combining this with $\Delta p_y = et\Delta B_\perp$ we find

$$\frac{t\Delta y\Delta B_\perp}{2} = \frac{\Delta\Phi}{2} \geqslant \frac{h}{2e} = \Phi_0.$$

This criterion tells us that quantum mechanical contrast effects are expected to become important when the flux difference between neighboring regions (the "flux inhomogeneity") is of the order of the flux quantum. If a precise measurement of the Lorentz deflection angle θ_L (or, equivalently, of the momentum transfer) is made, then it is impossible to measure also, on the same electron, where in the interval Δy the electron came from [47]. This limits the spatial range over which flux inhomogeneities (and therefore domain wall positions) can be measured by a single electron. One can use successive measurements on many electrons, however, to make measurements with any desired accuracy in both y and $\Delta B_\perp$; a detailed discussion of the relation between the "flux resolution" and the image exposure time can be found elsewhere [47].

2.3.2 Quantum Aspects of Fresnel and Foucault Observation Modes

The most readily observable quantum effect in Lorentz microscopy is the appearance of interference fringes at convergent wall images in the Fresnel mode (see Fig. 4). Experimental observations indicate that the number of fringes increases with increasing defocus, but the fringe spacing is constant. We now show that there is one fringe per fluxon $h/2e$ in the image, independent of defocus, magnification, and electron energy.

Consider again a film of thickness t, with an infinitely narrow domain wall centered at $y = 0$ (Fig. 10). A point source S at a distance s from the sample illuminates the area containing the domain wall. In Fresnel mode the image plane is located at a distance Δf below the sample. The Lorentz deflection of electrons passing through the domain on the left is $-\theta_L$, and at the image plane the electrons appear to originate from an effective source S', located at a distance $s\theta_L$ from the actual source (measured at the source plane). Similarly, electrons deflected by the angle $+\theta_L$ on the right-hand side appear to originate from the point source S''. The two sources will give rise to a region containing interference fringes with width δ_i. The angle subtended by the two sources at the image plane is equal to $\gamma = 2_s\theta_L/(s + \Delta f)$. The width of a single fringe in the image plane is then given by

$$\begin{aligned}\delta_\mathrm{i} &= \frac{\lambda}{\gamma} \\ &= \frac{h(s + \Delta f)}{2etsB_\perp},\end{aligned}$$

where we have made use of Eq. (7). Next we refer this fringe width back to the width δ_s in the sample plane:

$$\begin{aligned}\delta_\mathrm{s} &= \delta_\mathrm{i}\frac{s}{s + \Delta f} \\ &= \frac{h}{2etB_\perp} \\ &= \frac{\Phi_0}{tB_\perp}.\end{aligned}$$

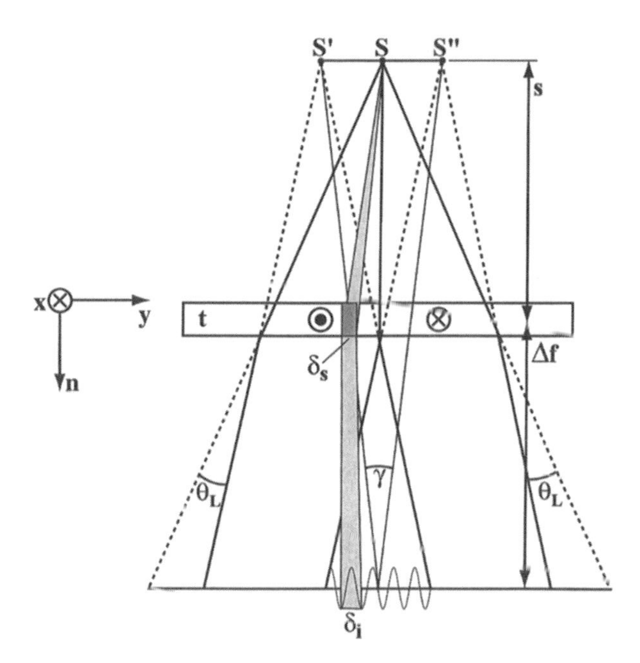

FIG. 10. Schematic illustration of the relation between a single fringe period in a Fresnel image (defocus Δf) and the corresponding region in the sample (gray rectangle). The flux through this region is equal to one flux quantum. (Based on a figure in Ref. [47].)

We find that $\delta_s t B_\perp = \Phi_0$, or that the flux contained in a region in the sample corresponding to a single projected fringe width in the image is equal to one flux quantum. The fringe width depends inversely on both film thickness and magnetic induction.

A similar argument can be made for the contrast fringes that appear in Foucault images. For an infinitely narrow wall the fringe spacing is again related to the flux quantum. For more realistic domain wall profiles the image contrast for both the Fresnel and Foucault imaging modes becomes more complex, and explicit numerical simulations are required to predict the intensity profiles of the corresponding interference fringes. An example of such a computation is discussed in Section 2.3.5. Before we enter this discussion we must first analyze how the microscope itself affects the image details.

2.3.3 Image Formation

In this section we describe the standard image formation theory for high resolution TEM work, on the basis of the microscope transfer function formalism. The description is somewhat superficial, and we refer the interested reader to one of the many excellent textbooks and review articles on the subject for more detailed information (e.g., Refs. [53, 55, 56]).

Consider a planar thin foil with thickness t, oriented normal to the electron beam. The incident beam is represented by a plane wave with wave vector $\mathbf{k} = \mathbf{n}/\lambda$, with λ being the relativistic electron wavelength. The thin foil sample modifies both the amplitude and phase of the incident wave, and the resulting wave function is

$$\psi(\mathbf{r}_\perp) = a(\mathbf{r}_\perp)e^{i\phi(\mathbf{r}_\perp)}, \tag{14}$$

where $a(\mathbf{r}_\perp)$ is the amplitude and $\phi(\mathbf{r}_\perp)$ is the phase of the wave function at the (in-plane) position $\mathbf{r}_\perp$. In the back focal plane of the objective lens (see Fig. 11), denoted by the in-plane reciprocal space frequency vector $\mathbf{q}_\perp$, the wave function is given by

$$\psi(\mathbf{q}_\perp) = \mathscr{F}[\psi(\mathbf{r}_\perp)]\mathscr{T}(\mathbf{q}_\perp),$$

and $\mathscr{T}$ is commonly known as the *microscope transfer function.* The symbol $\mathscr{F}$ represents the direct Fourier transform operator.

The transfer function can be decomposed into three separate factors (e.g., Ref. [55]):

$$\mathscr{T}(\mathbf{q}_\perp) = A(|\mathbf{q}_\perp|)e^{-i\chi(\mathbf{q}_\perp)}e^{-g(\mathbf{q}_\perp)} \tag{15}$$

where $A(\mathbf{q}_\perp)$ is the aperture function, $e^{-i\chi(\mathbf{q}_\perp)}$ is the phase transfer function, and $e^{-g(\mathbf{q}_\perp)}$ is the damping envelope. The characteristics of these factors are as follows:

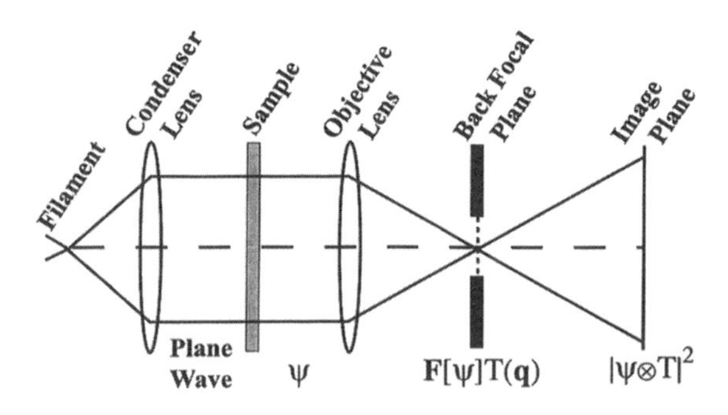

FIG. 11. Schematic illustration of the image formation process in a TEM.

1. The *aperture function* $A(\mathbf{q}_\perp)$ is equal to 1 inside the aperture and vanishes outside. For in-focus images one can vary the radius and position of the aperture through the aperture function $A(\mathbf{q}_\perp)$ to simulate the imaging conditions for the Foucault observation mode.
2. For the *phase transfer function* $e^{-i\chi(\mathbf{q}_\perp)}$ the phase shift is determined by the microscope defocus Δf and the spherical aberration C_s of the main imaging lens, and it is given by

$$\chi(\mathbf{q}_\perp) = \pi\lambda\Delta f\,|\mathbf{q}_\perp|^2 + \tfrac{1}{2}\pi C_s\lambda^3|\mathbf{q}_\perp|^4. \tag{16}$$

   It is assumed that objective lens astigmatism can be completely corrected. For Lorentz observations we can make use of the fact that the deflection angle is small compared to typical Bragg angles (i.e., $\theta_B/\theta_L \approx 100$). This in turn means that the contribution of spherical aberration, often the dominant contribution for high resolution TEM observations, is of no importance for Lorentz images (for a given C_s the phase shift owing to spherical aberration is roughly 10^{-6} times as large for a Lorentz deflected beam than for a Bragg diffracted beam). In fact, Lorentz lenses may well have spherical aberration coefficients in the range of hundreds of millimeters (or, indeed, several meters) without negatively affecting the image. The only factor remaining in the phase transfer function is then the microscope defocus Δf, which is the main variable for the Fresnel imaging mode.
3. The *damping envelope* $e^{-g(\mathbf{q}_\perp)}$ is determined mostly by the stability of lens currents and the accelerating voltage, and the main contributing factor (keeping only terms in $|\mathbf{q}_\perp|^2$) is due to the beam divergence θ_c:

$$g(\mathbf{q}_\perp) \approx \frac{(\pi\theta_c\Delta f)^2}{\ln 2}|\mathbf{q}_\perp|^2. \tag{17}$$

   The beam divergence angle is the angular half-width of the normalized Gaussian intensity distribution describing the electron source. Small values of θ_c describe more coherent electron sources.

We conclude that the microscope transfer function appropriate for Lorentz image simulations is given by

$$\mathcal{T}_L(\mathbf{q}_\perp) = A(|\mathbf{q}_\perp|)e^{-i\pi\lambda\Delta f|\mathbf{q}_\perp|^2}e^{-[(\pi\theta_c\Delta f)^2/\ln 2]|\mathbf{q}_\perp|^2}. \tag{18}$$

From the aberrated wave function in the back focal plane the image intensity can be derived by an inverse Fourier transform:

$$I(\mathbf{r}_\perp) = |\mathcal{F}^{-1}[\psi(\mathbf{q}_\perp)]|^2 = |\psi(\mathbf{r}_\perp) \otimes \mathcal{T}_L(\mathbf{r}_\perp)|^2, \tag{19}$$

where $\otimes$ indicates the convolution product and the function $\mathscr{T}_L(\mathbf{r}_\perp)$ is the *point spread function* appropriate for Lorentz microscopy.

2.3.4 Image Simulations: Example I

In this and the following section we illustrate the image simulation process using two examples of magnetization configurations: the uniformly magnetized sphere and a domain configuration typically observed in the magnetostrictive compound Terfenol-D. For the first example we make use of the explicit A–B equation to analytically compute both electrostatic and magnetic phase shifts. The second example makes use of a numerical method for the computation of the phase shifts. Furthermore, the first example addresses the application of Lorentz modes to the study of magnetic nanoparticles, an area of active research (e.g., Refs. [57–59]). The second example deals with a "bulk" magnetic system. The main ingredient for the computation of Lorentz images is the total A–B phase shift; once the phase shift is known, it is a trivial matter to compute Fresnel and/or Foucault images.

Before embarking on an experimental program involving Lorentz microscopy analysis of nanoparticles, one should attempt to answer the question, How small a magnetic moment can one detect with conventional Lorentz techniques? To address this question we use a simple model system: the uniformly magnetized sphere. The magnetic vector potential $\mathbf{A}$ of the system is available, and the simple geometry should also permit direct analytical evaluation of the electrostatic phase shift, thus, a closed-form expression for the strong phase object electron wave function can be used. More detailed computations can be found elsewhere [59].

The geometry of the problem is shown in the inset of Fig. 12a. The magnetic vector potential for the uniformly magnetized sphere can be found in the literature (e.g., Ref. [54, p. 197]), and, in a Cartesian reference frame, it is given by

$$\mathbf{A}(\mathbf{r}) = \frac{4\pi a^3}{3(r^3 > a^3)} M_0[y(\hat{\mathbf{x}} \sin\theta + \hat{\mathbf{z}} \cos\theta) - z\hat{\mathbf{y}}], \tag{20}$$

where $r = \sqrt{x^2 + y^2 + \ell^2}$ and the notation $(a > b)$ indicates that the larger of a and b should be used. The sphere with radius a and magnetization $\mathbf{M} = M_0(\sin\theta\hat{\mathbf{x}} + \cos\theta\hat{\mathbf{z}})$ is placed in the origin of the reference frame. The electron trajectory needed for the A–B phase shift integral is parameterized by the variable ℓ, and the trajectory intersects the plane $\ell = 0$ at the point $(x, y, 0)$.

Introducing normalized coordinates $\bar{r} = r/a$, $\bar{x} = x/a, \ldots,$ and using $M_0 = \frac{3}{8\pi}B$, with B being the saturation induction, we find after substitution

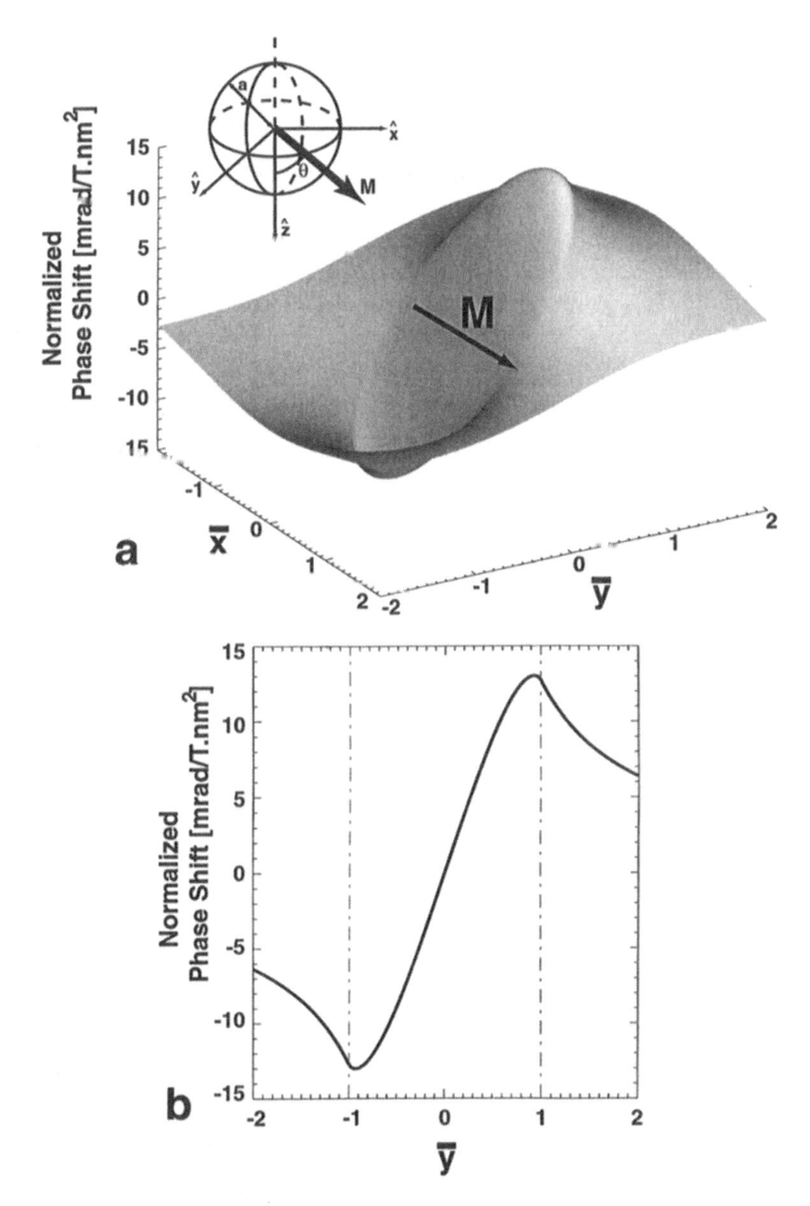

FIG. 12. (a) Magnetic phase shift for the uniformly magnetized sphere shown in the inset. (b) Profile across the phase shift surface (see text). (Reproduced with permission from Ref. [66], © the Royal Microscopical Society.)

into Eq. (10):

$$\bar{\phi}_{\mathrm{m}}(\mathbf{r}_\perp) = \frac{\phi_{\mathrm{m}}(\mathbf{r}_\perp)}{B_\perp a^2} = \frac{e}{\hbar}\frac{\bar{y}}{\bar{r}^2}\left\{1 - \beta\left(\bar{r}, \frac{3}{2}\right)\right\}, \tag{21}$$

where $\bar{r}^2 = \bar{x}^2 + \bar{y}^2$, $B_\perp$ is the in-plane component of **B**, and

$$\beta(r, q) = [1 - (r^2 < 1)]^q.$$

The function $\bar{\phi}_{\mathrm{m}}(\mathbf{r}_\perp)$ is shown as a shaded surface in Fig. 12a. A line trace along the line $(0, \bar{y})$ is shown in Fig. 12b. The trace is antisymmetric in the $\bar{y}$ coordinate. Note that this magnetic phase distribution includes contributions from the magnetic field outside the particle, the so-called fringe or demagnetization field, with a $1/r^3$ decay. If one would naively ignore this contribution, and only take into account the magnetization inside the spherical particle, then the resulting phase shift owing to the magnetic contribution would be too small by a factor of 2 at the center of the projected sphere, and by increasingly larger factors toward the outside of the sphere.

In addition to the magnetic A–B phase shift the transmitted beam electron will also experience a phase shift caused by the electrostatic mean inner potential V_s:

$$\bar{\phi}_{\mathrm{e}}(\mathbf{r}_\perp) = \frac{\phi_{\mathrm{e}}(\mathbf{r}_\perp)}{a\sigma V_s} = 2\beta(\bar{r}, \tfrac{1}{2}), \tag{22}$$

where σ is the interaction constant defined in Eq. (9). This function is radially symmetric around $\bar{r} = 0$.

If the spherical particle is embedded in a nonmagnetic medium of uniform thickness $2t$ and mean inner potential V_{m}, then the electrostatic phase shift would be given by

$$\bar{\phi}_{\mathrm{e}}(\mathbf{r}_\perp) = 2\left[\frac{V_{\mathrm{m}}}{V_s}\bar{t} + \left(1 - \frac{V_{\mathrm{m}}}{V_s}\right)\beta\left(\bar{r}, \frac{1}{2}\right)\right], \tag{23}$$

and for a particle lying on top of a support film of thickness $2t$, the phase shift would be

$$\bar{\phi}_{\mathrm{e}}(\mathbf{r}_\perp) = 2\left[\frac{V_{\mathrm{m}}}{V_s}\bar{t} + \beta\left(\bar{r}, \frac{1}{2}\right)\right]. \tag{24}$$

Assuming a pure strong phase object, the resulting exit wave function is

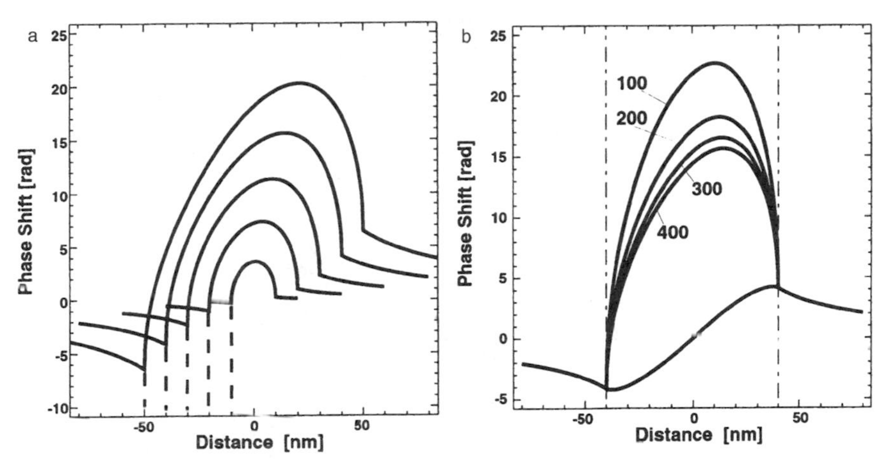

FIG. 13. (a) Total phase shift as a function of particle radius (other parameters as stated in the text). (b) Total phase shift versus microscope accelerating voltage for a 40-nm-radius, uniformly magnetized cobalt sphere. (Reproduced with permission from Ref. [66], © the Royal Microscopical Society.)

given by (using the quantum-mechanical sign convention)

$$\psi(\mathbf{r}_\perp) = e^{i[B_\perp a^2 \bar{\phi}_m(\mathbf{r}_\perp) + a\sigma V_s \bar{\phi}_e(\mathbf{r}_\perp)]}. \tag{25}$$

Note that the magnetic phase shift varies with the square of the particle radius, whereas the electrostatic contribution is linear in a. This behavior is illustrated in Fig. 13a, for the following cobalt material parameters: $V_0 =$ 29.46 V, 400 kV accelerating voltage, and particle radii from 10 to 50 nm. The saturation induction of cobalt is taken to be $B = 1.7$ T. The dependence on the microscope accelerating voltage is clearly displayed in Fig. 13b, for the same material parameters and a particle radius of 40 nm. Again, only the electrostatic part of the phase shift depends on the microscope accelerating voltage.

Next we can combine the analytical wave function [Eq. (25)] with the Lorentz transfer function [Eq. (18)] to compute Fresnel and Foucault images for a range of material parameters and particle sizes. Figures 14 and 15 show, respectively, Fresnel and Foucault image simulations for a set of five spherical cobalt particles of radii 5, 10, 15, 20, and 25 nm. The Fresnel images were computed for a 200 kV field emission microscope ($\theta_c = 10^{-5}$ rad) and the indicated defocus values. The particles in Fig. 14 were suspended in vacuum (left column), supported on a 20-nm-thick amorphous carbon film (center column), and embedded in a 40-nm-thick

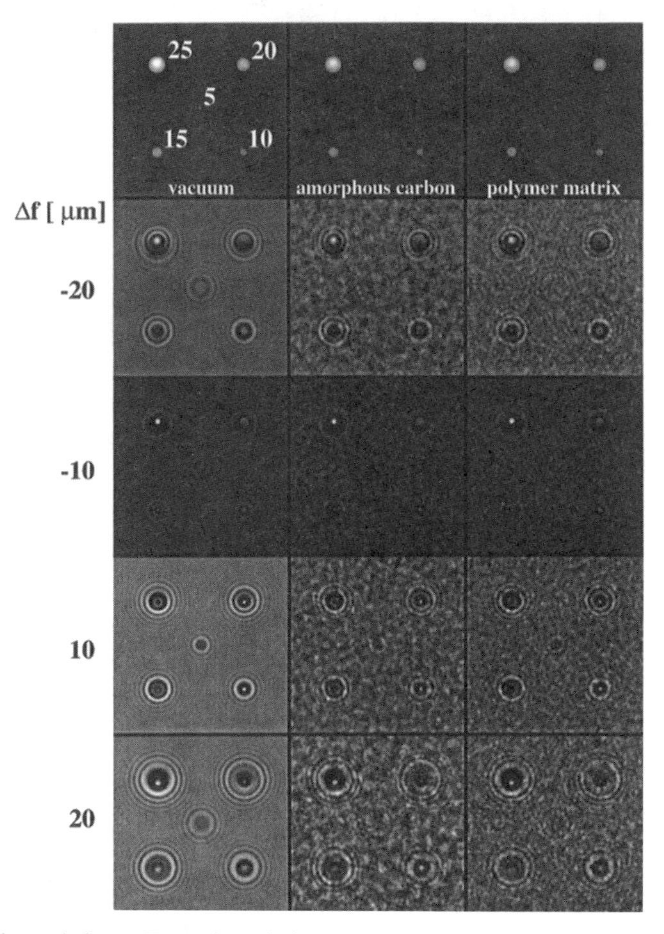

FIG. 14. Through-focus Fresnel mode images simulated for a 200 kV field emission microscope for five spherical cobalt particles with radii 25, 20, 15, 10, and 5 nm suspended in vacuum (left), on a 20-nm-thick amorphous carbon support (center), and embedded in a 40-nm-thick thermoplastic matrix (right). Other parameters are described in the text. (Reproduced with permission from Ref. [72].)

polymer matrix (right column) (e.g., Ref. [60]). The top row shows the calculated phase shift for all three cases. Both support media (amorphous carbon and thermoplastic matrix) have a density of about 2 g/cm^3 (i.e., about 100 carbon atoms per nm^3). Each atom position was randomly generated, and the atom's contribution to the electrostatic potential V_m was represented by a Gaussian disk. For a 5-nm film this implies the random generation of 131 million carbon atom locations for a field of view of

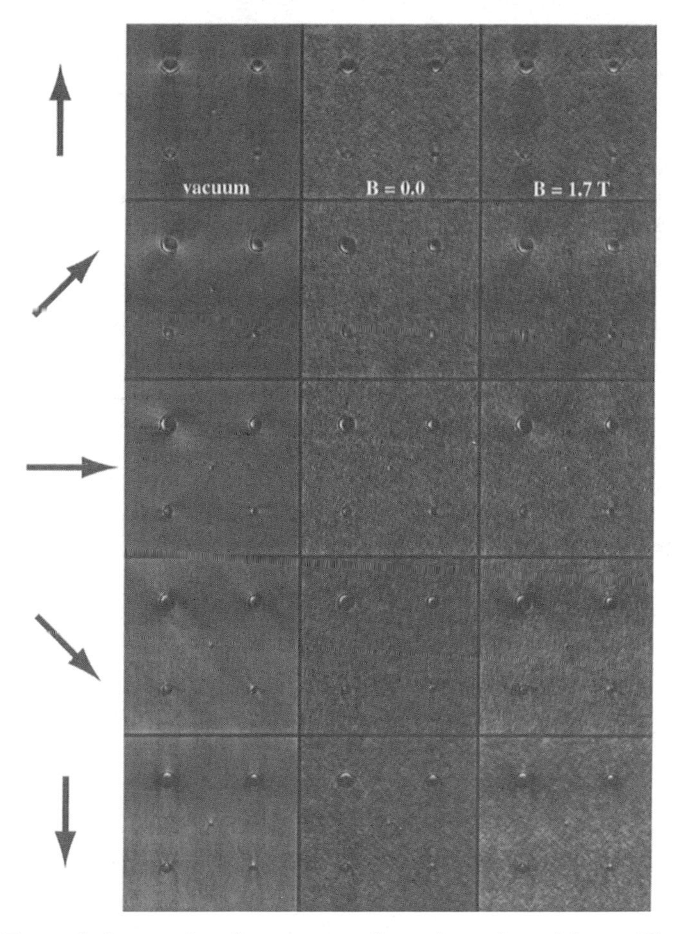

FIG. 15. Foucault images for the same configuration of particles as Fig. 14. The arrow indicates the direction of the aperture shift. The left column is for particles in vacuum; the other two are for a 40-nm thermoplastic embedding matrix. The center column does not have a magnetic contribution. (Reproduced with permission from Ref. [72].)

$512 \times 512\,\text{nm}^2$. The average potential of both films was set to the mean inner potential of graphite. Although the atom positions in the amorphous film were random (random walk with a step size of 0.15 nm), the thermoplastic model was generated by limiting the angular spread of bonds to 2° and using periodic boundary conditions.

The vacuum particles exhibit a clear off-center bright feature for all defocus values shown. This asymmetry is the signature of a magnetic contribution; a purely electrostatic potential would give rise to the same set of fringes, but symmetric around the center of the particle. For the smallest particles the asymmetry is only visible for much larger defocus values. For the particles suspended on or embedded in a support film, the contrast from the matrix itself overlaps with that of the particles, making it more difficult to separate out the magnetic component of the image contrast. The asymmetry is barely visible in the computed images of the 20-nm particle. Increasing the defocus to larger values will not be useful in this case, because the contrast from the support matrix will also become coarser.

For the Foucault images shown in Fig. 15 the defocus Δf is set to zero, so that the image contrast is entirely determined by the size and position of the aperture. For all simulations the radius was taken to be $0.3\,\mathrm{nm}^{-1}$, the same as the aperture size used for the Fresnel image simulations. The magnitude of the aperture shift is such that the aperture nearly touches the central undeflected beam, similar to what one would do experimentally. The image contrast consists of a bright off-center feature, which is mostly due to the electrostatic component (as can be seen from the center column, which has no magnetic contribution), and a typical double-lobe contrast feature outside the particle, which is associated with the magnetic information. The magnetic component of the Foucault contrast for the smallest particles is difficult to discern, even for the unsupported particles. In the presence of an embedding matrix the magnetic component of the image contrast may be clearly defined for the larger particles, but it becomes subtle or disappears completely for the smaller particles. All simulated images are astigmatism free; it is easy to show that the presence of residual image astigmatism can introduce contrast features which are rather similar to those due to the magnetic contribution. One must exercise extreme caution in interpreting both Fresnel and Foucault images of the smallest particle size range. It is also important to note that all simulations ignore the contribution of elastic scattering; in other words, all simulations use the *empty crystal approximation*, for which the crystal has only a mean inner potential. The image contrast owing to Bragg scattering of the lattice planes in the crystallites may be significant and may need to be taken into account in the simulations as well.

From Fig. 14 we find that there is a lower limit to the particle size for which the Fresnel imaging mode can produce useful magnetic image contrast. Although it is difficult to state a particular value for this lower limit (the actual value depends somewhat on the microscope parameters and on the support or embedding medium), it is not unreasonable to state that the lower limit is probably in the range of 10–20 nm particle radius for the cobalt particles used in the image simulations. For particles with a higher saturation induction, magnetic image contrast may be present down to smaller sizes, although the contrast from the support/embedding medium will always be present. One should be cautious in assigning magnetic origins to Fresnel type image contrast observed for particles in this size range or smaller. A similar remark holds for the Foucault observation mode, which is less reproducible than the Fresnel mode. Depending on the embedding medium and the magnitude of the in-plane component of the magnetic induction, one may be able to define a lower size limit. Most likely this limit will also be in the range of 10–20 nm radius particles. In Chapter 4 off-axis electron holography will be used to directly measure the phase distribution around small spherical cobalt particles (see Figs. 12–15 in Chapter 4). Electron holography combined with the theoretical modeling described in this section may be used to directly measure the electrostatic and magnetic image contributions. It can be shown [59] that electron holography can be reliably used for particle sizes where Fresnel and Foucault observation modes would fail.

2.3.5 Image Simulations: Example II

Magnetization configurations in thin films rarely have an analytical representation of either the vector potential or the magnetic induction. For such cases the magnetic phase shift imparted on the beam electron may be calculated by assuming that the magnetization has in-plane periodicity. It is usually not too difficult to create a periodic continuation of any given magnetization pattern; one can either pad the region of interest with a zero magnetization edge (making sure that the magnetization goes to zero smoothly as the edge is approached), or, if the spatial derivatives of the magnetization vanish at the edge, then one can readily use the pattern as a single unit cell. We will hence assume in this section that a two-dimensional (2-D) magnetization unit cell can be found, and to each pixel (i, j) in the cell we assign a three-dimensional (3-D) magnetization vector $\mathbf{M}(i, j)$. The magnetization is assumed to be constant along the beam direction.

It has been shown by Mansuripur [61] that for such a periodic configuration the magnetic A–B phase shift may be written as a 2-D Fourier series. We refer the reader to the original paper for the proof of this statement, and we continue by stating the explicit result. If the discrete Fourier components of the magnetization over a unit cell of $P \times Q$ pixels with pixel spacing D are represented by $\mathbf{M}_{mn}$, with $m = 1, \ldots, P$ and $n = 1, \ldots, Q$, then the A–B phase shift is given by the discrete 2-D Fourier transform:

$$\phi(\mathbf{r}_\perp) = \frac{e}{\hbar} \sum_{m=0}^{P} \sum_{n=0}^{Q}{}' \, \mathrm{i} \frac{t}{|\mathbf{q}_\perp|} G_{\mathbf{p}}(t, \mathbf{q}_\perp)(\hat{\mathbf{q}}_\perp \times \hat{\mathbf{z}}) \cdot [\mathbf{p} \times (\mathbf{p} \times \mathbf{M}_{mn})] e^{2\pi \mathrm{i} \mathbf{r}_\perp \cdot \mathbf{q}_\perp} \quad (26)$$

where the prime on the summation indicates that the term $(m, n) = (0, 0)$

does not contribute to the summation, $\mathbf{q}_\perp = \frac{m}{P}\mathbf{e}_x^* + \frac{n}{Q}e_y^*$ is the frequency vector, $\mathbf{e}_x^*$ and $\mathbf{e}_y^*$ are the reciprocal unit vectors, $\mathbf{p}$ is the beam direction expressed in the orthonormal reference frame, t is the sample thickness, a ^ indicates unit vectors, and $G_\mathbf{p}(t, \mathbf{q}_\perp)$ is given by

$$G_\mathbf{p}(t, \mathbf{q}_\perp) = \frac{1}{(\mathbf{p}\cdot\hat{\mathbf{q}}_\perp)^2 + (\mathbf{p}\cdot\hat{\mathbf{z}})^2} \frac{\sin\left(2\pi t \dfrac{\mathbf{p}\cdot\mathbf{q}_\perp}{\mathbf{p}\cdot\hat{\mathbf{z}}}\right)}{2\pi t \dfrac{\mathbf{p}\cdot\mathbf{q}_\perp}{\mathbf{p}\cdot\hat{\mathbf{z}}}} \tag{27}$$

For normal beam incidence ($\mathbf{p} \parallel \mathbf{z}$) the function $G_\mathbf{p}(t, \mathbf{q}_\perp)$ takes on the value 1. Numerical implementation of this equation is straightforward and, using fast Fourier transforms for a 512×512 array, takes less than 10 s on a Compaq Alpha 666 MHz platform.

Once the phase shift is known, the remainder of the image simulation proceeds along the same path as explained in the previous section, with the exception that the spatial frequencies at which the Lorentz transfer function $\mathscr{T}_L(\mathbf{q}_\perp)$ need to be evaluated are determined by the array spacing D. If the magnetization is not constant along the beam direction, then the above formalism can still be used, but one has to resort to a type of multislice method, where the entire thin foil is divided into thin slices normal to the incident beam; in each slice the Mansuripur equation is used to compute the phase, and a standard Fresnel propagator can be used to move from one slice to the next. This procedure has the added advantage that the phase shift arising from fringing or demagnetization fields may also be taken into account.

As an example of the simulation of Lorentz images for a periodic magnetization pattern, we shall use the compound Terfenol-D ($Te_{0.7}Dy_{0.3}Fe_2$), which has the cubic Laves phase structure, with $\langle 111 \rangle$ soft magnetic directions at room temperature. The magnetic domain walls in this system are 180° walls and also 71° walls, separating regions with different $\langle 111 \rangle$ oriented magnetizations. The magnetic domains coincide with rhombohedral distortion variants, and minimization of the magnetoelastic energy gives rise to regions (laminates) consisting of alternating lamellae of two magnetization directions, as described in detail by James and Kinderlehrer [62]. A typical magnetization pattern is shown in Fig. 16a; the region measures 512 nm on the side (sampled at 1 pixel per nm). The 71° domain wall (vertical wall) has a width of $\delta = 10$ nm, and the 180° zig-zag walls are Bloch walls of width $\delta = 16$ nm. The saturation induction of Terfenol-D is around 1 T, and a $t = 100$ nm foil was used for all simulations. For standard Fresnel and Foucault simulations the beam direction $\mathbf{p}$ is along the z-axis,

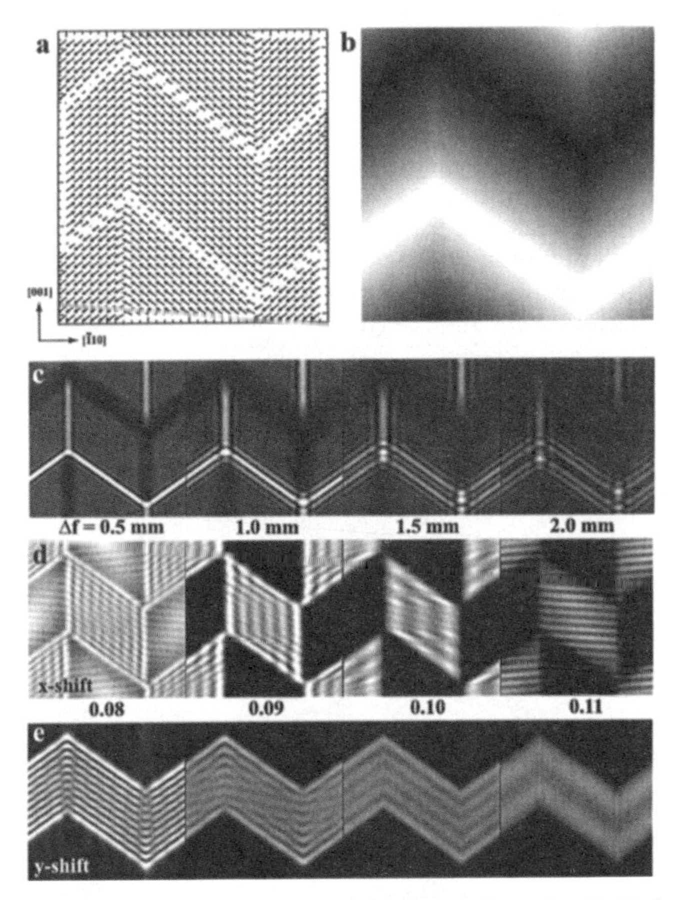

FIG. 16. (a) Schematic of a typical magnetization configuration in Terfenol-D. (b) Corresponding magnetic phase shift for a 100-nm-thick foil with a saturation induction of 1 T. (c) Simulated Fresnel focal series for a 400 kV microscope, with a beam divergence of 5×10^{-6} rad and a focal step size of 500 μm. (d,e). Simulated Foucault images for aperture shifts along the [$\bar{1}$10] (d) and [001] (e) directions. The aperture radius for all simulations is 0.1 nm^{-1}.

normal to the 2-D unit cell. All simulations were carried out for a microscope accelerating voltage of 400 kV. A uniform electrostatic potential was assumed, resulting in a constant phase shift across the unit cell; this electrostatic phase shift does not contribute to the image contrast and was ignored in all simulations. This is a good approximation inside the magnetic domains, but it may not be very accurate in the domain wall region; the discontinuity in the electrostatic potential at the domain wall (even for a wall of vanishingly small thickness) contributes to the image contrast in Fresnel mode and should be taken into account [48].

The first simulation step involves computation of the phase shift $\phi(\mathbf{r}_\perp)$ with Eq. (26). The resulting phase shift map is shown in Fig. 16b; the phase shift difference between the darkest and brightest pixels amounts to 30.0 rad ($\sim 9.55\pi$). Next we simulate a Fresnel focal series, for a 400 kV microscope with a beam divergence angle of 5×10^{-6} rad. The defocus step size is 500 μm, and the resulting images are shown in Fig. 16c. The fringes that appear for increasingly large defocus values have a constant spacing, as derived in Section 2.3.2, and the intensity profile agrees qualitatively with that observed for the Permalloy thin film in Fig. 4.

Foucault images for the same magnetization configuration are simulated by setting the defocus to zero and modifying the position of the diffraction aperture. The images shown in Fig. 16d,e correspond to aperture shifts along the $[\bar{1}10]$ and [001] directions, respectively. The magnitude of the aperture shift (in nm^{-1}) is noted. The aperture radius is 0.1 nm^{-1}. Note the presence of fringes in the bright domain regions; for a larger beam divergence angle, the fringes will disappear. The contrast depends strongly on the precise position of the aperture.

2.4 Magnetic Induction Mapping Methods

In Section 2.2.4 we saw that the Lorentz deflection owing to the in-plane magnetic induction component can be imaged as a spot splitting in the diffraction pattern. The patterns shown in Fig. 7 were obtained with a "regular" electron beam, and they contain the information of a rather large sample region. If one were to use a fine electron probe with a diameter in the nanometer range, then a measurement of the beam deflection as a function of probe position on the sample would provide a direct measure of the product $B_\perp t$. There are a few techniques capable of producing direct maps of the (in-plane) magnetic induction in a thin foil. Differential phase contrast (DPC) microscopy in a STEM [63] is perhaps the most successful method, and we briefly describe the technique here; a more detailed discussion can be found in Chapter 7. We also discuss a method equivalent to DPC that can be carried out on a regular TEM.

In a scanning TEM, the electron beam is focused into a fine probe that is scanned across the sample by means of scanning coils; these coils introduce a beam deflection that is superimposed on any Lorentz deflection caused by the sample. *Descanning coils* after the specimen (Fig. 17) can be aligned such that any tilt of the beam arising from the scanning coils is

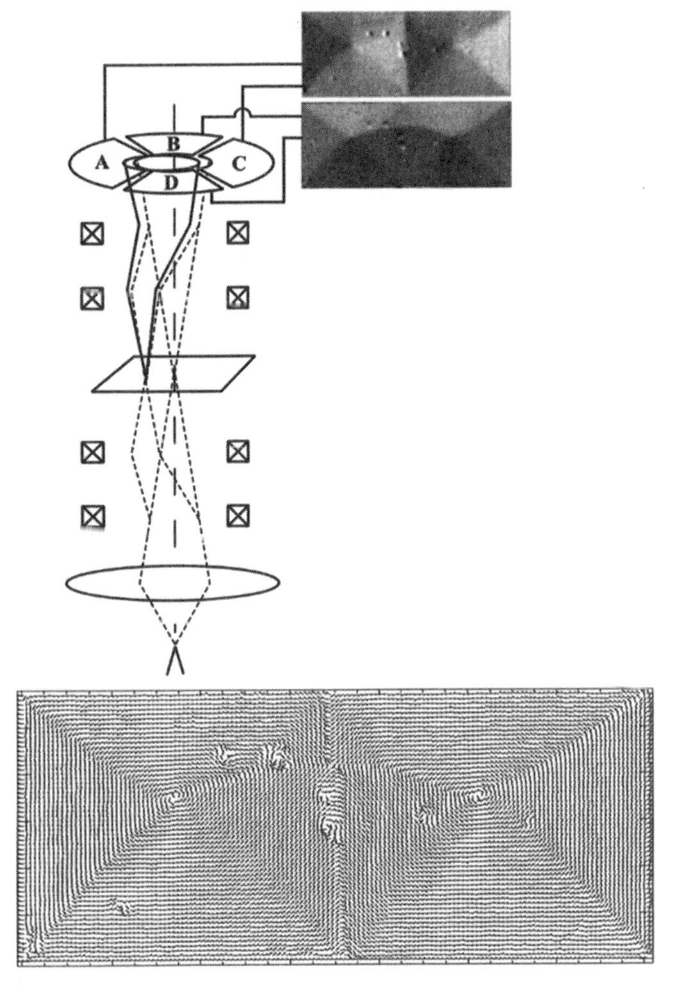

FIG. 17. Schematic representation of the DPC technique. The inset shows two experimental DPC images (courtesy of S. McVitie, University of Glasgow), and the corresponding vector map is also illustrated.

compensated for at the plane of the detector (dashed lines); any remaining shift of the beam over the detector is then due to the Lorentz deflection (solid lines). The detector typically consists of four quadrants with a central hole, and the difference signal measured between opposite quadrants is a direct measure of the Lorentz deflection. More complex detector geometries using eight quadrants have also been reported [64].

Figure 17 shows schematically how the descanning coils bring the incident probe (dashed line) back to the optical axis and through the hole in the detector, regardless of the position of the probe on the sample. The difference signals between the detector quadrants $A-C$ and $B-D$ are then monitored as a function of probe position. The insets near the top of Fig. 17 are actual experimental DPC images of a Permalloy element (82.5%Ni, 17.5%Fe), which has in-plane dimensions of $4.0 \times 1.5\ \mu m^2$. The DPC images were taken on a Philips CM20 as part of a magnetizing sequence, with an approximately zero component of in-plane field and an out-of-plane field of 110 Oe. The vector map was derived from the two DPC images by normalizing the intensities to a constant induction magnitude; two closure domains and a cross-tie wall in between are clearly visible. Because the thickness of this element is constant across the image, the DPC signal is directly proportional to the in-plane magnetic induction.

Magnetic induction mapping in the TEM was first introduced by Daykin and Petford-Long [65] and adapted to the Jeol 4000EX plus GIF combination by Dooley and De Graef [44]. A converged incident beam, with beam divergence angle θ_c, illuminates the region of interest on the sample. A small aperture truncates the transmitted beam in the back focal plane of the objective minilens. When the beam is tilted (using decoupled postspecimen image shift and tilt coils), the diffraction disk moves over the aperture. Images obtained with consecutive tilts are aligned and added together (approximating an integral with respect to beam tilt), and they constitute the equivalent of the quadrant detector signals in the STEM implementation of DPC.

Figure 18 shows four series of zero-loss Foucault images (slit width 15 eV), for two orthogonal tilt directions $\pm U$ and $\pm V$. The reader is referred elsewhere [66] for a full description of the calibration of the tilt directions with respect to the microstructure. After aligning the Foucault images in each tilt series, they are summed on a per-pixel basis. The sums are shown in Fig. 19a, along with the difference maps. It can be shown analytically [65] that the intensity in the difference image between two summed tilt series is proportional to the in-plane induction component in the orthogonal direction.

Using the Mansuripur algorithm, one can again compute the phase shift due to the magnetization pattern shown in Fig. 16a and vary the incident beam direction $\mathbf{p}$ to simulate the beam tilts used in the experiment. This is followed by a calculation of the corresponding Foucault images for a centered aperture, and the images are summed together for increasing beam tilt angles. The resulting sums and their differences are shown in Fig. 19b. There is good qualitative agreement between the outlined area in the calculated pattern and that in the experimental images. The simulation

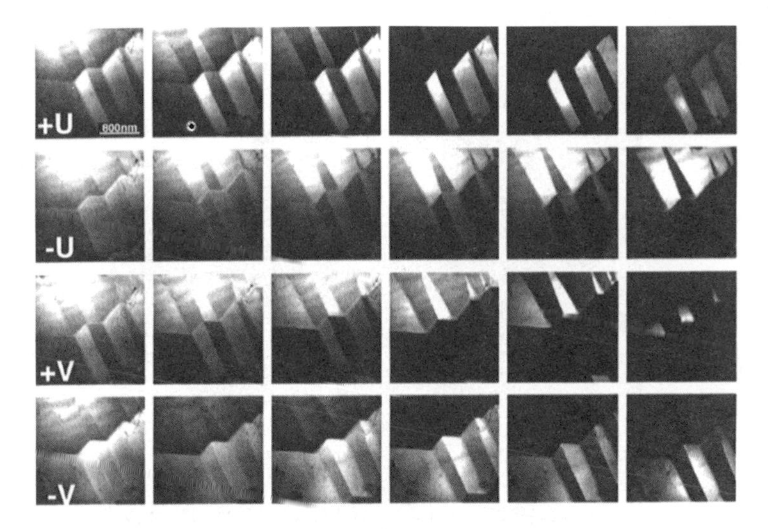

FIG. 18. Foucault images as a function of postspecimen beam tilt for two orthogonal tilt directions $\pm U$ and $\pm V$. (Reproduced with permission from Ref. [44], © Cell Press.)

parameters were identical to the ones used in Section 2.3.5, with $\theta_c = 1$ mrad, a beam tilt increment of 3.5 μm rad, aperture radius of 0.05 nm^{-1}, and an incident beam energy of 400 kV. Tilt directions in the simulation were taken to be the same as in the experiment.

2.5 The Transport-of-Intensity Equation

It is clear from the discussion in the preceding sections that the phase of the electron wave is the relevant quantity. Experimental determination of this phase is a nontrivial problem, as the phase information is lost in the image formation process. Once the phase function is known, then the magnetic induction can be determined simply by a gradient operation [Eq. (13)]. Direct determination of the phase information is thus highly desirable. In Chapter 4 McCartney *et al.* describe how the phase can be reconstructed through electron holographic methods. Barty, Paganin, and Nugent describe in Chapter 5 how noninterferometric techniques based on the so-called transport-of-intensity equation (TIE) can be used to accomplish the same. In this section we show how the TIE formalism relates to the transfer function formalism introduced in Section 2.3.3.

We start from the Lorentz transfer function [Eq. (18)]. For small deflection angles (i.e., small $|\mathbf{q}_\perp|$), and small defocus values Δf, both the

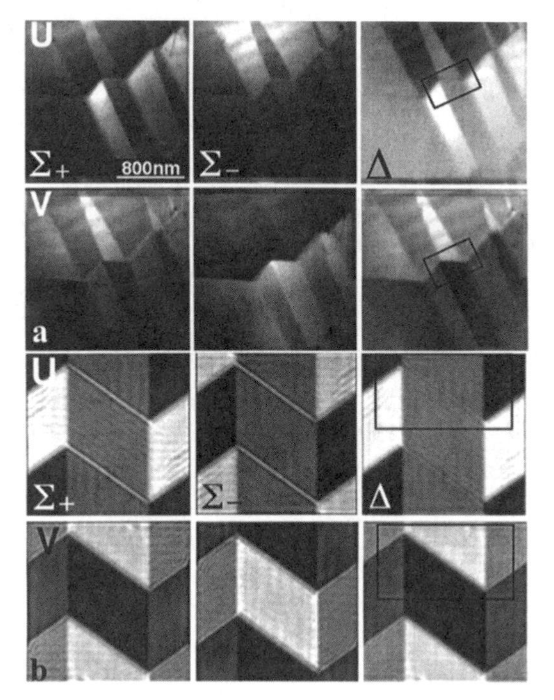

FIG. 19. (a) Experimental sums of the tilt series in Fig. 18. The difference images are shown in the rightmost column. (b) Calculated summed Foucault tilt series for the parameters stated in the text. The rectangular region outlined in the difference images should be compared with the rectangular area outlined in (a). (Reproduced with permission from Ref. [44], © Cell Press.)

phase transfer function and the damping envelope may be expanded in a Taylor series, and if we truncate the expansion after the quadratic term in $|\mathbf{q}_\perp|$ we find the following for the "paraxial" wave function:

$$\psi(\mathbf{q}_\perp) = \mathscr{F}[\psi(\mathbf{r}_\perp)](1 - z|\mathbf{q}_\perp|^2),$$

where $z = z_r + iz_i$ and

$$z_r = \frac{(\pi\theta_c\Delta f)^2}{\ln 2};$$

$$z_i = \pi\lambda\Delta f.$$

The wave function in the image plane is given by the inverse Fourier transform of this expression, which is [dropping the argument $\mathbf{r}_\perp$ in the exit wave function of Eq. (14) and writing $|\mathbf{q}_\perp| = q$]

$$\psi = ae^{i\phi} - z\mathcal{F}^{-1}[\mathcal{F}(ae^{i\phi})q^2].$$

Writing $\mathcal{F}[ae^{i\phi}] = f(\mathbf{q}_\perp)$ we have

$$\begin{aligned}
\mathcal{F}^{-1}[\mathcal{F}[ae^{i\phi}]q^2] &= \iint q^2 f(\mathbf{q}_\perp)e^{2\pi i\mathbf{q}_\perp\cdot\mathbf{r}_\perp}\, d\mathbf{q}_\perp \\
&= \frac{-1}{4\pi^2}\iint f(\mathbf{q}_\perp)\nabla^2 e^{2\pi i\mathbf{q}_\perp\cdot\mathbf{r}_\perp}\, d\mathbf{q}_\perp \\
&= \frac{-1}{4\pi^2}\nabla^2[\mathcal{F}^{-1}[f(\mathbf{q}_\perp)]] \\
&= \frac{-1}{4\pi^2}\nabla^2[ae^{i\phi}],
\end{aligned}$$

and therefore

$$\psi = ae^{i\phi} + \frac{z}{4\pi^2}\nabla^2[ae^{i\phi}].$$

The image intensity is then given by the modulus squared, and after a straightforward computation (ignoring terms of order λ^2) we find for the image intensity in the paraxial approximation and for small defocus values:

$$I = a^2 - \frac{\lambda\Delta f}{2\pi}\nabla\cdot(a^2\nabla\phi) + \frac{(\theta_c\Delta f)^2}{2\ln 2}[a\nabla^2 a - a^2(\nabla\phi)^2]. \tag{28}$$

Equation (28), which is valid for small defocus values, states that for a uniform background intensity (i.e., $\nabla^2 a = 0$), magnetic contrast will be visible only in those regions where the gradient of the phase $\nabla\phi$ is not constant, i.e., at the domain walls (in the absence of magnetization ripple). This justifies the use of extrapolations to zero defocus of divergent wall image widths.

The intensity a^2 is the in-focus intensity, which we represent by $I(\mathbf{r}_\perp, 0)$; the out-of-focus image is then given by $I(\mathbf{r}_\perp, \Delta f)$. Next we can derive an equation for the phase ϕ. Consider an underfocus and overfocus image, for the same defocus magnitude $|\Delta f|$:

$$I(\mathbf{r}_\perp, |\Delta f|) = I(\mathbf{r}_\perp, 0) - \frac{\lambda|\Delta f|}{2\pi}\nabla \cdot I(\mathbf{r}_\perp, 0)\nabla\phi]$$

$$+ \frac{(\theta_c|\Delta f|)^2}{2\ln 2}[\sqrt{I(\mathbf{r}_\perp, 0)}\,\nabla^2\sqrt{I(\mathbf{r}_\perp, 0)} - I(\mathbf{r}_\perp, 0)(\nabla\phi)^2],$$

$$I(\mathbf{r}_\perp, |\Delta f|) = I(\mathbf{r}_\perp, 0) + \frac{\lambda|f|}{2\pi}\nabla \cdot [I(\mathbf{r}_\perp, 0)\nabla\phi]$$

$$+ \frac{(\theta_c|\Delta f|)^2}{2\ln 2}[\sqrt{I(\mathbf{r}_\perp, 0)}\,\nabla^2\sqrt{I(\mathbf{r}_\perp, 0)} - I(\mathbf{r}_\perp, 0)(\nabla\phi)^2].$$

Subtracting the second from the first, and rearranging terms, we have

$$-\frac{2\pi}{\lambda}\frac{I(\mathbf{r}_\perp, |\Delta f|) - I(\mathbf{r}_\perp, -|\Delta f|)}{2|\Delta f|} = \nabla \cdot [I(\mathbf{r}_\perp, 0)\nabla\phi],$$

and in the limit of vanishingly small defocus we finally arrive at the so-called *transport-of-intensity equation* (TIE) [67–70]:

$$\nabla \cdot [I(\mathbf{r}_\perp, 0)\nabla\phi] = -\frac{2\pi}{\lambda}\frac{\partial I(\mathbf{r}_\perp, 0)}{\partial z}. \tag{29}$$

The name of this equation was first coined by Teague [71], who showed that this formalism could be applied to phase retrieval. Note that the beam divergence term is even in the defocus, which means that it cancels out in the difference between underfocus and overfocus images. The solution to this equation can be derived by introducing a new variable Ψ, such that

$$\nabla\Psi = I(\mathbf{r}_\perp, 0)\nabla\phi.$$

This turns the TIE into a standard inhomogeneous Poisson equation, which can be solved by means of Fourier transform methods:

$$\nabla \cdot \nabla\Psi = \nabla^2\Psi = -\frac{2\pi}{\lambda}\frac{\partial I}{\partial z}.$$

The gradient of the phase is then given by

$$\nabla\phi = \frac{\nabla\Psi}{I(\mathbf{r}_\perp, 0)} = -\frac{e}{\hbar}[\mathbf{B}(\mathbf{r}_\perp) \times \mathbf{n}]t(\mathbf{r}_\perp), \tag{30}$$

where $\mathbf{n}$ is the unit normal in the beam direction and t is the local sample thickness. The formal solution to the transport-of-intensity equation is derived in Chapter 5 [see Eqs. (18) through (22) in Chapter 5].

More details on this phase reconstruction method can be found in Chapter 5. The method is readily applied to any set of Fresnel images. Consider the square area outlined in Fig. 3c. From the corresponding areas in Fig. 3a,d and knowledge of the defocus step size Δf, we can compute the partial z-derivative and numerically solve the TIE [via Eq. (22) in Chapter 5] for the phase $\phi(\mathbf{r}_\perp)$. The resulting phase map is shown in Fig. 20a (gray

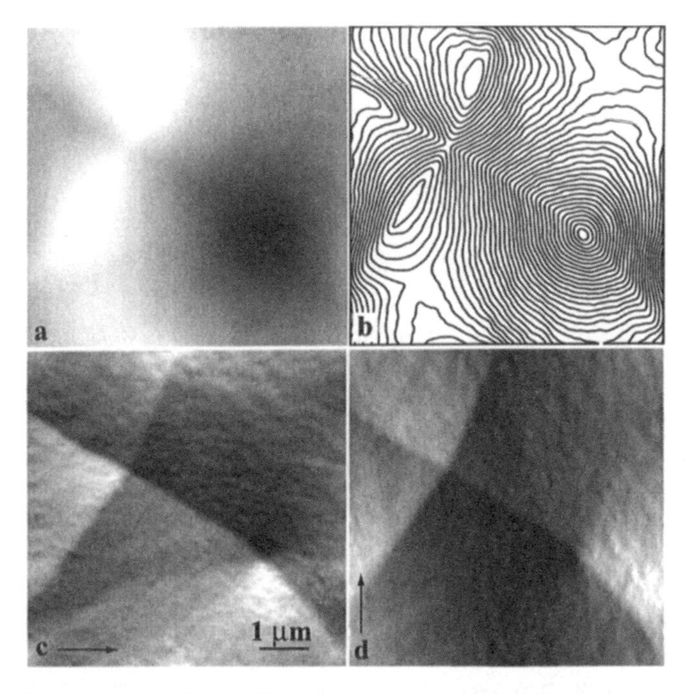

FIG. 20. (a) Reconstructed phase for the square area outlined in Fig. 3c. (b) Contour plot of the phase $\phi(\mathbf{r}_\perp)$; the vortex is clearly visible at the lower right. (c) B_x and (d) B_y maps, obtained by numerical differentiation of the phase. The maps are only qualitatively correct, because the microscope defocus was not calibrated.

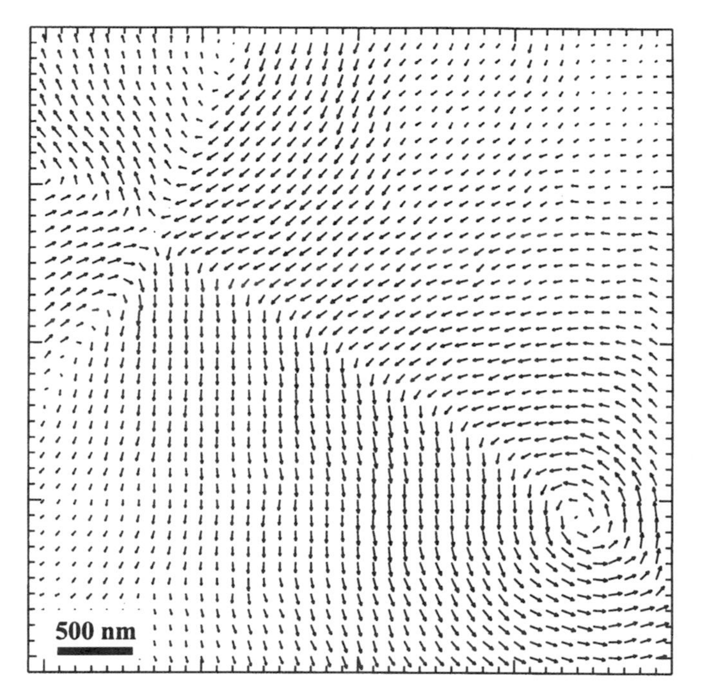

FIG. 21. Vector map derived from the component images in Fig. 20c,d. Each vector is an average over a 4×4 pixel square region (or $100 \times 100\ \text{nm}^2$).

scale) and Fig. 20b (contour map). Since the defocus was not calibrated, the phase map is only qualitatively correct. The vortex shows up clearly in the contour plot (Fig. 20b) as a conical feature, whereas the cross-tie wall corresponds to a saddle point in the phase function.

The in-plane magnetic induction can then be derived from the phase map by numerical differentiation, or from Eq. (30). The resulting maps are shown as gray scale images in Fig. 20c,d. The two maps can then be combined into a vector map (Fig. 21), where each arrow represents the average over a 4×4 pixel region. The vortex on the bottom right is clearly resolved, and the divergent nature of the cross-tie wall to the upper left is also apparent. Along the Bloch wall connecting the vortex to the cross-tie wall the vectors are somewhat shorter than away from the domain walls, reflecting the presence of an out-of-plane magnetization component along the Bloch wall.

Acknowledgments

The author would like to acknowledge stimulating discussions and/or electronic conversations with J. Dooley, N. T. Nuhfer, D. L. Laughlin, M. E. McHenry, J. Bain, C.-M. Park, J. H. Scott, Y. Zhu, D. Paganin, M. R. McCartney, J. Chapman, P. Aitchison, B. Ferrier, S. McVitie, and K. Kirk. This work was partially supported by the National Science Foundation under Grants DMR 9403621 and DMR 9501017 and by Rhône-Poulenc. Portions of this work have appeared as original papers: the author would like to acknowledge the Royal Microscopical Society for permission to reproduce Figures 1, 2, 4, 6, 12, and 13 from Ref. [66]; Cell Press for permission to reproduce Figures 1, 3, 4, and 5 from Ref. [44]; and the Materials Research Society for permission to reproduce Figures 3 and 5 from Ref. [72].

3. Recent Advances in Magnetic Force Microscopy: Quantification and *in Situ* Field Measurements

Romel D. Gomez

Department of Electrical and Computer Engineering
University of Maryland
College Park, Maryland
and Laboratory for Physical Sciences
College Park, Maryland

3.1 Introduction

Since its invention in 1987 magnetic force microscopy (MFM) has emerged as one of the most important imaging tools for surface magnetism. Its superb resolving capability combined with relatively lenient requirements of operation helped to make it a standard fixture in microscopy laboratories worldwide, and MFM has significantly impacted the diverse areas of magnetic recording technology, microelectronics, microelectromechanicals (MEMs) development, geology, chemistry, materials science, and biomedical applications. Excellent review articles, which thoroughly discuss the development of MFM and the relevant theoretical underpinnings, can be found in the literature [73–75]. Most of these articles are based on in-house fabricated instruments and thus contain the seminal ideas that led to today's implementation. The technique has undergone continuous refinement in both implementation and theoretical modeling in the 1990s, and commercial instruments have more or less converged to a standard practice. The typical MFM probe, for example, is a thin-film-coated silicon or silicon nitride cantilever, and the mechanical deflection is measured using a reflected optical light beam. Similarly, with the probe magnetized normal to the sample, the images are regarded as the distribution of magnetic charges on the sample.

The research trends in MFM instrument development have been toward quantification, resolution, and sensitivity enhancement and toward the adaptation to allow sample inspection in the presence of some external stimuli. These developments are fairly recent, and the primary aim of this chapter is to highlight this progress. To make this chapter self-contained, a summary of the relevant models is also presented as background material, along with the specific applications of imaging in the presence of a magnetic field.

EXPERIMENTAL METHODS IN THE PHYSICAL SCIENCES
Vol. 36
ISBN 0-12-475983-1

ISSN 1079-4042/01 $35.00

3.2 Review of Magnetic Force Microscopy

Magnetic force microscopy produces a two-dimensional image of the derivatives of the local magnetic field from the surface. The lateral resolution is typically 100 nm [76], although 25 nm and smaller magnetic features have been reported under certain conditions [77, 78]. MFM uses a cantilever probe containing a magnetic tip, which is oscillated near its resonance frequency and scanned at a constant height above the sample. The variation of the magnetic interaction at different regions of the sample changes the resonance frequency of the cantilever, which is manifested as an amplitude or phase change of the cantilever oscillation. The magnetic image is obtained by recording one or more of these parameters point by point as a function of the lateral position of the tip. With the aid of computer graphics, the raw data array is converted to a gray-scale or some pseudo 2-D plot to generate a two-dimensional rendition of the surface magnetic features. The contribution from the surface topography is often eliminated by first measuring it on the first pass and retracing the contour on the second pass as the MFM data are acquired [79].

Despite the simplicity of MFM, many factors influence the magnetic contrast, and one can obtain very different images of the same surface, depending on the height of the tip, the orientation of the probe and strength of the probe, the topography mapping, and so on. The key in accurate interpretation of the images is the precise understanding of the contrast mechanism as well as the anticipation of the various effects. A goal of this chapter is to provide an overview of magnetic force microscopy and to highlight some of the most useful models in image interpretation.

3.2.1 Surface Forces and the Force–Distance Curve

Let us begin by describing the interaction between the probe and the sample. As the tip is brought in close proximity to the sample, it experiences a variety of forces [80]. Effective on the atomic scale are ionic and Pauli repulsion, metallic adhesion, chemisorption, and physisorption, whereas on the nanometer length scales the interactions are dominated by van der Waals, capillary, and coulombic forces. For extended charge distributions, the long-range electrostatic and magnetostatic forces couple the tip and sample up to distances of hundreds of nanometers.

In general, the interaction is quite complex and depends on the specifics of the system such as the tip and sample composition, environment, contact area, tip–sample spacing, and electric and magnetic potential differences. The exposition of these effects is beyond the scope of this work, and the reader is referred to the excellent discussion by Israelachvili [81]. Nevertheless, for the purpose of understanding magnetic force microscopy, a reasonably good starting point is the empirical Lennard-Jones (L-J) potential

$$V(z) = \frac{a}{z^{12}} - \frac{b}{z^6}, \tag{1}$$

which yields

$$F(z) = -\frac{\partial V}{\partial z} = 12\frac{a}{z^{13}} - 6\frac{b}{z^7}, \tag{2}$$

where z is the distance between the tip and the surface. The second term dominates for large separation distance z, and the force in this regime is negative and vanishes asymptotically. This implies that the surface exerts an attractive interaction on the tip, which of course diminishes with distance. A plot of Eq. (2) is shown in Fig. 1, where we arbitrarily chose $a = b = 1$. At closer tip–sample spacing, the strength of the attractive force goes through a maximum (corresponding to the valley in the plot) and reverses direction. At even closer distances, the force crosses zero at $z_c \approx 1.24(a/b)^{1/6}$, and a further reduction in z leads to large positive values of the force, indicating stiff repulsion.

A straightforward experiment can be used to elucidate the model and to determine the parameters a and b in Eq. (1) for a specific system. In the force–distance experiment, the deflection of the cantilever is continuously measured as the tip is pushed against and pulled away from the surface. A typical result in air is shown in the lower part of Fig. 1. The abscissa is the amount of travel of the tip, z_t, while the ordinate is expressed either as the amount of deflection, z_d, or, equivalently, as the force of interaction since the spring constant is known. Starting from the right, the cantilever suffers no appreciable deflection for large tip–sample distances. However, as the tip approaches it bends toward the sample, which corresponds to a negative deflection. It reaches a maximum negative deflection at $z_t^{\max}$. However, further reduction in tip–sample spacing begins to deflect the tip in the opposite direction. The small dip represents the maximum attractive force in the L-J type potential, whereas the linear term represents the repulsive contribution at small distances. It is perhaps surprising at first that the tip appears to extend into the sample by as much as 80 nm from its value at the maximum of the attractive force. However, we note that z_t is the amount of tip travel and not the tip–surface distance, z. As the tip contacts the surface, it sees the huge L-J potential barrier. To relate the measurement with the model, one needs to make the translation $z = z_t - z_d$, and to do a best fit with the data using a and b as the adjustable parameters. We shall leave this an exercise for the interested reader; here we merely note that the slope of the linear term is nearly equal to the cantilever spring constant.

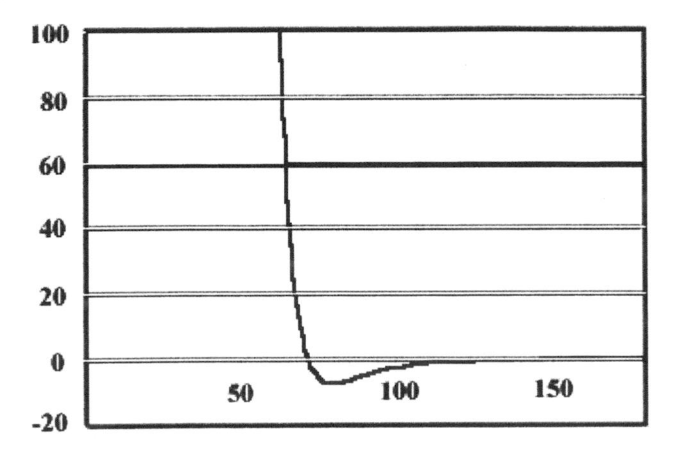

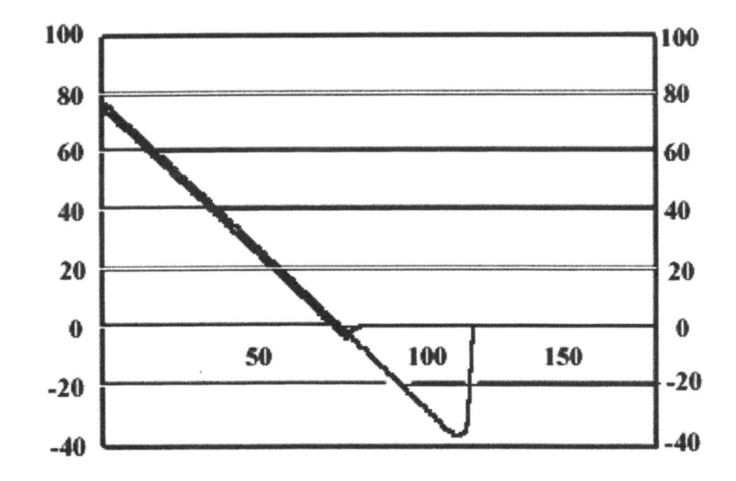

FIG. 1. Lennard-Jones potential and force–distance measurement.

An interesting artifact occurs when the tip is retracted. The retract curve continues to exhibit an attractive force far beyond z_t^{max}. The tip has to be substantially retracted before it breaks free of the attractive force and snaps back into the undeflected state. This behavior indicates that the tip has somehow adhered to the surface after making contact, which is possibly due to the adhesion force of the water layer. The liquid layer itself is on the order of a monolayer or several nanometers, but the capillary action on tip contact causes the liquid to form a meniscus at the tip. In this particular case, the tip has to retract by as much as 45 nm, corresponding to a force of about 21 nN to break free of the meniscus forces.

The above analysis has significant implications for performing MFM imaging. For instance, for any given height above the surface lower than 45 nm, the initial deflection of the probe and hence the resonance frequency are double valued and dependent on the manner in which the tip was placed at this height. Any quantification schemes, therefore, would have to be cognizant of this effect and should properly take them into account. An especially severe case happens when the height is fortuitously set at the snap point. In this case, the images would exhibit large variations that have nothing to do with magnetic forces. In some instruments, the user can define the initial height prior to the MFM scan, which makes it possible to choose one of the two branches for the entire image.

In the presence of magnetic or electrostatic interaction, the potential in Eq. (2) is modified to include these interactions. These coulombic forces vary as $1/z$, which is much longer range than the L-J potential. Similarly, at large separations, the z-components of the force gradient of the long-range forces (i.e., df/dz) are much more dominant. For these reasons, electrostatic force microscopy (EFM) and MFM measurements are normally operated with tips that are far (i.e., tens of nanometers) from the sample surface.

3.2.2 Topography and Force Gradient Measurement

Let us take a closer look at atomic force microscopy. A block diagram of a typical setup is shown in Fig. 2 (see color insert). The essential components are the cantilever probe, the deflection detector [82], the piezo actuators for sample lateral positioning (xy), the tip–sample fine position actuator (z), the tip–sample coarse positioner, the probe oscillator, and the control electronics for feedback and raster control. The atomic force microscope (AFM) can be used as a surface profilometer or force gradient mapper.

One way to map out the surface topography is by the so-called constant height mode. The surface undulations are obtained by dragging the tip across while in contact with the surface, and the interaction is repulsive. The recorded deflection of the probe constitutes the topographic image. As there is no feedback loop, the images can be acquired fairly rapidly (i.e., up to 30 frames per second). The disadvantage, however, is that the absence of an active control of the vertical position of the tip limits the usage to relatively smooth surfaces. Smooth surfaces do not produce cantilever excursions beyond the harmonic range, and they reduce the possibility of the tip crashing into a surface protrusion. An alternative method, known as the constant force mode, uses a feedback loop that adjusts the tip height to keep a preset deflection of the cantilever. The reference deflection or set point is equivalent to the amount of the loading force that the sensor is applying locally to the surface. Depending on the value of the set point, the AFM can be operated in the repulsive regime, where the tip is pushed against the surface, or in the attractive regime, where the tip is several tens of nanometers away from the surface.

Regardless of the execution, these techniques directly measure local forces. In principle, the same techniques can be used to measure the magnetic forces by using a magnetized probe [73, 83]. The difficulty, however, is that magnetic forces are much weaker than the adhesion forces, so that images are heavily contaminated by the topography. An alternative technique, using scanning tunneling microscope (STM) with a magnetic coated, rigid [84] or flexible freestanding thin film probe, improves the magnetic signal relative to the topography well enough to show distinct magnetic features on the surface [85, 86]. Although electrostatic and magnetostatic forces are weaker than the van der Waals type, their gradients are much larger [87]. Following the suggestion of the inventors of the AFM [88], this property was later exploited by others [89–93] by developing the tools for local force gradient detection. Force gradient detection has become the de facto method for MFM implementation, as it does not require conducting samples and the images are more immune to surface topography.

3.2.3 Force Gradient Detection

As mentioned earlier, an excellent alternative to mapping the long-range electrostatic and magnetostatic forces is the measurement of the derivative of the force or the force gradient. The technique is often referred as the alternating current (AC) "slope detection" [89, 90]. The cantilever is vibrated normal to the surface by means of a piezoelectric oscillator attached to the fixed end of the lever. The motion of the cantilever is detected as before, and the AC amplitude, frequency, or phase of the oscillation are measured. As force imaging, the instrument can be operated in the constant height or "constant force gradient" modes. The difference between the two is the presence of an active feedback loop in the constant gradient mode. In systems where the topography can be measured concurrently with the force gradient, the constant height mode is preferred. It has the advantage of faster data acquisition, without the detrimental effects of tip crash as the tip is held at a fixed height above the surface. The images, in this case, are made up of either the amplitude, frequency, or phase shift as a function of tip position.

The relationship between the local force gradients and the change in either of the oscillation parameters can be understood in terms of the equation of motion for a one-dimensional forced coupled harmonic oscillator with damping [89, 93, 94]:

$$m\ddot{z} + \gamma\dot{z} + \left[\frac{1}{m}\frac{\partial F(x,\ y)}{\partial z_0} - K\right] z = F_0 \cos \omega_d t, \tag{3}$$

where m is the cantilever effective mass, γ is the damping coefficient, F_0 is the amplitude of the forced oscillation, and ω_d is the driving frequency. The terms enclosed in the brackets, namely, the natural cantilever spring constant K and the effective spring constant of the tip–surface interaction, can be regarded as the equivalent spring constant of the system. The steady-state solution of Eq. (3) is well known:

$$z(x,\ y,\ t) = D \cos(\omega_R t - \delta), \tag{4}$$

with

$$\omega_R = \sqrt{\frac{K}{m} - \frac{1}{m}\frac{\partial F}{\partial z_0}} = \sqrt{\frac{K - F'}{m}}, \tag{5}$$

$$D = \frac{f_0/m}{\sqrt{(\omega_R^2 - \omega_d^2)^2 + 4\omega^2\beta^2}} = \frac{F_0/m}{\sqrt{(\omega_R^2 - \omega_d^2)^2 + \omega_R^2\omega_d^2/Q^2}}, \tag{6}$$

$$\delta = \cos^{-1}\left[\frac{\omega_R^2 - \omega_d^2}{\sqrt{(\omega_R^2 - \omega_d^2)^2 + \omega_R^2\omega_d^2/Q^2}}\right]. \tag{7}$$

The dependence on the lateral coordinates $(x,\ y)$ enters via the tip–sample interaction. When the system is driven at resonance, the amplitude and phase, respectively, reduce to

$$D_0 = \frac{F_0 Q}{m\omega_R^2}, \tag{8}$$

$$\delta_0 = \frac{\pi}{2}. \tag{9}$$

Under this condition, the force gradient can be derived by differentiating Eq. (5) and assuming that $K \leqslant \partial F/\partial z_0$:

$$\frac{\partial \omega_R}{\partial F'} = \frac{1}{m\omega_R} \approx \frac{\omega_R}{2K},$$

or

$$\delta\omega \approx \frac{\omega_R}{2K}\,\delta F'. \tag{10}$$

If one chooses to measure amplitude change, Martin *et al.* [89] have shown that the highest sensitivity measurements would yield

$$\Delta D = -\frac{2Q}{3\sqrt{3}K}\Delta F' D_0. \tag{11}$$

Similarly, oscillation phase can be related to the force gradient as

$$\Delta\delta = \frac{Q\Delta F'}{K}. \tag{12}$$

In arriving at Eq. (12), we recognize that the quality factor Q is of the order of 100–1000 and that the resonance frequency is in the tens of kilohertz, making the argument of the arcosine in Eq. (7) small. Equations (10)–(12) are the central equations in force gradient mapping, as they relate the measured quantities to the specific change in force gradients.

3.2.4 Theory of Magnetic Force Microscopy

Up to this point, the origin of the tip–surface force gradient F' is arbitrary. It can arise from electrostatic, magnetostatic, or even frictional forces. If the tip is magnetic, then the force on the lever is given by the negative of the gradient of the magnetostatic energy,

$$\mathbf{F} = -\nabla E = -\mu_0 \nabla \int_{V'} \mathbf{M}(\mathbf{R}') \otimes \mathbf{H}(\mathbf{R} + \mathbf{R}') dV'. \tag{13}$$

R is the vector from the origin of the field to the center of the magnetization and $\mathbf{R}'$ is the vector relative to the center of the tip magnetization vector $\mathbf{M}(\mathbf{R})$. Note that the mutual interaction is assumed to have no effect on the magnetic properties of either the probe or the sample. Strictly speaking this assumption is wrong by virtue of the hysteretic properties of magnetized bodies. Nevertheless, this is adequate as long as the local coercivity of the sample and the probe are 100 Oe or higher.

In component form, Eq. (13) is equivalent to

$$F_k = \mu_0 \int_{V'} \frac{\partial}{\partial x_k} M_i H_i \, \mathrm{d}V' = \mu_0 \int_{V'} M_i \frac{\partial H_i}{\partial x_k} \mathrm{d}V', \tag{14}$$

where we adopt the convention that the sum is over the repeated index and it runs from 1–3, denoting the Cartesian coordinate axes. The second equality is true since **M** does not depend on the unprimed coordinates. The components of the force gradient are obtained readily by differentiation with respect to x_l,

$$F'_{k,l} = \mu_0 \int_{V'} M_i \frac{\partial^2 H_i}{\partial x_l \partial x_k} \, dV'. \tag{15}$$

Equations (14) and (15) yield explicit expressions for the force and the force gradients along any given axis. They show that, in general, the cantilever will experience forces in all three directions, and the gradients contain mixed derivatives with respect to the coordinate variables.

Apart from special requirements of compactness and extra immunity to tip-crashes [95], most MFM applications vibrate the probe nominally along the z-axis, which means that only the z-component of the force gradient is sampled. The modeling situation is thus considerably simplified. Further, the cantilever is constrained to deflect only along the z-axis, which restricts the measured force to only be the z-component. In this case, the relevant equations reduce to [96]

$$F_z = \mu_0 \int_{V'} M_i \frac{\partial H_i}{\partial z} \, dV' = \mu_0 \int_{V'} \left(M_x \frac{\partial H_x}{\partial z} + M_y \frac{\partial H_y}{\partial z} + M_z \frac{\partial H_z}{\partial z} \right) dV', \tag{16}$$

$$F'_{z,z} = \mu_0 \int_{V'} M_i \frac{\partial^2 H_i}{\partial z^2} \, dV' = \mu_0 \int_{V'} \left(M_x \frac{\partial^2 H_x}{\partial z^2} + M_y \frac{\partial^2 H_y}{\partial z^2} + M_z \frac{\partial^2 H_z}{\partial z^2} \right) dV'. \tag{17}$$

Even with considerable simplifications, Eqs. (16) and (17) show that the force and force gradients are mixtures of the contributions of the z-derivatives of the x-, y-, and z-components of the surface field. The force is proportional to the first derivative, and the force gradient is proportional to the second derivatives of the individual components of the field. The relative contribution from each component is determined by the orientation of the magnetic moment of the probe.

Again, in practice, the probe is magnetized along a definite axis. This allows us to further simplify the force equations by setting two of the components of the magnetization to zero. In the case of z-axis alignment, the force gradient becomes

$$F'_{z,z} = \mu_0 \int_{V'} M_z \frac{\partial^2 H_z}{\partial z^2} \, dV'. \tag{18}$$

Equation (18) is the most widely used relation for MFM contrast, and it does a very good job of accounting for the qualitative image features.

However, this relation is far too simplistic to be used for quantitative analysis. In practice, there is no micromagnetic reason to support the assertion that the magnetization of the probe is uniform and directed exclusively along the z-axis. Likewise, the special condition of oscillation purely along the z-axis is seldom satisfied. Indeed, owing to instrument design restrictions, the probe is mounted at some angle α, typically 10° off the vertical axis. It is particularly troublesome that these contributions produce similar perturbations on the images, making it harder to develop unique image interpretation models. The analysis of Schonenberger and Alvarado [97, 98] has revealed some interesting consequences in three dimensions.

To illustrate this point, let us consider a two-dimensional model where we assume that the magnetization is solely along the z-axis, but the cantilever is tilted by an angle α. The force component normal to the axis of the cantilever is given by

$$\begin{aligned} F_{\rm n} &= F_x \sin\alpha + F_z \cos\alpha \\ &= \sin\alpha \int_{V'} \mu_0 \frac{\partial}{\partial x}[M_z H_z]\,{\rm d}V' + \cos\alpha \int_{V'} \mu_0 \frac{\partial}{\partial z}[M_z H_z]\,{\rm d}V'. \end{aligned} \tag{19}$$

Similarly, the force gradient operator along this axis is found by

$$\hat{\mathbf{n}} \times \nabla = \sin\alpha \frac{\partial}{\partial x} + \cos\alpha \frac{\partial}{\partial z}. \tag{20}$$

Operating on Eq. (19) and expanding the trigonometric terms for small angle α, one obtains

$$F_z' \approx \int_{V'} \mu_0 M_z \left[(1 + 2\alpha)\frac{\partial^2 H_z}{\partial z^2} + 2\alpha \frac{\partial^2 H_z}{\partial z \partial x}\right] {\rm d}V'. \tag{21}$$

But the field components are Hermitian pairs $\partial H_x/\partial z = \partial H_z/\partial x$, which allows us to write

$$F_z' \approx \int_{V'} \mu_0 M_z \left[(1 + 2\alpha)\frac{\partial^2 H_z}{\partial z^2} + 2\alpha \frac{\partial^2 H_x}{\partial z^2}\right] {\rm d}V'. \tag{22}$$

In comparison with Eq. (17), Eq. (22) produces the same force gradient as the case when the probe is vibrated exclusively along the z-axis, but with the magnetization components given by $M_z = (1 + 2\alpha)M_0$ and $M_x = 2\alpha M_0$. Hence, in general, quantitative analyses that attempt to characterize subtle MFM features would have to be sophisticated enough to discriminate between these effects.

3.2.5 Models for Probe Magnetization

For the force gradient expressions in the preceding section to be useful, it is necessary to have precise knowledge of $\mathbf{M(r)}$. A wide variety of MFM probes are in use today, including etched ferromagnetic wires, thin-film-coated Si tips, and individual magnetic clusters attached at the tips. To date, the descriptions of the magnetization for these tips are open questions, since direct measurements of the distributions have yet to be performed. For our purposes, it will suffice to consider a few widely used probe magnetization models and to understand the gross effects on the images.

3.2.5.1 Point–Dipole and Field Interaction Model. The simplest and most common model for the tip is the so-called point–dipole model [99]:

$$\mathbf{M(R)} = \mathbf{m}_0 \delta(\mathbf{R}), \tag{23}$$

where one assumes that the tip moment $\mathbf{m}_0$ is localized into a singular point somewhere in the tip. In this case, the integration over the volume of the tip becomes trivial, and Eqs. (16) and (17) simplify to

$$F_z = \mu_0 \left(M_x \frac{\partial H_x}{\partial z} + M_y \frac{\partial H_y}{\partial z} + M_z \frac{\partial H_z}{\partial z} \right), \tag{24}$$

$$F'_{z,z} = \mu_0 \left(M_x \frac{\partial^2 H_x}{\partial z^2} + M_y \frac{\partial^2 H_y}{\partial z^2} + M_z \frac{\partial^2 H_z}{\partial z^2} \right). \tag{25}$$

The point–dipole model yields a particularly intuitive way to interpret MFM images, as the images can be regarded as a mixture of the second derivatives with respect to z of the individual field components. Furthermore, it can be shown that near the surface the normal component of the magnetic field H_z, and its derivatives with respect to z, resemble $\nabla \times \mathbf{M}$ or $\mathbf{M} \times \mathbf{n}$. In other words, if the probe were exclusively along the z-axis, then the MFM images would correspond to the distribution of the magnetic charges on the surface. On the other hand, the in-plane field component H_x mimics the distribution of an in-plane magnetization. This is seen easily by modeling a typical longitudinal pattern with arctangent transitions [74]:

$$\mathbf{M}(x) = -\left(\frac{2M_r}{\pi}\right) \tan^{-1} \frac{x}{a}, \tag{26}$$

with a being the transition length parameter. The resulting fields for a single transition are

$$H_x = 4M_r \left[\tan^{-1}\left(\frac{x(\delta + z)}{x^2 + a^2 + a(\delta + z)}\right) - \tan^{-1}\left(\frac{xz}{x^2 + a^2 + az}\right)\right], \tag{27}$$

$$H_z = 2M_r \ln \left| \frac{x^2 + (\delta + z + a)^2}{x^2 + (z + a)^2} \right|. \tag{28}$$

We consider only two dimensions for brevity without loss of generality. For a repetitive pattern, the total field is a superposition of fields from an alternating pattern and is given as

$$\mathbf{H}_{\text{total}}(x, z) = \sum_n (-1)^n \mathbf{H}(x - nd, z). \tag{29}$$

Plotting Eqs. (27) and (28) is left as an exercise, and the reader is urged to demonstrate that indeed the assertions above are well justified.

This model was successfully used by Mamin *et al.* [100] to analyze their MFM images of periodic longitudinal recording patterns. The profiles they obtained were in excellent agreement with the data, provided that canting of the tip was properly accounted for and that the tip–sample distance was much greater than the experimental conditions. Later work by the same group resolved the inconsistency by accounting for the finite size of the probe.

3.2.5.2 Extended Tip Model: Dipolar and Monopolar Interaction. The main prediction of the point–dipole model is that the images correspond to the second derivative of the field. For extended tips, however, we shall see that this conclusion is not always the case. Various geometries have been studied in detail in the literature, including triangular, truncated pyramidal, and spherical structures, and they have been compared with experimental results. The models were highly specific for their probes, and the solutions were obtained numerically. As such, it is somewhat hard to derive a general physical intuition on the basis of these models. For the purposes of illustrating the effect of finite probes, we consider a simple case of a single-domain cylinder of length L, uniformly magnetized along the z-axis, and used to image a two-dimensional periodic longitudinal pattern. The scalar magnetic potential for the region above the medium is a solution of Laplace's equation [101] and must be of the form

$$V_m(x, z) = \sum_k [A_k \cos kx + B_k \sin kx] e^{-kz}, \tag{30}$$

where $k = 2\pi/\lambda$ and λ is the characteristic wavelength of the pattern. The fields are then given by

$$H_x(x, z) = -\sum_k k[A_k \sin kx - B_k \cos kx] e^{-kz}, \tag{31}$$

$$H_z(x, z) = -\sum_k k[A_k \cos kx + B_k \sin kx] e^{-kz}. \tag{32}$$

The coefficients A_k and B_k depend on the specifics of the transition model.

Using Eq. (18), the force can be written as

$$F_{z,z} = -\mu_0 \iint_{x',z'} M_z \frac{\partial}{\partial z} \sum_k k[A_k \cos k(x + x') + B_k \sin k(x + x')]e^{-k(z+z')}\,dx'\,dz'. \tag{33}$$

Integrating over z' from the tip–sample spacing (taken to be zero) to L (the domain length), we obtain

$$F_{z,z} = -\int_{x'} M_z \sum_k k[A_k \cos k(x + x') + B_k \sin k(x + x')]e^{-kz}(e^{-kL} - 1)\,dx'. \tag{34}$$

The characteristics of Eq. (34) change dramatically with the ratio of the characteristic wavelength of the pattern and the domain length L. In the short wavelength limit (i.e., $kL \gg 1$), one finds that

$$\begin{aligned} F_{z,z} &= \mu_0 \int_{x'} M_z \sum_k k[A_k \cos k(x + x') + B_k \sin k(x + x')]e^{-kz}\,dx' \\ &= -\mu_0 \int_{x'} H_z(x + x')\,dQ'_m. \end{aligned} \tag{35}$$

Q'_m of course is the effective charge on the apex of the tip. Following a similar analysis, the force gradient can be written as

$$\begin{aligned} F'_{z,z} &\approx \mu_0 \int_{x'} M_z \sum_k k^2[A_k \cos k(x + x') + B_k \sin k(x + x')]e^{-kz}\,dx' \\ &= -\mu_0 \int_{x'} \frac{\partial H_z(x + x')}{\partial z}\,dQ'_m. \end{aligned} \tag{36}$$

Thus, in the limit that the patterns have characteristic dimensions much less than the length of the tip domain, the force is proportional to the field, and the gradient is proportional to the derivative with respect to z. This implies that the tip appears as a point charge. Qualitatively, this is reasonable because the magnetic field of a periodic pattern of wavelength λ ($=2d$) decays as $e^{-2\pi z/\lambda}$. The field at the far end of the tip domain is reduced to $e^{-2\pi}$ relative to the end near the sample. Thus, the force is primarily due to the charges close to the sample. Figure 3 is a plot of the MFM profile using this model and using the data of Rugar *et al.* [74]. It shows the excellent agreement of the "point–charge" model for both the line charge approximation and the arctangent transition models. More importantly, we find excellent agreement for the effective separation, which is 0.8 μm, a realistic distance.

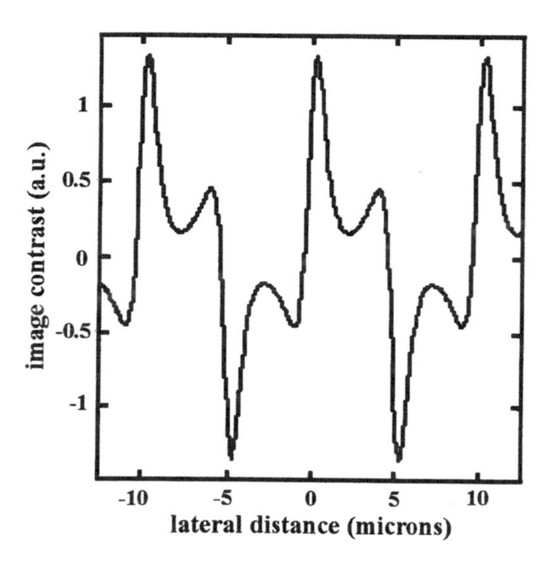

FIG. 3. Comparison of the point–charge model with the data of Rugar *et al.* [74].

Now, let us consider the other extreme, namely, the low frequency limit with $kL \ll 1$. In this case, e^{-kL} can be expanded in a Taylor series. To first-order expansion, Eq. (33) becomes

$$F_{z,z} = \mu_0 \int_{x'} M_r \sum_k k^2[A_k \cos kx + B_k \sin kx]e^{-kz}\,\mathrm{d}x', \tag{37}$$

and the force gradient is

$$F'_{z,z} = \mu_0 \int_{x'} M_r \sum_k k^3 L[A_k \cos kx + B_k \sin kx]e^{-kz}\,\mathrm{d}x'. \tag{38}$$

These equations are precisely those of the point–dipole model, where the force is proportional to the first derivative of the local field and the force gradient proportional to the second derivative.

In summary, the models provide a straightforward explanation of why MFM images of the same pattern can look very different when observed using different probes. The most relevant parameter is the effective length of the domain of the probe, which defines whether the patterns are in the high or low frequency regimes. In the high frequency regime the contrast arises from a monopole type interaction, whereas a dipolar interaction is expected for low frequency patterns. This result was first pointed out by Hartmann [102], who argued that in the high spatial frequency limit, when the probe dimensions are not much larger than the characteristic variation lengths of the microfield, the image formation is quite insensitive to the actual geometrical and micromagnetic fine structure of the probe. The predominant factor in image formation depends on the characteristic dimensions of the probe along the *x*, *y*, and *z* axes. Consequently, the quality of image simulation does not depend on the choice of more realistic, and often more complicated, models. Rather, it has been suggested that a model with simplified geometries should be employed. Conversely, in the low spatial frequency limit, the image contrast crucially depends on the geometrical configuration of the probe. This is very crucial in identifying and distinguishing small micromagnetic features such as Bloch or Néel domain walls.

A number of researchers have modeled various tip geometries that they believe accurately describe the specific probes they use. There are three basic classes of MFM probes, namely, electrochemically etched ferromagnetic wires, thin-film-coated nonmagnetic tips, and mechanically attached magnetic structures. Schonenberger and Alvarado [97] calculated the response function for both a constantly magnetized tip and a flux-conserving truncated cone and found that the high frequency roll-off is determined by the tip radius, and sets in appreciably for $k > 2R$. The single-domain tip acts like a band-pass filter with a low frequency cutoff that depends on L. In contrast, the flux-conserving tip of the same aspect ratio exhibits a flat response up to the lowest frequencies. Wadas and Grutter [103] considered the case of perfectly sharp transitions of vertically magnetized thin film imaged using a truncated pyramidal probe of interior taper angle α. They found that the image strongly depends on the orientation and that the highest resolution is not related to the tip with the smallest tip radius. For any specific height above the surface and for any aspect ratio, there exists a critical angle α_c for achieving the highest resolution, which could vary by up to a factor of 3 for different values of α.

3.2.5.3 Resolution and Sensitivity Limits. From the preceding analysis, it is clear that estimates of resolution and sensitivity are system dependent. However, without resorting to specifics, it is possible to derive some intuitive general conclusions.

1. Resolution and sensitivity cannot be maximized simultaneously. Sensitivity requires a large interaction force, which requires a large interacting volume. Resolution is inversely proportional to the active volume. A corollary follows: the best resolution can be obtained only for systems that have large gradients.
2. Although a point–dipole tip, in principle, provides better resolution than a point–charge tip, a true point–dipole tip cannot be realized. The dipole field falls off as R^{-3}, whereas a point–charge falls off as R^{-2}. This is

responsible for the enhanced resolution. However, by its very nature, the dipole approximation is valid as long as the separation between the charges is small in comparison with the distance of observation. Rugar *et al.* [74] numerically analyzed the resolution of etched wires with conical tips as a function of the tip length/tip–sample distance and the radius of the tip. They found that for point dipoles, the ultimate full width at half-maximum is one-half the tip–sample spacing, and this could be achieved by using very short tips ($L_t/z_0 < 0.01$). In typical MFM applications, the tip–sample spacing is of the order of 25–50 nm, so that, strictly speaking, the point–dipole response is realizable only for $L_t < 5$ Å. A magnetic cluster of this size would most likely be superparamagnetic.

3.2.5.4 Estimate of the Extent of the Probe Field.

A final point that deserves some discussion is an estimate of the field from the probe. For this purpose, it is convenient to regard the tip as a uniformly magnetized cylinder of length L, radius R, and at a distance z from the surface. The field strength at the axis is given by

$$H = \frac{M_0}{2}\left[\frac{L}{\sqrt{L^2 + R^2}} - \frac{z}{\sqrt{z^2 + R^2}}\right]. \tag{39}$$

For a nickel tip of saturation magnetization $M_0 = 450$ emu/cm^{-3} (Ni), with a radius of 50 nm and $L \gg z$, the stray field is about 160 Oe at distance of 30 nm from the surface. If a thin film were used instead, one can estimate the field as a superposition of two concentric cylinders of opposite magnetization and varying only in R and z by an amount d, the film thickness. In this case, the equation becomes

$$H = \frac{M_0}{2}\left[\frac{z + d/2}{\sqrt{(z + d/2)^2 + R^2}} - \frac{z - d/2}{\sqrt{(z - d/2)^2 + R^2}}\right]. \tag{40}$$

Taking the film thickness as 10 nm and using $M_0 = 500$ emu/cm^3 (CoPtCr), then the field reduces to 21 Oe at the same conditions. For thin films, there is an added reduction in the intensity with lateral distance, as the thin film is effectively a dipole layer and thus varies as x^{-3}. More accurate numerical calculations using similar values show that the field from a thin film tip had diminished to less than 2 Oe for a lateral distance of 200 nm, whereas a conical wire probe produces in excess of 70 Oe at 200 nm and 20 Oe at 2.5 μm [104]. Numerical calculations of Bryant *et al.* for axially symmetric geometry showed similar characteristics between solid and thin film probes, although the values they obtained were roughly 10 times larger than the previous estimates [105]. Streblechenko *et al.* [106] used electron holography techniques to derive the field profile and magnitudes near conventional MFM tips. From the reconstructed images, they were able to calculate the various multipole terms. The calculated field was of the order of 620 Oe at the tip surface and 310 Oe at 10 nm from the tip.

The field distributions from tips have been studied by several groups using electron microscopies. Frost *et al.* [107] used off-axis electron holography to reveal the magnetic lines of force of commercial pyramidal thin-film-coated tips. The reconstructed images dramatically show the return lines at the tip edges and lines that emanate from the apex of the tips. Earlier, Sueoka *et al.* [108] used Lorentz transmission electron microscopy (LTEM) to obtain a comparative study of etched and coated tips. They found that the etched CoFeSiB tips were single domain and produced fields generated by a monopole. With Permalloy-coated tips, on the other hand, they showed complex magnetic structures with small domains. The etched tips had the advantage of very high aspect ratios that facilitated the formation of large domains. This was carried further by Ruhrig *et al.* [109], who in 1996 formed 150-nm-wide needles using electron beam deposition (EBD) and coated them with CoNi alloy. They showed by using Foucault LTEM imaging a monopolar stray field distribution at the apex. More importantly, they successfully used these tips to obtain high resolution MFM images. Work by others [110] on these so-called EBD tips yielded resolution claims of the order of 35 nm [111].

By realizing that the field from the probe is nonzero, one is confronted with the particularly difficult issue of probe-induced sample magnetization distortion. A dramatic example of this effect was demonstrated in NiFe islands using a high moment iron probe [112]. The domain walls were displaced as the tip was rastered across the surface, and the effect was aggravated as the tip–sample distance was reduced. Many others observed similar phenomena as manifested by abrupt Barkhausen jumps in the image [113], spontaneous contrast reversal in the images [114], or enhanced contrast in the dissipation image [115]. To fully appreciate the influence of the stray field from the tip, one has to model the interaction along with the instrument response. In general, the field from the tip is of the order of several tens of Oersteds, and the effect would consequently be most prominent for soft samples [116].

3.2.6 Image Reconstruction

The next step in interpreting MFM images is to reconstruct the actual distribution of the magnetization of the surface on the basis of the force or force gradient measurements. Several approaches have been reported in the literature, and here we provide a brief summary. Hartmann was the first to show a semiquantitative analysis of the microfield of a Bloch wall [99]. His approach was to start with the simplified microfield components of a Bloch wall, which formed the basis for calculating the "compliance" or force gradient response. He further pointed out that the microfield profiles can be determined, in principle, if the magnetic probe configuration is known in detail. Since that time, well-defined MFM probes have been developed, and work has been performed under the specific assumptions of the probe. Chang *et al.* [18] treated MFM images as essentially distributions of the magnetic charges, which are convoluted with the instrument response function. The instrument response function was obtained by the inverse Fourier transformation of the image of a long, narrow nickel strip, which was assumed to represent localized point charges at the ends. Mayergoyz *et al.* [116a] realized that the MFM by its very nature was sensitive to the free producing or curl-free component of the magnetization. They argued that the divergence-free component is essentially undetectable by any force-sensing probe. However, by assuming that the magnitude of the individual local dipole moments is constant and that the vectors are planar, they found an expression to recover the divergence-free part. They employed a two dimensional Fourier transform to model the instrument response function and imaged a small cobalt island to extract it. The magnetization vectors of magnetic recording tracks were successfully reconstructed by their routine, including the transverse components at the track edges. More recently, Hug *et al.* [116b] presented an extensive transfer function approach to calculate the force of the MFM tip and the stray field that is due to a perpendicularly magnetized medium having an arbitrary magnetization pattern. Unlike the earlier methods, the extended magnetization of the tip was incorporated in the analysis. They calculated the two-dimensional Fourier transform of the surface charges on the top and bottom surfaces of the thin film, which are the so-called field transfer functions. The microfields can be obtained from the inverse transforms of these functions. They then considered the force as the convolution of the surface and volume charge densities of the tip with the microfields from the surface.

3.3 Quantification Techniques: Models and Experiments

The previous sections give us some relations that provide physical intuition and qualitative interpretation of MFM images to deduce the magnetic state of the sample. The MFM images can be used in conjunction with established models of the sample magnetization to derive quantities such as transition lengths of recording patterns, Neel or Bloch domain wall widths [102], and anisotropy. Because these quantities depend only on the spatial coordinates, it is unnecessary to know the absolute magnitude of the interaction force.

The difficulty, however, arises when one attempts to quantify parameters, such as the absolute magnitude of the field gradients, that depend on the accurate measurement of the force gradients. For the relations in the preceding section to be useful for quantification, the intrinsic spring constant of the cantilever, the quality factor Q, and the effective moment of the probe must be known. Several research groups propose calibration schemes or provide estimates for the tip–sample interaction force. Several researchers propose the imaging of a standardized system, such as a metal strip [117–120] or single-crystal surfaces [121], to determine the magnetic moment of the probe; they also use standard formulas to estimate the spring constant of the cantilever and use the resonance (amplitude versus driving frequency) curve to measure the quality factor Q. Heydon *et al.* [122] used torque magnetometry to measure the moment of the probe, and Proksch *et al.* [123] used a modified flux–gate type probe to directly measure the magnetic field of the samples. There are, however, no methods that prescribe a self-contained calibration procedure to determine the mechanical and magnetic characteristics of the probe by utilizing only the measurements of the instrument itself.

In this section a straightforward method for calibrating the MFM is described [124]. It is particularly instructive because the approach also demonstrates electrostatic force microscopy (EFM). On the practical side, the procedure lays out a method for distinguishing electrically continuous or broken connections in narrow electrical strips, which are used, for example, in VLSI circuits or in electromigration failure analysis.

The procedure is as follows. Consider a nonmagnetic conducting strip, which is electrically wired so that it can either be biased relative to the probe or allow current to flow along the strip. In the former case, the interaction force is purely electrostatic; in the latter, the force is predominantly magnetic [125]. Figure 4 (see color insert) shows two force gradient scans of the same 10-μm-wide strip. The only difference between them is the electrical connection. On the left-hand image, a direct current (DC) of 11 mA is allowed to flow along the strip, and the probe senses the resulting magnetic field gradients. Premagnetization of the probe along the normal direction, as discussed in the previous section, yields images that are sensitive to the normal component of the field. This explains the confinement of the contrast along the edges of the strip, which is visible from the average line profile. On the right-hand image in Fig. 4, one of the electrodes was removed from the strip and connected to the probe. A potential difference was established between the tip and the sample. The image shows dark contrast above the strip, because the electrostatic force is purely attractive.

3.3.1 Determination of the Mechanical Parameter Using EFM

Regardless of the connection, we use Eq. (12), which we rewrite in the form

$$\Delta\phi = K_p \frac{\partial F_z}{\partial z}. \tag{41}$$

We combined the ratio of the quality factor and the spring constant in one parameter K_p. This is valid as long as the scanning probe microscope (SPM) is operated in a linear regime.

To determine K_p, an EFM image of the test sample is obtained by biasing the metal structure at a voltage V relative to the probe at ground. Note that a regular MFM probe has been used, because the magnetic coating is also electrically conducting. The attractive electrostatic force is generated on the tip from the coulombic attraction of the charge on the sample and its image charge on the other side of the grounded metallic probe.

The electrostatic force is a function of the bias voltage V and capacitance C and is given by

$$\frac{\partial F_z^e}{\partial z} = -\frac{V^2}{2}\frac{\partial^2 C}{\partial z^2}. \tag{42}$$

At any given voltage V, the spatial variation depends only on the capacitance. Moreover, Eq. (42) specifies that the force gradient, and thus the image contrast, should vary as V^2. Figure 5 (see color insert) shows the acquired images at a constant height above the surface at different bias voltages. The contrast is artificially reversed to emphasize the changing magnitude of the force gradient. From the images, a plot of the maximum phase contrast versus V is derived and shown at the bottom of Fig. 5. The quadratic dependence of the contrast with the voltage is readily apparent from this plot. A least squares fit to a second-order polynomial yields the quadratic coefficient equal to 0.17. This number, which we denote as $(\Delta\phi/V^2)_{\mathrm{meas}}$, is one of the calibration constants. By using Eqs. (41) and (42), we find that it is equal to the product $K_p/2 \times \partial^2 C/\partial z^2$.

The next step in finding K_p is to obtain a relation for the capacitance of the system as a function of z. For this purpose, we recognize that the capacitance between the probe tip and the surface can be regarded as a finite area plate capacitor. One can, of course, solve the capacitance numerically, using a variety of tools for any given geometry, and subsequently obtain a definite value for the second derivative of C with respect to z, evaluated at the tip–sample height. However, it is instructive to use the Schwartz transformation, a standard technique used to solve finite square capacitance. The solution for this problem is well known [126], and twice differentiation with respect to z and evaluating at $z = h$, the tip–sample height, yields

$$\frac{\partial^2 C}{\partial h^2} = 2\varepsilon_0 A\left(\frac{1}{h^3} + \frac{2}{\pi^{3/2} R h^2}\right), \tag{43}$$

where A is the effective cross-sectional area of the tip and R is the tip radius. The h^{-3} dependence comes from the $1/h$ "infinite area" capacitance term, whereas the h^{-2} term comes from a logarithmic term from the fringe field contribution. An additional set of measurements that supports the validity of Eq. (43) was the dependence of the contrast image with the tip height, h. The log–log plot demonstrated the 2 power law dependence on h.

We can then combine this expression with Eqs. (41) and (42) to extract an expression for K_p:

$$K_p = \frac{2(|\Delta\phi|/V^2)_{\text{meas}}}{2\varepsilon_0 A(1/h^3 + 2/\pi^{3/2} R h^2)}. \tag{44}$$

Using the effective area $A = \pi R^2$, together with the dimensions of the probe obtained from scanning electron microscopy (SEM) measurement, $R = 40$ nm, and the predefined lift height, $h = 50$ nm, we find the mechanical calibration for this probe to be $K_p = 369$ degrees/N/m.

3.3.2 Determination of the Probe Moment Using MFM

Using the mechanical proportionality constant, we can calculate the effective probe moment along the z-direction. For simplicity, we assume that the probe is oriented along the z-axis and so is the cantilever deflection. Using the point–dipole approximation, we obtain

$$m_z = \frac{\partial F^m/\partial_z}{\mu_0(\partial^2 H_z/\partial z^2)} = \frac{|\Delta\phi_{\text{magn}}|}{K_p \mu_0(\partial^2 H_z/\partial z^2)}. \tag{45}$$

The next step is to obtain a specific calibration for the magnetic contrast. To do this, we image the same strip at a specific current and measure the change in phase. The data shown in Fig. 5 can be used for this purpose. We measure the maximum phase change, which occurs at the edges of the strip, as expected. This second calibration constant yields the value of $|\Delta\phi_{\text{magn}}|$ in Eq. (45). To complete the equation for the moment, we derive an expression for the derivatives of the field. Again, because of the simplicity of the geometry, we easily solve for an analytical expression for the field at the edge of the strip. The answer is

$$H_z(0, h) = \frac{I_0}{4\pi L t}\left\{h \ln\left(\frac{L^2 + h^2}{h^2}\right) - (h - t)\ln\left\{\frac{L^2 + (h-t)^2}{(h-t)^2} + 2L\left(\tan^{-1}\frac{h}{L} - \tan^{-1}\frac{h-t}{L}\right)\right\}.\right. \tag{46}$$

The field derivatives can be calculated from Eq. (46) and evaluated for the specific parameters used in the experiment (i.e., $h = 50$ nm and $I_0 = 11$ mA). Substitution into Eq. (46) yields 4.4×10^{13} kA/m^3. Using this value and the measured phase of 0.045 degrees in Eq. (45), we obtain the effective moment for the probe, $m_z = 2.22 \times 10^{-15}$ A m^2 (2.22×10^{-12} emu). This value is of the same order of magnitude as, but somewhat lower than, other estimates of similar probes.

It is interesting to note that even with lower values, the magnitude of the moment is large enough to cast some doubts on the actual validity of the point–dipole model. It turns out that in order to obtain this value of the magnetic moment, consistent with the bulk magnetization value of the magnetic film on the tip, the active volume would have to be quite large and of the order of (100 nm)^3. Furthermore, if the film thickness were taken to be 40 nm, with magnetization of 600 emu/cm^3 for CoCr, as specified for this probe, then magnetic material would have to extend by distance of 200 nm from the apex of the tip [124]. In order words, the tip occupies a very large volume, and it is far from a point–dipole. Thus, a better model should take into account the finite size of the tip.

We recall the discussion in Section 3.2 for extended tip models. The characteristic length scale of the field variation of our strip is of the order of 100 nm, which we derived from the smallest magnetic feature in Fig. 4. Hence, the criterion $kL \approx 1$ is satisfied, which means that the monopolar interaction model, given by Eq. (36), is more appropriate in this case. Assuming that the magnetization is more or less constant, then we arrive at the following Eq. for the magnetic charge:

$$Q_m = \frac{|\Delta\phi_{\text{meas}}|}{K_p \mu_0 (\partial H_z / \partial z)}, \tag{47}$$

from which we obtain (using the same geometrical parameters) the field gradient value as 5.4×10^9 A/m^2 and the magnetic charge as 1.81×10^{-8} A m. This value is roughly an order of magnitude lower than the estimates of Kong and Chou [119] for their 65-nm one-sided coated cobalt tips. If we assume that the charge density is of the form $\mathbf{M} \times \mathbf{n}$, then the effective area is given by Q_m/M. Using the magnetization for CoCr films, we obtain an effective area of about 3 nm^2, which is consistent with the point–magnetic charge model.

The procedure described is highly simplified but nevertheless provides reasonable estimates of the parameters, and it highlights the adequacy of using a specific model for the probe. The simplification was deliberate in order to emphasize the methodology rather than numerical calculations. This can therefore be implemented more accurately by using finite element modeling for calculating the capacitance of the tip–sample system as well as the distribution of the field from the current strip for all space. Similarly, it might be judicious to use some deconvolution techniques that accurately reflect the off-normal components of the oscillating cantilever and the possible spatial variation of the magnetization of the tip.

3.4 Advanced Techniques: Imaging in the Presence of an External Field

The MFM images taken at zero field or in the ambient magnetic field of the earth provide information about the remanent or demagnetized states of the sample. To address important questions, such as the mode of magnetization reversal, switching field distributions, magnetic pinning mechanisms, and so on, it is necessary to be able to acquire MFM images in the presence of an external field [127, 128]. These measurements, together with the results of bulk magnetometry, provide unique insights in our understanding of local magnetization processes, which ultimately lead to new perspectives in the design of magnetic devices and the tailoring of properties for optimized performance.

The implementation of MFM in the presence of a field requires a straightforward addition of a variable magnetic field source that is compatible with the physical restrictions of the microscope hardware [127, 129]. Naturally, the significant design requirements are the direction of the applied field and the capability to generate highly controlled and sufficiently high fields with minimal adverse effects in the images [128]. An electromagnet design to produce an in-plane or horizontal field is shown in Fig. 6a (see color insert). The dimensions were constrained by the available space of a *Nanoscope Multimode II* platform. The material for the yoke and other coil specifications were chosen to optimize the magnetic field strength generated at the gap location compatible with an existing power supply and subject to a maximum temperature of 40°C. Heat is particularly hazardous because it could cause drifts that distort the images, alter the susceptibility of the yoke, and, in severe cases, depolarize the piezoelectric scanners.

In this design, the deep-gap magnetic flux intensity generated is well known and is given by $B_g \approx \mu_0 NI/L_g$, where N is the total number of turns, I is the current, and μ_0 is the permeability of free space. In our design the relevant parameters are $N = 4700$ turns and $L_g = 0.76$ mm. Ideally, it is expected to produce $B_g = 2330$ Gauss at 30 mA and to reach a saturation at $B_g =$ 4500 Gauss at around 50 mA. The performance is shown in Fig. 6b. The horizontal component of the field was measured using a calibrated miniature Hall probe (*LakeShore HT 2100*). The magnetic field varies linearly with current from 0 to 25 mA, which provides the assurance that within this range the iron yoke is more or less in the reversible regime of its magnetization curve. High frequency fluctuations in the field were of the order of about 0.5%. The temperature was measured by spot welding a calibrated thermocouple near the gap region. It shows that a direct current of 25 mA elevates the temperature to about 40°C and produces a field of about 1500 Oe. This is roughly 64% of the ideal deep-gap value, which we expect because the pole pieces are narrow, the permeability of the core is not infinite, and losses occur from the other air gap where the two core parts meet. It is rather difficult to increase the field significantly beyond 1500 Oe, corresponding to a power input of 15 W. With the *Nanoscope II* multimode, there is little freedom to incorporate any kind of cooling other than convection. In certain cases, we obtained fields as high as 3000 Oe, by reducing the gap to 0.35 mm.

We have fabricated an electromagnet using an improved design that utilizes the larger space and fixed sample scanning scheme in the large platform SPMs. It includes a Peltier water cooling system and can handle as much as 50 W. It is capable of up to 3.5 kOe at room temperature with a gap of nearly twice the previous design.

Apart from the design details, an important consideration that must be considered is the direction of the magnetic field on the sample. If the sample were to be placed deep inside the gap, then the field would predominantly lie along the horizontal direction. However, since the MFM cantilever is nearly horizontal, it necessitates that the sample be positioned flush with the pole piece plane. This means that the field is reduced considerably from the deep-gap estimate, and the direction of the field vector varies as a function of the horizontal distance from the pole piece. A very good model to describe the vector field is given by Bertram in his book *Theory of Magnetic Recording* [130]. He modeled the field from a write head with a geometry very similar to that of our electromagnet. The horizontal and vertical components of the magnetic field are expressed as a modified Karlqvist equation:

$$\frac{H_x(x, y)}{H_0} = \frac{1}{2\pi}\left(\tan^{-1}\frac{g+2x}{2y} + \tan^{-1}\frac{g-2x}{2y}\right) + \frac{g}{2\sqrt{2\pi}}\,\frac{\{\sqrt{[x^2-y^2-(g/2)^2]^2+4x^2y^2} - x^2 + y^2 + (g/2)^2\}^{1/2}}{\sqrt{[x^2+y^2-(g/2)^2]^2+y^2g^2}},$$

$$\frac{H_y(x, y)}{H_0} = -\frac{1}{4\pi}\log\left[\frac{(2x+g)^2+4y^2}{(2x-g)^2+4y^2}\right] - \mathrm{Sgn}(x)\,\frac{g}{2\sqrt{2\pi}}\,\frac{\{\sqrt{[x^2-y^2-(g/2)^2]^2+4x^2y^2}+x^2-y^2-(g/2)^2\}^{1/2}}{\sqrt{[x^2+y^2-(g/2)^2]^2+y^2g^2}}, \tag{48}$$

where H_0 is the deep-gap field, x is the in-plane coordinate measured from the center of the gap of length g, and y is the transverse component. In the case of $y = 0$, that is, when the sample is at the same level as the pole piece, we can calculate the out-of-plane/in-plane component ratio as a function of the x-axis position in the gap. The general trend of this ratio is a one-to-one correspondence for $x \leqslant 0.25g$, and it dramatically rises for $x > 0.25g$. In practice, this implies that if the imaged area is exactly centered at the gap and the scan range is, say, $\pm 0.1g$, then the normal component varies from zero at the center to roughly 10% contribution at the scan edges. However, close to the pole pieces, $x \pm g/2$, the normal component is nearly equal to the horizontal component. The message, therefore, is that one would have to be perfectly aware of the position of the sample inside the electromagnet and be cognizant of the fact that there will always be some out-of-plane component of the field.

3.4.1 Probe-Induced Effect: Component-Resolved Imaging and Probe Hysteretic Effects

In *in situ* field experiments, both the sample and the probe experience the same applied field, so that there is potential ambiguity in ascribing any image change as exclusively due the characteristics of one of the components. This confusion is further aggravated in situations where the coercivities of the sample and probe are very similar. It is thus quite important to first ascertain the probe-induced changes in the images. Fortunately, the effects can be anticipated by using the force gradient models that we developed in the last section. For magnetic patterns whose characteristic wavelength is long in comparison with the characteristic length of the active moment of the probe, the point–dipole model or Eq. (25) is appropriate. In either case, the image is regarded as proportional to a mixture of the z-derivatives of the different components of the field. The relative contribution of each component of the sample field is weighted by the component of the probe moment. A suitably large in-plane external field, for example, can reorient the probe moment in its direction. This can dramatically alter the weights of the contributions of the sample field, and consequently change the image contrast. Furthermore, the probe could exhibit hysteresis effects, so that MFM at any given field could depend on the history of field application.

The magnitude of the effect is, of course, a function of the specifics of the experiment, and the ambiguity is minimized if the coercive fields of the probe and sample are far apart. Nevertheless, we will illustrate the probe-induced effects by considering a specific commercially available thin film probe and a longitudinal recording thin film hard disk medium. The results are presented in Fig. 7 (see color insert). The sequence was started by premagnetizing the probe along the vertical direction and acquiring the initial image (Fig. 7A). The horizontal field was cycled by slowly raising it up to $+500$ Oe, ramping back down to -500 Oe, and again reversing direction to up to $+700$ Oe. Images were acquired at each field increment all along the cyclic path, and the chronological order is denoted by the labels **A** to **H** in Fig. 7. The images show the qualitative changes for two types of patterns with different wavelengths. The line profiles shown in the middle of each image correspond to the averages of the line scans of the long wavelength pattern, which were derived by averaging the line scans of the upper (long wavelength) track. Because the images were obtained from the same area and because the external field was low in comparison with the coercivity of the medium ($H_c > 1600$ Oe), the magnetization component of the probe along the track direction (m_x) can be estimated. This can be done by fitting each of the experimental line shapes with the modeled response given by Eq. (25) and extracting the relative coefficients m_x and m_z. However, because we know that the z-component contribution is confined at the transition regions, we simply can derive the m_x contribution for each image by measuring the average contrast levels at the midpoints of the "bits." The resulting plot clearly exhibits the hysteretic property of the probe. Some arbitrarily chosen points along the magnetization loop of the probe are shown in images **A** through **H** in Fig. 7. They depict how changing the probe magnetization alters the image formation.

Assuming that the magnetization of the probe is unaffected by the stray field of the sample, Fig. 7A corresponds to the tip moment nominally oriented along the z-direction. Thus, the contrast arises predominantly from $\partial^2 H_z/\partial z^2$. By direct differentiation of Eqs. (27)–(29) with respect to z, it is straightforward to verify that these derivatives are localized in the regions of large divergence of the magnetization. In Fig. 7B, on the other hand, the external field of 450 Oe has reoriented the probe moment in the direction of the applied field, which caused a nonnegligible m_x contribution in the force gradient. This introduced a term proportional to $\partial^2 H_x/\partial z^2$ or **???** in the force gradient. It can easily be shown by direct substitution of $m_z = m_y = 0$ and $m_x =$ constant in Eq. (25) for the point–dipole, or by substituting H_x for H_z in Eq. (36) for the point–charge model, that indeed the derivatives mimic the magnetization pattern. This is manifested as the contrast spreading from the transition regions and onto the entire track. The line profile shows the emergence of contrast in the central regions of the magnetization. It is worthwhile to note, however, that although the horizontal component becomes more prominent, the contribution from the z-component is still quite significant. Indeed, further increase in the applied field beyond 500 Oe does not appreciably alter the images. This implies that the probe has a strong anisotropy favoring magnetization along the z-direction.

On the return path, we observe that the image at zero field (Fig. 7C) differs only slightly from the saturated state. This indicates that horizontally realigned domains have high remanence and maintain their in-plane orientation. This large remanent magnetization can be beneficial in being able to premagnetize the probe along the horizontal direction (instead of the customary vertical direction) to increase its sensitivity in detecting the local horizontal field contribution. An intriguing possibility thus exists to selectively image specific components of the field by appropriately orienting the probe magnetization direction. In other words, by a judicious choice of probe magnetization direction or sample rotation, it may be feasible to individually map out the z-derivatives of the three Cartesian components of the local surface magnetic field vectors. An attempt to determine the different components of the field has been reported in the literature, and it was specifically applied to the cross track components of the recording pattern.

The onset of m_x reversal occurs at roughly 400 Oe as depicted in Fig. 7D. This also identifies the in-plane coercive field for this specific probe. Note, however, that although $m_x \approx 0$ the image is significantly different from that in Fig. 7A. We observe the emergence of an additional weak contrast on the left of the bright transition areas. The profile clearly shows this effect. This suggests that the tip cannot be modeled as having a uniform moment that rotates with the applied field. If this were the case, then the image in Fig. 7D would be identical to that in Fig. 7A. Instead, we can think of the tip as breaking up into many domains, some of which can be rotated by the field. The unusual profile, namely, the emergence of a negative pulse at the bright edges, can be modeled by assuming that the moment of some domains on the tip changes as the tip is scanned across the surface. In these domains, the instantaneous moment of the tip follows the field and consequently produces a purely attractive force. Experiments by others using superparamagnetic tips produce uniformly dark contrast images for recording patterns. Thus, the contrast arises as a combination of the domains that are exclusively in the z-direction and those that are nearly paramagnetic. The z-directed domains produce the sharp bright/dark pattern as in Fig. 7A, whereas the other type produces the uniform dark contrast. The superposition yields the image in Fig. 7D. As the field is increased further to 450 Oe, image Fig. 7E exhibits the contrast reversal. The contrast is almost the exact complement of Fig. 7B. It is also quite interesting that the contrast at the transition region, which we attribute to the z-component of the field, has also reversed. This implies that the magnetization along the z-axis has also changed sign at 450 Oe. This is reasonable because, according to Eq. (48), the applied field also has a significant out-of-plane component that may be responsible for the z-axis moment reversal. More significantly, the fine structure around the transition observed in Fig. 7D has completely vanished. Presumably, the external field has pinned the moment in the x-direction, and it can no longer be instantaneously rotated by the sample field.

On removal of the field, the resulting image (Fig. 7F) shows very little difference from Fig. 7E, except for the somewhat weaker vertical component. At +400 Oe, Fig. 7G is analogous to Fig. 7D, with the exception that the dark/bright fine structure occurs at both transition edges. As in Fig. 7D we speculate that because the probe is in a demagnetized state, the influence of local fields at the dark regions induces transient switching of some magnetic domains of the probe, which produces the fine structure. The reason for the existence of the fine structure in both transition regions is as follows. Bright-to-dark contrast reversal of course occurs as the out-of-plane moments are reversed, whereas the ancillary dark strips are due to those domains that switch with the sample field. That this occurs in Fig. 7G but not in Fig. 7D may be due to the asymmetry between the two cases, as the tip itself is misoriented by 10° relative to the normal axis.

For the sake of completeness, we mention that a very similar situation exists for perpendicularly magnetized samples. The predictions of the contrast change as the mixture of the derivatives of the field still hold. The interpretation that the z-component imaging is sensitive to the distribution of magnetic charges still holds. One distinction, however, is that in this case the volume charge density vanishes, and the discontinuity of magnetization at the surface–air interface produces the surface magnetic charges. Hence, the z-component images are proportional to M_z. From Laplace's equation, it follows that the x-component images must be proportional to the variation of the surface charges along the x-direction. In other words, the x-component images are sensitive to the domain walls [86].

3.5 Applications

3.5.1 Erasure Process of a Thin Film Medium

Having understood the response of the probe to the field, let us apply the *in-situ* field to investigate the microscopic behavior of a sample undergoing a gradual DC erasure. A typical result is shown in Fig. 8 (see color insert). The plot at the center is the descending branch of the measured magnetization curve, and the images are labeled according to the positions on the curve. The curve represents the magnetization component being reversed by the applied field.

The evolution, as evident in the data, can be described as follows. Figure 8A is the initial, unperturbed state at zero field. Prior to imaging, the probe was magnetized along the z-axis. Following our intuitive interpretation, the contrast variation mimics the distribution of the magnetic charges from the patterns. These are confined along the transition areas, where the magnetization changes direction, leading to a large $-\nabla \times \mathbf{M}$ contribution. The bright regions correspond to positive or "head-to-head" transitions, whereas the dark regions are the reverse.

The field of 200 Oe is too low to perturb the medium, but it causes an orientation change of the moment of the probe. As discussed in the previous section, the image in Fig. 8B is sensitive to the z-derivative of H_x. For longitudinal media, because H_x closely mimics the magnetization distribution, it follows that Fig. 8B can be regarded as the magnetization distribution corresponding to the charge distribution in Fig. 8A. At higher fields, we anticipate that the probe would simply be increasingly diverted along the x-direction and will merely enhance the sensitivity to the magnetization distribution. As the field is increased, no significant variation nor contrast degradation was observed up to 600 Oe, consistent with the M–H loop, which shows that the magnetization was little changed from its saturation value for small external fields.

At 696 Oe, however, changes in the patterns became apparent. The subtle expansion of the bright areas and the distinctive roughening of the transition and interior regions signify the initial stages of magnetization reversal. The increase in transition edge irregularity suggests that the leading edges of the favorably magnetized ripples have extended in response to the applied field into areas previously magnetized in the opposite direction. Direct comparison of the bit lengths in Fig. 8B and Fig. 8D confirms that transition roughening intensified and that the bright regions have extended by an average of about 1 μm, accompanied by a corresponding remission of the dark contrast areas. Likewise, the increase in "roughening" of the dark areas indicates that the specific areas in the interior regions switch in the direction of the applied field. This is in qualitative agreement with the results of micromagnetic simulations at various states along the hysteresis loop. The observed roughening may be attributable to the formation of a large number of vortex structures. As the field approached the bulk coercivity, adjacent growing domains began to break through the entire length of the unfavorable domains. In certain areas, large neighboring transition edge ripples have coalesced from both sides, which causes the breakup of the unfavorably magnetized areas. This can be seen in Fig. 8D where light-intensity streaks appear in most of the short-bit-length dark contrast areas. In sections where this occurred, the brightness intensity at the transition region has been diminished. This is a consequence of the disappearance of free magnetic charges at the transition regions when neighboring identically magnetized ripples combine. Streaks have similarly appeared on the wider dark areas at 926 Oe and expanded laterally in some locations.

Note, however, that as soon as the percolation occurred, the average location of the transition boundaries remained stationary. This implies that the new channel to reverse magnetic moments, namely, to expand the percolated areas, is energetically more favorable than the displacement of the transition boundary. Indeed, the erasure of the pattern considerably hastened beyond 926 Oe. With an additional 100 Oe, at 1027 Oe, most parts have been reoriented in the direction of the field. What remains after this point are isolated clusters. These pinned areas are quite stable and require fields that are significantly higher than coercivity to switch. The terminal distribution of these clusters appears to be uncorrelated with the recording pattern, and their random distribution may reflect the intrinsic local properties of the medium.

In summary, we conclude that the erasure process in thin film media occurs in several steps. The precursor is the track width reduction and enhanced transition edge variation. The second phase is the percolation of reversed islands and their rapid lateral expansion. The final phase is the switching of the pinned areas at fields much higher than the nominal coercity of the medium. This work highlights the particulate nature of the medium. Pattern erasure does not involve the motion of domain walls but rather the reversal of the individual grains.

3.5.2 Magnetic Characteristics and Wall Motion of Small NiFe (Permalloy) Elements

Understanding the characteristics of patterned NiFe films at nanometric length scales is an area of immense technological and fundamental scientific importance. Permalloy is used in magnetic sensing applications such as magnetic recording, in MEM development including flexure arms in micro-actuators, and in magnetoelectronics as spin-filtering electrodes. The optimization of film properties for these applications is crucially dependent on the understanding of its magnetic properties at the microscopic level. Moreover, as the elements diminish in size, the total energy, whose minima determine the magnetic configuration of the system, becomes dominated by the specific geometry of the system. It is thus meaningful to systematically study the magnetic behavior of islands that have similar size and dimensions as their intended applications.

From a scientific point of view, Permalloy is well known to have low crystalline and magnetoelastic energy contributions, which makes it an excellent system to understand domain motion driven primarily by the balance of magnetostatic, exchange, and anisotropy energies. Several sophisticated numerical micromagnetic calculations, combined with advanced fabrication and diagnostic techniques, are making it possible to predict and engineer the magnetic properties of submicron-sized islands. Smyth *et al.* [131] have investigated the role of particle size, aspect ratio, and interparticle spacing on the hysteretic properties of the ensemble and compared them with *ab initio* calculations, whereas Zhu *et al.* [132] have modeled the micromagnetics of exchanged biased 10-μm sized islands and demonstrated remarkable agreement between the model and the measurements. More recently, Schrefl *et al.* [133] have systematically studied the influence of edge shapes on the switching dynamics of submicron particles using Lorentz transmission electron microscopy (LTEM) and micromagnetics to demonstrate how the end shapes affect the switching field. Prior to those studies, Fredkin *et al.* [134, 135] had developed micromagnetic codes that showed complex edge and multivortex structures, depending on the exact manner in which the magnetic field was applied, and the findings were later confirmed by Hefferman *et al.* [136] using Lorentz electron microscopy. The MFM technique is naturally suited to complement these investigations.

3.5.2.1 Classification of Island Magnetization Patterns. We investigated arrays of islands prepared by several groups in Europe and the United States. Figure 9 (see color insert) shows the magnetic configuration of each of these islands after applying a 200 Oe field in the direction of the arrow and imaging at zero applied field [137]. The additional images in the inset of Fig. 9 were observed after applying an in-plane field to roughly 200 Oe and imaging at zero field.

The islands were prepared by K. J. Kirk and J. N Chapman at the University of Glasgow using electron beam lithography and thermal evaporation in ultrahigh vacuum conditions. The patterns were prepared by electron beam lithography on silicon-based substrates [138]. The material is an evaporated soft $Ni_{80}Fe_{20}$ alloy ($B_s = 1.2$ T), 26 nm thick. The substrate was diced to about 2 mm and was placed inside the gap of an electromagnet. The setup allows magnetic force microscopy images [128] to be acquired while a well-controlled continuously variable in-plane magnetic field was applied. Low moment tips were selected to minimize tip-induced perturbations. The pattern has been layed out to cover a range of lateral aspect ratios from 1 to 16 covering micron to submicron dimensions. Despite the variation in geometry, it is interesting that the patterns can be conveniently classified into seven unique configurations. Three of the classes are "solenoidal." Closure domains are formed at zero field in these cases, and the islands individually exhibit no net magnetic moment. These are classified as follows [137]:

Type A: Four domain closure, exemplified by the 3 μm by 3 μm patterns.
Type B: Four-domain closure with four 90° walls and one 180° wall, as in the 2 μm by 3 μm island.
Type C: Four-domain closure with four 90° walls and one 180° wall with cross-ties and Bloch lines, as in the 3 μm by 2 μm island.
Type D: Seven-domain closure pattern, as the two lower islands in the inset of Fig. 9.

Similarly, the nonsolenoidal patterns exhibit three unique configurations, namely,

Type E: uniformly magnetized central region with closure domains at the ends, such as the 1 μm by 4 μm pattern and most of the islands of width 0.5 μm.
Type F: single domain, with no observed fine structure at the ends, such as all of the 0.25-μm-wide islands.
Type M: distinct net moment as exhibited by charge separation at the ends, but with nonuniform and complex internal structures, such as the 4 μm by 2 μm, 4 μm by 3 μm, and 4 μm by 4 μm islands.

A cartoon of the configurations is given in Fig. 10.

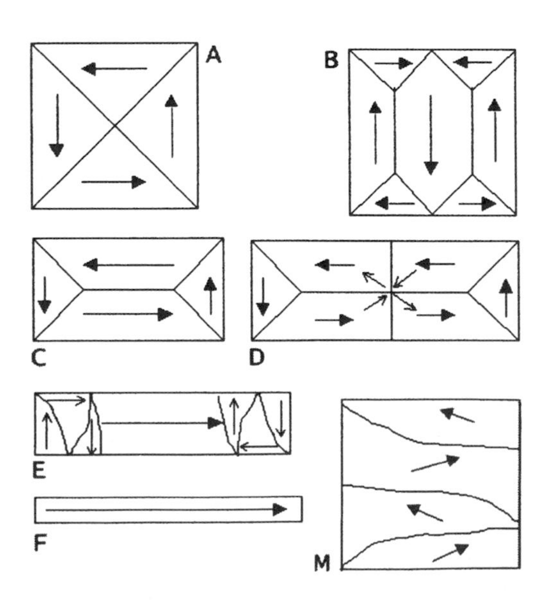

FIG. 10. Classification of observed domains in small NiFe elements.

The transition from solenoidal to nonsolenoidal, with the exception of the complex pattern M, occurs at about an aspect ratio of 3. It was also verified, from subsequent magnetizing measurements, that the direction of the applied field determines the choice. Specifically, if the field is applied along the shape easy axis (i.e., the long direction of the islands), then the resulting configuration will be nonsolenoidal; the converse is true for field applied along the short axis, hard direction. The four-domain and seven-domain configurations (see, e.g., Runge *et al.* [139]) at remanence appear to be energetically quite similar, as the probability of finding either configuration in any specific island is roughly 50% in different magnetizing cycles. Cross-tie and Bloch walls appear at an aspect ratio slightly above 1 and depend on the direction of the applied field. The appearance of additional cross-tie inclusions is related to the aspect ratio, and from our limited sample size, it appears that new inclusions are formed every 400 nm on the 180° walls. As far as the nonsolenoidal patterns are concerned, the complex type M is most ubiquitous for larger islands and for low aspect ratios. In most cases, this pattern changes into one of the solenoidal configurations after applying a slight reverse field of about 10 to 15 Oe. Note that, although rather complex, the patterns from different islands share common features such as the crisscrossed curved domains in the interior as well as the splitting of a domain wall as it extends to the other edge. Thus, it is reasonable to infer that the type M domain is truly a result of energy minimization, rather than a fortuitous structure from some stochastic phenomenon. The type E pattern, which if not for the closure pattern at the ends would have been single domain, is perhaps one of the most stable of all the domains. It is found from aspect ratios of 3 to 8 as long as the island width is greater than 500 nm. In most applications to date, the size and aspect ratios are in the range where the type E pattern is expected. At these dimensions, a true single domain is exhibited when the aspect ratio is greater than or equal to 4 and the width is less than 500 nm.

3.5.2.2 High Resolution Imaging: Bloch, Neel, Cross-tie Walls. Now let us consider a high resolution image of one of the interesting type C islands in Fig. 11 (see color insert). Since the MFM detects the divergence of magnetization or the magnetic charge distribution, the contrast comes primarily from the domain boundaries and not from the domains themselves. The divergence of the local magnetization at the domain boundary produces the negative and positive charges which, in the image, appear as dark and bright contrasts along the walls. On this basis, the distribution of charges at all the domain walls and the magnetization of the domains can be qualitatively discerned. The four 90° walls oriented at 45° from the corners are clearly visible. They converge to two separate points at the midsection of the island, where they connect with a 180° wall along the horizontal axis. A magnetization vector map, obtained using a public micromagnetic code developed at The National Institute for Standards and Technology (NIST) [140], is included in Fig. 11 as well, with the divergence of the vector field at each point shown in gray scale. The images show qualitative agreement of the model with the MFM images. The requirement that the magnetostatic energy be a minimum implies that the free charges ($\mathbf{M} \times \mathbf{n}$) vanish at the edges, making the magnetization vector lie parallel to the edges. The only sources of magnetic charges are the walls themselves and the cross-tie inclusions. At the intersection of the 90° and 180° walls are vortex structures, forming a curling magnetization vector. The region near the vortex is divergence free but contains a singularity at the center. The micromagnetic model shows distinct fine structures at the center of the vortices, although the singularities are perhaps below the resolving power of the MFM and hence unobserved. It is also quite intriguing that the micromagnetic model predicts that the width of a 90° wall is somewhat larger than that of the 180° wall, and this is more or less confirmed in the MFM data. The intense bright/dark contrast at the center of the image is due to a cross-tie structure. These arise from lowering of the magnetostatic energy of the 180° wall by breaking up into two opposite magnetized regions. Unlike the vortices, the sense of circulation of those moments is opposite to the magnetization of the main domains. Hence, a large divergence is consequently formed, which extends from the 180° wall and transverse to the 180° wall and into the main domains. The model predicts that the cross-ties would have somewhat weaker contrast than the walls themselves, which is in apparent disagreement with the data. One possible reason is that the MFM samples are integrated over a rather larger region than the idealize model, and thus the contrast is intensified.

The magnetization vectors for the configurations can also be interpreted on the basis of the proposed model, albeit the end closure patterns for type D and the internal structure of type M are more complex in that they exhibit domain boundaries and uniformly magnetized domains.

3.5.2.3 Magnetizing Experiments: Domain Wall Motion and Reversal Processes. The micromagnetic evolution for any given island is a function of its geometry, configuration type, and the direction of the applied field as well as island-specific defects and local stoichiometric variations. Additionally, hysteretic effects are also quite prevalent, so that the behavior could be different between different magnetizing cycles. Thus, thorough understanding will invariably involve a huge number of samples of various parameters. Nevertheless, despite this intricacy, we would like to develop general qualitative insights into the nature of local magnetization processes by considering the islands at our disposal. We will concentrate on the simplest case, where the field is applied along the shape-induced hard and easy axes.

In the case of the solenoidal patterns, applying an external field has the initial effect of increasing the size of the domains parallel to the applied field at the expense of the unfavorably oriented domains. This is true for both the four- and seven-domain closure patterns, where the domain expansion is accomplished by the movement of the 90° and 180° walls. An example of this is illustrated in Fig. 12 (see color insert), which shows a 3×4 pattern magnetized along the hard and easy axes. At zero field, the patterns exhibit four-domain closure arrangement similar to the square islands, but with the 90° walls curved to compensate for the 1.33 aspect ratio. At low fields, the favorable domains increase in proportion to the applied field. We further observed, by cycling the field within this small range, that the patterns were reversible and that we can regain the specific configuration independently of the magnetizing history. The onset of irreversibility occurs when the vortex breaks up into another wall. The field required is dependent on the direction: 40 Oe for the easy axis and 70 Oe for the hard axis, respectively. In both cases, a 180° wall is formed where the vortex used to be and as the left and right domains, which are magnetized antiparallel to each other, begin to coalesce. Raising the field causes the length of the newly formed wall to increase, and the wall subsequently divides into several segments through the formation of Bloch lines and cross-ties. In the hard axis case, only one cross-tie is formed, as opposed to two in the easy axis magnetization. This is of course reasonable because the easy axis field allows longer 180° walls to form. As the field is further increased, the system undergoes another irreversible transition marked by the complete disappearance of the unfavorable domain. We label this field at H_{s1}, which happens at roughly 70 Oe for the easy axis case and at 90 Oe for the hard axis field. This transformation appears to takes place abruptly once a critical length of the 180° wall is reached. Beyond this point, the pattern then becomes a large uniformly magnetized domain bounded by charged triangular segments at both ends. The charges at the ends are indicated by the bright and dark contrast at both ends and imply that either the end domains are nonuniformly magnetized (i.e., divergence of **M** is nonzero) or there is slight out-of-plane alignment of the magnetization at these regions. Unfortunately, MFM alone cannot distinguish between these two possibilities. This configuration corresponds to the technical saturation, beyond which the effect of higher field is to slowly reduce the area of the charged region as more of the atomic moments are forced to align with field.

There is very little change observed from the saturated configuration as the field is lowered. In fact, the saturated state persists even when the field falls below H_{s1}. The transition back into the closure domain pattern occurs at a field denoted as H_{s2}, and in the case of the 3×4 island magnetized in the hard axis, it corresponds to about 15 Oe. However, for the easy axis magnetization, the transition back to a solenoidal pattern occurs at a slightly negative field of 10 Oe. Often, the islands form the complex domain type M in lieu of the closure pattern and require a slight negative field to revert back to the closure formation. This is generally the situation for islands magnetized along the easy axis. The converse is also true for islands magnetized along the hard axis. In these cases, H_{s2}—the formation of a closure domain—is close to H_{s1}, albeit always lower. The difference is in inverse relation to the aspect ratio.

Next we focus on the behavior of the nonsolenoidal islands, types E and F, and attempt to understand the mechanism of moment reversal when the field is applied along the easy axis. The behavior of the end domains in a type E pattern is typified by a 1×3 island shown in Fig. 13 (see color insert). The mechanism of reversal involves the gradual migration of the end domains toward the center of the islands, as shown in the 20 Oe and 40 Oe scans. At the slightly higher field of 50 Oe, we observe that the contrast on the top edge has changed from dark to bright, from which we conclude that a spontaneous reversal of the magnetization has occurred. Furthermore, it is interesting that the actual reversal has taken place while scanning. Note that in the beginning of the bottom-to-top scan at 50 Oe, no switching has yet occurred, as the lower edge features are virtually identical with those at 40 Oe. We suspect that the slight field from the probe has added to the external field to induce the reversal. The richness of the reversal mechanism of the type E patterns is in stark contrast with the true single domain, type F islands. For these islands, the pattern is unchanged except at the switching field, which induces an inversion of the contrast at the opposite ends.

In conclusion, the behavior of the domain structures of even the simplest thin film elements is complex and depends crucially on the geometry and the direction of the applied field. By using magnetic force microscopy, each of the islands in Fig. 9 was analyzed, and the details are reported in the thesis of V. T. Luu at the University of Maryland [141]. Despite the large variation in geometry and size, and despite the difference in the specific values of the critical fields, the behavior is quite predictable. One can conveniently classify the behavior of the islands on the basis of local magnetization loops first introduced by Hefferman *et al.* [136]. The types are shown in Fig. 14. Type I is the classic M–H loop expected for a Stoner–Wohlfarth (SW) single particle [142], and it occurs for high aspect ratio islands that are nearly single domain and magnetized along the easy axis. Type II describes a process where the magnetization is reversible for small fields, undergoes a transition to a saturated state at H_{s1}, retains a net moment at zero field on the descending branch, and reverts to a reversible state at H_{s2}. Complex domains are formed at remanence. Type II loops govern the reversal mechanism of low aspect ratio islands that are magnet-

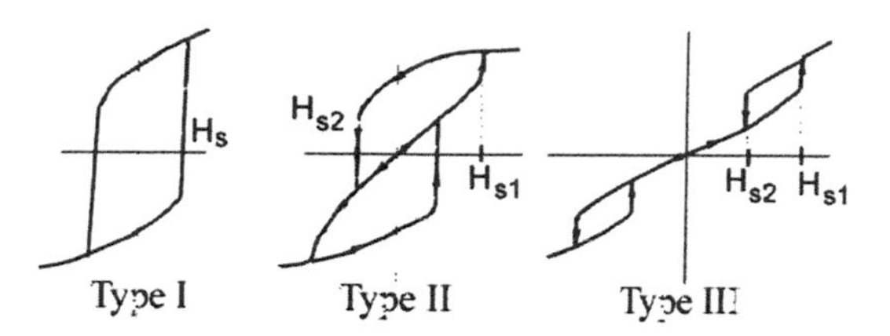

FIG. 14. Classification of microhysteresis loops.

ized along the easy axis. Type III exhibits reversible behavior up to relatively high field. It undergoes a transition to a saturated state at H_{s1} and reverts back to a reversible configuration at H_{s2}. The type III loops describe the behavior of low aspect ratio islands.

In the preceding section, we mentioned that the true single-domain characteristics for rectangular Permalloy islands depended on both the aspect ratio and the width of the particles. Koo *et al.* [143] studied a large number of similar particles with nominal thickness of 25, 35, and 55 nm to verify the formation of single-domain particles. They found that a minimum width of about 500 nm is required to impede the formation of end closure patterns and thus form a true single-domain particle. Their measurements, which spanned aspect ratios from 1 to 16, showed that the critical width is independent of the aspect ratio provided that it is larger than 2.4. For the smaller aspect ratios, however, they discovered that the onset of single-domain behavior for a given aspect ratio depended on the width. In particular, islands having a width larger than 350 nm were multidomains up to an aspect ratio of 2 : 1, whereas the 310-nm-wide islands were single domain at a lower aspect ratio of 1.87. The multidomain character of the 350-nm-wide particles persists even as the length increases, but it eventually reverts back to single-domain behavior when the aspect ratio (AR) reaches a critical value, which depends on the thickness. These transitions occur at the following lengths: 0.97 μm (AR = 2.43) for 23 nm thickness, 1.34 μm (AR = 3.62) for 35 nm thickness, and 1.57 μm (AR = 4.13) for 55 nm thickness. The trend in forming single domains is toward increasing aspect ratio with increasing thickness.

The magnetization of the single-domain particles is bistable, and the MFM measurements with applied field merely show the reversal of the bright/dark contrast in the middle of the scan whenever a critical field is reached. More interestingly, by counting the number of domains that switched as a function of the field, we were able to form the switching field distribution of the islands. The switching distributions were symmetric and peaked functions. For the islands where measurements were made, namely,

1.34 μm × 0.37 μm (AR = 3.62) and 2.21 μm × 0.37 μm (AR = 5.9), the curves were centered at 40 and 120 Oe, respectively, with nearly identical spreads (full width at half-maximum) of about 40 Oe.

Finally, we comment on the dependence of the switching fields on the domain configurations. Our results show that the switching field of single-domain particles is strongly dependent on the aspect ratio. Nearly square or circular single-domain particles have switching fields of several Oersteds, whereas particles of 1 : 6 aspect ratio have switching fields of 120 Oe. Others have also reported switching fields of nearly 700 Oe for Permalloy particles of 1 : 10 aspect ratio [127, 131]. This could be explained on the basis of the coherent rotation model, wherein the shape-induced anisotropy term increases with aspect ratio. Others, who used *ab initio* micromagnetic calculations, also predict the strong dependence of the single-domain switching field with aspect ratio and the reduced width (= width/thickness) [131]. The numerical calculations also predicted a state with no remanence for an aspect ratio of 2.3, which was not experimentally observed for 0.41 μm × 0.175 μm × 0.05 μm, and the prediction for squareness deviated substantially with the experimental results at the lowest aspect ratios (<4) [131]. We regard the calculations as correct in describing the hysteresis of multidomain particles, which occurs for the larger particles. The prediction of zero remanence can be construed as being consistent with those particles that form solenoidal intermediate states. As far as the multidomain islands are concerned, our results show weak dependence on aspect ratio. Indeed for both of the aforementioned multidomain configurations, the switching field occurs at roughly the same value of about 50 Oe, and it occurs independently of the aspect ratio. This is reasonable because switching is induced by the movement of the closure domains. As the data suggest, the size of the end closure domains does not change with aspect ratio, so that the energy or the strength of the magnetic field to move them should be invariant with size as well. Similarly, those islands that form intermediate closure patterns are expected to have weak dependence on the aspect ratio, since the motion of 90° or 180° walls is independent of the aspect ratio. One possible factor that could affect the switching field is the number of cross-ties or other inclusions that appear on the domain walls. These features, of course, depend on the size and geometry of the islands.

3.5.3 Nanostructured Cobalt and Iron Islands

Driven by their important applications in diverse areas of technology, the micromagnetic properties of nanostructured cobalt and iron islands are currently attracting considerable attention. For instance, patterned cobalt elements are regarded as possible candidates for ultrahigh density magnetic storage media [144]; some spin-dependent devices use nanoferromagnets as active layers, nanodots are being considered for quantum computation, and regular arrays of iron islands are used as flux-pinning centers to increase the critical superconducting current density, to name just a few. Their applications will undoubtedly increase dramatically in the coming years as initiatives in nanotechnology are being realized.

Numerous research groups have active programs to fabricate and control the properties of these elements [131, 145, 146]. In the case of patterned media applications, one approach is to form columnar structures with diameters as small as 25 nm and lengths of the order of 100 nm to 1 μm. Because of the high aspect ratios (height/diameter), these islands form stable magnetizations normal to the surface. This configuration, known as the perpendicular medium, potentially yields the highest possible packing fraction for very high areal densities. The other alternative is an in-plane medium, where rectangular bars lie flat on the surface and with thickness much smaller than the lateral axes. In this case the magnetization is primarily along the plane and oriented according to the interplay between the shape and crystalline anisotropy axes. The reader is referred to the literature elsewhere for the treatment of perpendicular media, as our focus here is exclusively on the in-plane islands. The reason is that MFM, being a surface probe technique, is ideally suited to study in-plane patterns because the magnetization along the entire island is accessible. In perpendicular islands, only one end is exposed, and it is difficult to ascertain the micromagnetic structure in the interior of the columns.

From the point of view of patterned media and other applications, the two most important considerations are (1) how to form single-domain islands and (2) how to control the switching process. New *et al.* [147] successfully fabricated submicron in-plane cobalt islands of various aspect ratios using electron-beam lithography. An example of these islands, with dimensions 200 nm × 400 nm × 20 nm, is shown Fig. 15 (see color insert).

It is evident that despite the similarity of the topography, the orientation of the moment is richly varied. New *et al.* [147] modeled the dispersion of the easy axis as a function of aspect ratio. Their model used a random-walk approach, which took into account the finite number of grains that comprise each island. Because of the finite number of grains per island, the sum of the individual grain anisotropies does not average to zero. As a consequence, there will be dispersion or spread of the easy axis centered about the long axis of the island. The width of the dispersion is directly related to the aspect ratio, and it is roughly inversely related to the number of grains. The result of the calculations is shown in Fig. 16 (see color insert) for several cases. Ensembles of these islands were imaged using the MFM, and the direction of the easy axis for each island was measured. The bar graph superimposed in Fig. 16 is the result of the MFM measurements. Considering that the calculations assumed single-domain particles, whereas the measurement showed a substantial fraction being multidomain, the agreement of the data confirms the validity of the dispersion model.

To study the interaction of the islands with an external field, we monitored the magnetic evolution of individual nanostructures as a function of applied field. The results are shown in Fig. 17 (see color insert). At the beginning of the experiment, we chose two neighboring islands that were in opposite directions at $H = 0$ Oe. The island on the left was nominally oriented along the direction of applied field. The moment had an initial angle of about 27° with the horizontal axis, which was the direction of the in-plane field. As the field was gradually increased, we observed the moment to rotate in the direction of the field, reaching to just above 5° off the horizontal axis at 960 Oe. The process of rotation, however, does not follow dependence on the field predicted by a classic Stoner–Wohlfarth particle. In the SW particle, the anisotropic energy is proportional to the square of the sine of the angle between the easy axis and the moment, α, whereas the magnetostatic energy is proportional to the cosine of the angle between the field and the moment $(\alpha - 27°)$. Thus, the ratio $\sin^2\alpha/\cos(\alpha - 27°)$ must be proportional to the field. We find, by inspection, that this is not the case and that some pinning effects thwart the moment rotation in the range from 360 to 660 Oe. Thus, we might conclude that despite the "quasi-single-domain" configuration of the island, it cannot be described simply by a SW model. This is perhaps justified, by taking into account that the bulk uniaxial crystalline anisotropy field in cobalt is 6300 Oe, which is much larger than the demagnetizing fields for this geometry (i.e., 500 Oe for the long axis and 1400 Oe for the short axis). Thus, we expect that the behavior of the islands would be dictated primarily by the strong crystalline anisotropy of the randomly oriented grains.

The island on the right in Fig. 17 shows the switching reversal process. It had an initial orientation of roughly 180° with respect to the field. The configuration remained unchanged up to 560 Oe, but it spontaneously switched at 660 Oe in the middle of a scan. The fact that the reversal occurred while scanning suggests that the switching field is probably slightly higher than 660 Oe, and the balance was provided by the field from the probe. We considered an ensemble of about 100 islands and found a rich variety of magnetic configurations as well as a broad switching field distribution. Roughly 30% of the islands exhibited multidomain structures at zero field. Most of these islands showed a multistep reversal process, where the individual domains within an island would switch at different fields. The rest of the islands can be considered quasi single domains, as these showed some variability not only in the orientation of the moment but also in the fine structure in interior areas. These islands switched spontaneously at various applied fields, yielding a peaked distribution centered at 750 Oe, with full width at half-maximum spread of 500 Oe. We observed no correlation between the multidomain characteristics and the switching field. Comparison with our earlier results in Permalloy shows that the governing parameter for the Co islands is the crystalline anisotropy. In the case of Permalloy, which has little or no crystalline anisotropy, the 2 : 1 aspect ratio for islands of comparable size forms well-defined and almost identical single-domain structures. The relative mean switching field is lower for Permalloy (in the range of 100 Oe), and it critically depends on the shape of the ends up to an 8 : 1 aspect ratio.

In recognition of this fact, new directions have emerged in fabricating single-domain cobalt particles. Our collaborators at Stanford have achieved breakthroughs in making patterned submicron Co islands on Si(110) and MgO(110) crystals [148]. The islands were grown epitaxially, and the cobalt overlayer followed the substrate lattice. Our preliminary MFM results on MgO(110) in Fig. 18 (see color insert) show that these islands were predominantly single domain, even if the shape and crystalline easy axes were oriented in perpendicular directions. The c-axis of the cobalt layer is oriented along the [001] direction, and the images show that the moment followed the crystalline easy axis regardless of the shape easy axis direction. The distinction between the two cases is the appearance, on about 15% of the islands, of bi-domain structures when the crystalline and shape anisotropy axes are orthogonal. Furthermore, the switching field distribution is single peaked (~2000 Oe) in the parallel case, and double peaked (1500 Oe to 2000 Oe) in the orthogonal case. The full width at half-maximum were about 500 Oe. These results are suggestive of two distinct reversal mechanisms: a direct single domain to single domain mode and a second one, through the formation of an intermediate bi-domain state. These observations, plus the fact that the switching fields were lower than the bulk crystal anisotropy fields, suggest either some lattice mismatch between the substrate and the film or switching occurs though a mechanism that we do not yet fully understand.

Acknowledgments

The author is indebted to I. D. Mayergoyz for his theoretical support, to his past and present students T. V. Luu, A. O. Pak, A. Anderson, A. Kratz, M. C. Shih and H. C. Koo, his collaborators, R. L. White, R. F. Pease, R. H. New, S. Ganesan, K. J. Kirk, J. N. Chapman and W. Egelhoff for various contributions. Funding support from LPS, NSF-CAREER Award, and NSF-MRSEC grant are gratefully acknowledged. This work is dedicated to the memory of my mother, Julita del Rosario Gomez.

4. ELECTRON HOLOGRAPHY AND ITS APPLICATION TO MAGNETIC MATERIALS

M. R. McCartney,[1] R. E. Dunin-Borkowski,[1,2] and David J. Smith[1,3]

[1]Center for Solid State Science, Arizona State University, Tempe, Arizona
[2]Department of Materials, University of Oxford, Oxford, United Kingdom
[3]Department of Physics and Astronomy, Arizona State University, Tempe, Arizona

4.1 Introduction

The transmission electron microscope (TEM) is widely used for characterizing the microstructure of materials, and there are several imaging modes that are sensitive to magnetic microstructure. All of these techniques take advantage of the Lorentz force, which involves sideways deflection of the incident high-energy electrons by the magnetic field of the sample. The most well-established techniques are variants of Lorentz microscopy, described in Chapter 2, that use highly defocused images to provide lines of dark and light contrast at magnetic features such as domain walls [149, 150]. Electron holography is an alternative TEM technique that provides access to the phase and amplitude of an electron wave after it has traversed through the sample [151, 152]. Because the phase of the electron wave is altered by magnetic (and electric) fields, electron holography can be used to study magnetic materials. Digital recording of holograms facilitates the quantitative measurement of the field in and around the sample with high sensitivity. Moreover, because any loss of resolution associated with defocusing is not involved, the technique is inherently capable of achieving a spatial resolution for magnetic microscopy that can approach or exceed 1 nm, depending on the optics of the particular microscope. In practice, these resolution limits have yet to be achieved, mostly because of limitations of the recording and/or processing methods. A further attraction of electron holography as a technique for studying magnetic materials is that it allows access to much smaller regions than are accessible by other magnetic imaging techniques. With the recent downscaling of dimensions for magnetic storage devices, the technique thus has great potential for contributing to important industrial problems, as well as advancing fundamental scientific research. Conversely, as electron holography usually requires the transmission of electrons through the specimen, it has the disadvantage that samples must be limited in thickness ($<0.5\,\mu$m). In this chapter, we first describe the development of electron holography and the basic principles of the technique, with particu-

EXPERIMENTAL METHODS IN THE PHYSICAL SCIENCES
Vol. 36
ISBN 0-12-475983-1

ISSN 1079-4042/01 $35.00

lar reference to magnetic imaging. We then survey some applications of electron holography, which include studies of hard magnets, magnetic media, and magnetic microstructures. Finally, we discuss prospects for further development of the technique, such as real-time viewing of dynamic events and the *in situ* application of variable external fields.

4.2 Development of Electron Holography

Electron holography relies on the interference of two (or more) coherent electron waves to produce a hologram, which must then be reconstructed in order to retrieve a complex electron wavefunction that carries phase and amplitude information about the sample. At least 20 forms of electron holography have been identified [153], and many of these have been realized in practice. The most commonly used is termed off-axis (or sideband) electron holography, which is described in the next section, and relies on the use of an electrostatic biprism developed by Möellenstedt and Düker [154] to overlap a vacuum or reference wave with the wave scattered by the object. Early electron-holographic experiments [155] were limited by the brightness of the available tungsten hairpin filament used as the electron source, which directly impacted on the coherence of the incident beam. The later development of the high-brightness, field emission gun (FEG) for the TEM [156] greatly facilitated the practical implementation of electron holography. The FEG has been used almost exclusively as the electron source for all subsequent electron holography applications reported in the scientific literature.

Traditionally, wavefunction reconstruction from electron holograms has been achieved off-line using optical methods [157]. With the advent of the slow-scan charge-coupled device (CCD) camera for digital recording [158], coupled with the rapid growth in computer speed and memory, digital processing of electron holograms has become widespread [159], paving the way to truly quantitative electron holography [160, 161]. The correction of microscope aberrations, as originally proposed by Gabor [162] as a means to surpass conventional microscope resolution limits, has been achieved using both optical [163, 164] and computer methods (e.g., Refs. [165, 166]). Digital recording also removes the need for (tedious) correction of the nonlinearity of photoplate optical density, thus enhancing the speed and reliability of the reconstruction process.

Off-axis electron holography has a long history of applications to electrostatic fields. Some noteworthy examples include the following. Reverse-biased p–n junctions were studied at low magnification ($\sim 2500\times$) by Frabboni *et al.* [167], and electrostatic fields associated with charged latex spheres were investigated by Frost *et al.* [168]. Depletion region potentials at a Si/Si p–n junction were reported by McCartney and Gajdardziska-Josiforska [169]. The two-dimensional electrostatic potential associated with deep-submicron transistors has been mapped with a resolution of about 10 nm and a sensitivity approaching 0.1 V [170]. The electrostatic potential and associated space charge across grain boundaries in $SrTiO_3$ has also been reported [171].

Historically, electron holography of magnetic materials has been limited to a spatial resolution of more than 10 nm because of the need to locate the sample in a region of low external field to prevent magnetization saturation. Important milestones have included imaging of the field distribution associated with magnetic recording tape [172], experimental confirmation of Aharonov–Bohm magnetic phase shifts at low temperature [52], and observations of vortex lattices in a superconductor [173]. In most TEM applications, the strong objective lens was usually switched off, and imaging was achieved using the remaining diffraction/projector lenses. A scanning TEM configuration, with an inherent potential for improved resolution, involved raising the sample out of the lens to a point at which the residual field was less than 50 G [174]. By using the objective lens in a long-focal length mode, the spatial resolution could still approach 1 nm. Special minilenses can also be inserted into the bore of the objective lens to achieve improved resolution without adversely affecting the remaining field at the specimen [175]. Several applications of electron holography to magnetic materials are described in detail below.

4.3 The Technique of Off-Axis Electron Holography

We focus our attention here on the technique of off-axis electron holography, since it is this holography method that has been used in most electron microscope studies of magnetic materials. The microscope geometry for off-axis electron holography in the conventional TEM is illustrated schematically in Fig. 1; equivalent configurations can also be achieved in the scanning TEM (STEM) using a stationary defocused probe [153, 176]. The sample is examined using defocused, coherent illumination, usually from a FEG electron source. It is positioned so that it covers approximately half the field of view, and an electrostatic biprism is used to overlap the vacuum or reference wave with the wave that has traveled through the sample to form an interference pattern. The biprism normally consists of a thin ($< 1\,\mu m$) metallic wire or quartz fiber coated with gold or platinum, which is biased by means of an external direct current (DC) power supply or battery, typically to a voltage of between 50 and 200 V for the examples

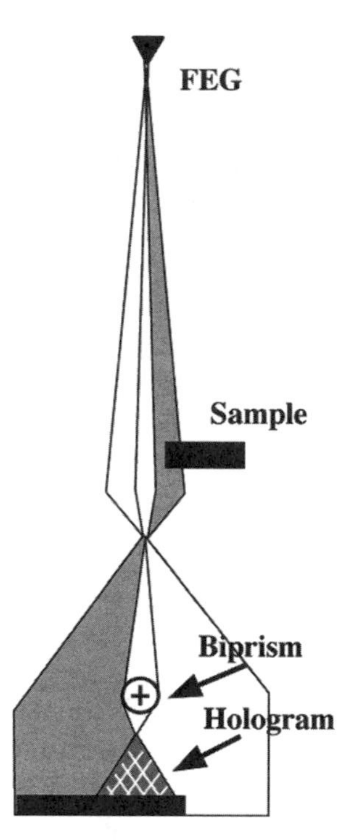

FIG. 1. Schematic ray diagram showing setup used for off-axis electron holography in the TEM. Essential components are the field emission gun (FEG) electron source used to provide coherent illumination and the electrostatic biprism that causes overlap of object and (vacuum) reference waves.

shown here. It is useful if the biprism is rotatable, as this allows for flexibility in aligning the interference fringes with respect to features of interest in the sample.

The biprism may be placed at several positions along the beam path. The most common arrangement is for the biprism to occupy one of the selected-area aperture positions. In this case, the interference pattern is formed at the first image plane, which is translated electron-optically to below the selected area plane by increasing the excitation of the diffraction or intermediate lens so that the image is located just below the biprism. The spacing of the interference fringes and the extent of their overlap are then referenced to the magnification of the image at this plane [161].

This magnification depends on the method used to ensure that the magnetic sample is located in a field-free region. For example, turning the normal objective lens off results in a low magnification and a correspondingly large field of view, but a relatively poor image resolution. Additional magnification may then be provided by using a postcolumn imaging filter. Alternatively, the sample can be located outside the field of the immersion objective lens, resulting in a far-out-of-focus image that can be corrected for defocus effects during the reconstruction process [176]. The use of a weak imaging lens below the normal objective lens, such as the so-called Lorentz minilens of the Philips CM200 FEG-TEM, allows magnifications of up to 70,000 × to be obtained, and it allows spatial resolution in the reconstructed phase image equal to the 1.4 nm information limit of the lens [177]. A special low-field (~5.5 G) objective lens for studying small magnetic particles has been reported to provide a maximum magnification of 500,000 × [178].

The biprism may also be placed in one of the condenser aperture positions. This configuration is equivalent to creating two closely spaced, overlapping plane waves incident on the sample [179]. An equivalent mode is obtainable using the stationary focused probe of the STEM [176]. If the observation plane is then defocused relative to the sample, the resulting interference pattern records the difference in phase shift between adjacent areas, resulting in a differential phase contrast (DPC) image. In this configuration, either a rotating biprism or a rotating sample holder is required, so both components of the in-plane magnetic field can be characterized. This approach is particularly attractive for holography applications in that it eliminates the need for a vacuum reference wave. However, it has the disadvantage that the hologram contains information about only one component of the magnetization.

For the quantitative analysis of phase shifts, a linear detector with a large dynamic range, such as a CCD camera or an imaging plate, should be used for recording the holograms [160, 180]. Digital acquisition also facilitates further computer processing. The sampling density of the recovered amplitude and phase is determined primarily by the effective pixel size of the hologram as referred to the sample, as well as by the size of the extracted sideband used for reconstruction (see below). The intensity distribution in the holographic interference fringe pattern using two coherent waves can be written in the form

$$\begin{aligned} I(x, y) &= I\Psi_1(x, y)|^2 + |\Psi_2(x, y)|^2 + |\Psi_1(x, y)||\Psi_2(x, y)|[c^{i(\phi_1-\phi_2)} + e^{-i(\phi_1-\phi_2)}] \\ &= A_1^2 + A_2^2 + 2A_1A_2\cos\Delta\phi, \end{aligned} \tag{1}$$

where Ψ is an electron wavefunction, ϕ is the electron phase, and the subscripts refer to the reference and object waves. The recorded hologram thus consists of a series of cosinusoidal fringes that are superimposed on a

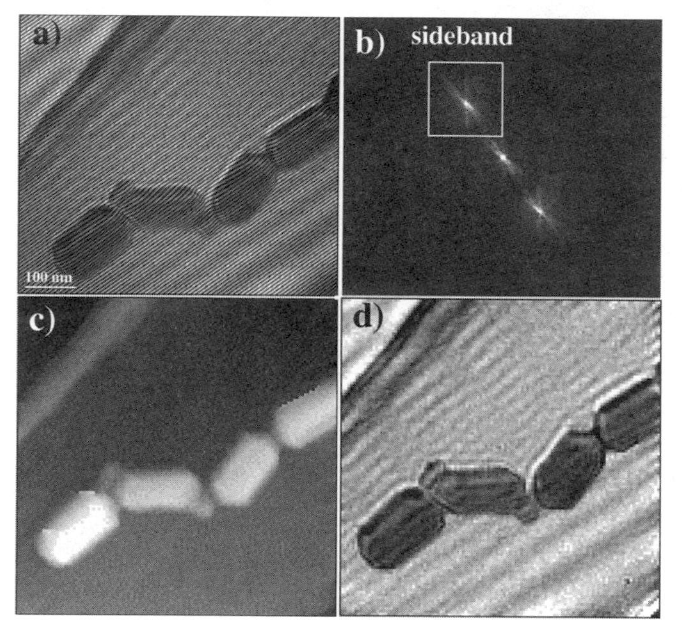

FIG. 2. (a) Off-axis electron hologram showing a chain of magnetosomes; (b) Fourier transform of (a) indicating sideband used in phase reconstruction; (c) phase image; (d) amplitude image.

normal TEM bright-field image. Changes in the positions of the fringes across the image record the relative phase shift of the electron wave after passing through the sample. Loss of contrast of the interference fringes owing to finite beam divergence (effective source size) and energy spread (temporal coherence) will directly affect the accuracy and precision of the reconstructed holographic phase.

To illustrate the technique, Fig. 2a shows an off-axis electron hologram of a chain of magnetite crystals from a single magnetotactic bacterium supported on a holey carbon film. A region of vacuum is located in the upper left-hand corner of the hologram. The large deviations in the spacing and angle of the fringes as they cross the crystallites record the phase shift of the electron wavefunction. The Fourier transform of the hologram shown in Fig. 2b provides a two-dimensional map of these changes in frequencies, which are responsible for the intensity distribution around the two sidebands that are centered on the fundamental cosine frequency. Each of the complex conjugate sidebands contains complete information about the phase and amplitude of the image wave, whereas this information is unavailable from the central autocorrelation function. The separation of these sidebands arises from the respective tilt of the object and reference waves and depends on the voltage applied to the biprism. The use of one of these sidebands in the reconstruction process gives rise to the term "off-axis" holography, and it provides a solution to the overlapping twin-image problem of in-line holographic techniques [153].

Reconstruction of the hologram proceeds by extracting and recentering one of the sidebands, followed by calculation of its inverse Fourier transform. The phase and amplitude of the resulting complex image are then

$$\phi = \tan^{-1}\left(\frac{i}{r}\right)$$
$$A = \sqrt{r^2 + i^2}, \tag{2}$$

where r and i are the real and imaginary data sets, respectively.

Figure 2c, d shows the reconstructed phase and amplitude, respectively, of the magnetosome hologram. The phase is usually calculated modulo 2π, which means that 2π phase discontinuities that are unrelated to particular specimen features appear at positions in the phase image at which the phase shift exceeds this amount. The phase must therefore be unwrapped carefully using suitable phase-unwrapping algorithms before reliable interpretation of the image features becomes possible [161]. It is also customary to record reference holograms in the absence of the sample so that artifacts associated with local irregularities of the imaging and the recording systems can be identified and excluded.

Neglecting dynamical diffraction effects, which can be significant for crystalline materials oriented close to zone-axis projections [181], the change in the phase of the electron wave after it has passed through the sample is given (in one dimension) by the expression

$$\phi(x) = C_E \int V(x, z)\,dz - \frac{e}{\hbar} \iint B_{\perp}(x, z)\,dx\,dz, \tag{3}$$

where z is the incident beam direction, x is a direction in the plane of the sample, V is the mean inner potential and $B_{\perp}$ is the component of the magnetic induction perpendicular to both x and z [182]. The constant C_E is given by

$$C_E = \frac{2\pi}{\lambda E}\frac{E + E_0}{E + 2E_0}, \tag{4}$$

where λ is the wavelength, E is the kinetic energy, and E_0 is the rest mass energy of the incident electron.

If neither V nor B vary with z, and neglecting magnetic and electric fringing fields outside the sample, this expression can be simplified to

$$\phi(x) = C_E V(x)t(x) - \frac{e}{\hbar}\int B_{\perp}(x)t(x)\,\mathrm{d}x. \tag{5}$$

Differentiation with respect to x leads finally to

$$\frac{\mathrm{d}\phi(x)}{\mathrm{d}x} = C_E \frac{\mathrm{d}}{\mathrm{d}x}\{V(x)t(x)\} - \frac{e}{\hbar} B_{\perp}(x)t(x). \tag{6}$$

In a sample of uniform thickness and composition, the first term of Eq. (6) disappears, and the in-plane induction is then proportional to the phase gradient. However, for many magnetic samples studied by TEM, the thickness may vary rapidly. Consequently, the mean inner potential term $V(x)t(x)$ may then dominate both the phase and the phase gradient, thus complicating any attempts to quantify the magnetization in the sample. Additional steps are then required during processing in order to extract the magnetization, and these will be demonstrated in the next section.

When a Lorentz lens is used for imaging, the spatial resolution of the reconstructed phase image is not limited by lens aberrations. Rather, the resolution for magnetic materials is limited by the strength of the magnetic signal and the phase sensitivity of the hologram (i.e., the signal-to-noise ratio in the phase image). The phase sensitivities that can be achieved depend on a number of experimental parameters [160], but values of $2\pi/100$ can routinely be achieved. Under these conditions, one can detect magnetic structures in thin films with details on a scale of about 5 nm producing a phase shift of $2\pi/100$ [183]. Smaller features would require thicker films, greater magnetization, or longer acquisition times.

4.4 Applications to Magnetic Materials

4.4.1 Hard Magnets

Because of the need for an electron-transparent sample, concerns about reliability are inevitable whenever a bulk magnetic material is thinned for TEM or holographic examination. For example, Fig. 3 shows underfocus Lorentz images of a sample of die-upset $Nd_2Fe_{14}B$ that was prepared for electron microscopy by standard dimpling and ion-milling procedures [184]. A variety of serpentine and Y-shaped domains, which extend to the sample edge in Fig. 3a, are outlined by lines of light and dark contrast at the domain walls. The shapes of these domains are similar to those previously observed at grain boundaries in sintered Nd–Fe–B [185]. In

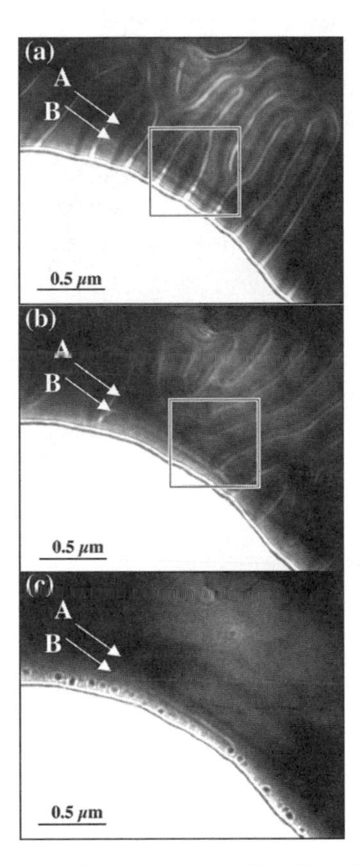

FIG. 3. Comparison of Lorentz images from a thin film of $Nd_2Fe_{14}B$: (a) recorded at room temperature; b) recorded at 300°C; (c) recorded at 400°C.

addition, two structural features can be seen, one having dark contrast and the other one bright, which are labelled A and B in Fig. 3. These lie parallel to the vacuum edge and intersect the domains that form the stems of the Y domains.

Heating the sample to a nominal temperature of 300°C and subsequent cooling to room temperature resulted in the domain rearrangement that is visible in Fig. 3b. Domain walls have been released from the thin edge of the sample, although some remaining domains interact with features A and B. Further heating to 400°C, which is above the Curie temperature of 312°C [186], resulted in complete disappearance of the domain structure, as shown in Fig. 3c. The persistence of the contrast at A and B indicates that these features are structural in origin. Feature A is revealed to be some sort of

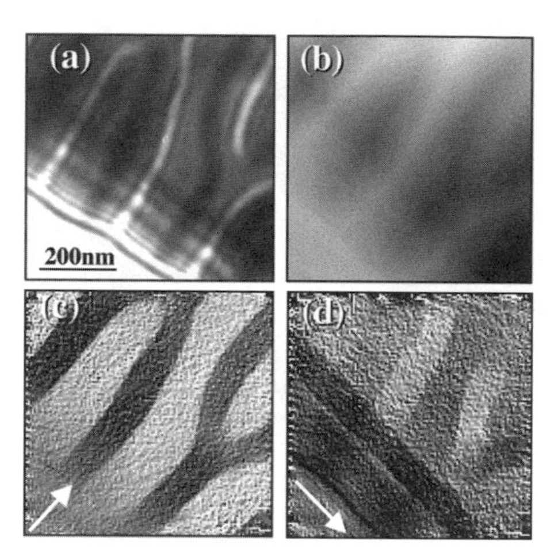

FIG. 4. (a) Enlarged Lorentz image of $Nd_2Fe_{14}B$ at room temperature; (b) phase image reconstructed from the hologram of the area shown in (a); (c, d) gradients of phase proportional to the magnetization component along the directions indicated.

planar defect, whereas B is a grain boundary that does not appear to strongly pin domain walls which are perpendicular to it, in agreement with other observations [187].

An enlargement from the Fresnel image of Fig. 3a is shown in Fig. 4a, and the corresponding phase image after holographic reconstruction is displayed in Fig. 4b. Bright ridges and dark valleys in the phase image clearly correspond to the positions of bright and dark lines in the underfocused Fresnel image, and they indicate the domain wall positions by changes in slope, predicted by Eq. (6). The width of the walls can be estimated by measuring the width of the region over which the slope of the phase perpendicular to the wall changes abruptly. Figure 5 shows an example of a single-pixel line scan across the bright ridge in the center of Fig. 4b. The small squares mark the phase data, and the lines are least-squares fits to the slopes of the phase components perpendicular to the wall. Although limited by the 5 nm pixel size of the reconstructed image, the line plot puts an upper limit of 10 nm on the wall width. This result agrees with theoretical values [186] (for details, see Section 8.4.4.3), especially given that no special care was taken to ensure that the walls were parallel to the electron beam path.

To map the magnetic induction within the sample, simple gradients of the phase image at $\pm 45^\circ$ angles were calculated according to Eq. (6), and these are shown in Fig. 4c, d. Figure 4c shows the induction component parallel to the sample edge (dark line at lower left), and this is primarily associated with the defect at A in Fig. 3. The grain boundary at B appears as a decrease

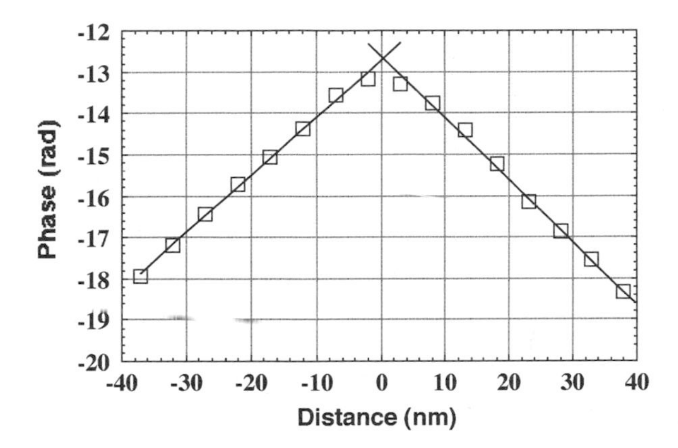

FIG. 5. Profile of phase cross a 180° domain wall in $Nd_2Fe_{14}B$. The phase slope is constant within domains on either side of the wall. The upper limit on the wall width is 7 nm.

in intensity over 15–20 nm. The induction component perpendicular to the edge is shown in Fig. 4d and it appears as striped domains with little disruption at A or B and diminishing in intensity as the sample becomes thinner near the edge. An estimate of the thickness profile can be made by processing the amplitude of the reconstructed hologram to yield the relative-thickness image [169]. This image indicates that the major thickness increase occurs in the grain nearest the sample edge and that the thickness becomes nearly constant at a distance of 150 nm from the vacuum. The effect of this thickness wedge is to enhance the apparent magnetization component parallel to the edge in the thinner part of the sample at the same time as the magnitude of the component perpendicular to the edge is decreasing. By assuming a mean free path for inelastic scattering of 60 nm, the absolute value of the thickness at the center of the image is estimated to be approximately 90 nm. This estimate is based on experimentally determined cross sections for Fe [188] and on prevailing experimental conditions, namely, medium effective acceptance aperture and weakly diffracting conditions.

By combining the two gradient images into a vector map, a direct image of magnetic structure within the domains is produced, as shown in Fig. 6a. The vector map, which has been superimposed on a low contrast image of the x-gradient, is divided into 20×20 nm^2 squares. The minimum vector length is zero (indicating out-of-plane induction), and the maximum vector length calculated from the modulus of the gradients near the center of the image corresponds to $4\pi M_s = B = 1.0$ T for a thickness of 90 nm. The vector

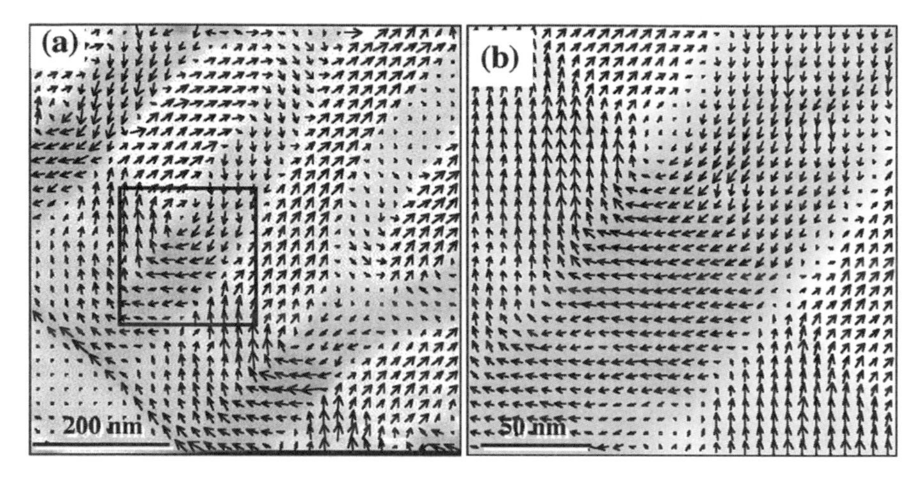

FIG. 6. (a) Induction map derived from phase gradients in Fig. 4; (b) enlargement of area indicated in (a) showing singularities and domain wall character.

map shows that the magnetizations of the domains in this region of the sample are oriented at approximately 90° to each other, rather than 180° which might have been expected from a material with such strong uniaxial anisotropy. The region of the sample between A in Fig. 3 and the vacuum edge comprises a series of domains that appear to serve to close the in-plane induction more or less parallel to the thin edge. Pairs of singularities at the intersection with the striped domains are reminiscent of cross-tie walls that contain periodic arrays of Bloch lines of alternating polarity [189]. The induction map can be interpreted only qualitatively in the thinnest region of the sample, given the additional complications of the thickness gradient described above.

The grain boundary has only a limited effect on the magnetization, and the magnetization across the two grains is coupled. Contributions to the phase at the grain boundary may arise not only from changes in the induction at this position but also from preferential thinning or changes in composition within the boundary that affect the mean inner potential.

Figure 6b shows a higher magnification map of the area, outlined in Fig. 6a, that encloses a pair of the singularities. This map is divided into 6.7×6.7 nm^2 squares, and the maximum vector length is again 1.0 T. Portions of the domain walls that run perpendicular to the sample edge show distinct out-of-plane character, and there is an apparent tendency for the induction in the center domain to rotate toward a 180° orientation near the vortices. The vortices show Blochlike character with vanishingly small vector length, indicating small in-plane components. The majority of the 90° walls show large in-plane components as expected for a thin sample. However, care is needed in interpreting details of the induction maps owing to undetermined effects of the fringing fields immediately above and below the sample surfaces.

A more intuitive picture of changes in magnitude and direction of the micromagnetic structure following thermal annealing for such a complicated region can be obtained by mapping the phase gradients onto a suitable color wheel, as shown in Fig. 7a (see color insert). The hue of the wheel indicates the direction of the in-plane induction, and the saturation value indicates the magnitude, which ranges linearly from zero at the center to a maximum of 0.8 T for Fig. 7b and 0.5 T for Fig. 7c. In these color maps, large-scale variations and local differences in the magnetic structure become strikingly apparent. The coupling between the grains can be identified in Fig. 7b by the similarity in the blue and green hues across the grain boundary marked by the thin magenta line.

Figure 7c is a color induction map of the same area (see Fig. 3b) after heating to 300°C and subsequent cooling to room temperature. The well-defined domains on the left-hand side have disappeared, and reduced phase gradients indicate that the remaining structure has a large out-of-plane component. The domains that had previously been oriented perpendicular to the edge of the sample have rotated by approximately 90°. These are now approximately 180° domains as indicated by their alternating red and green striped appearance. The magnetizations on either side of the grain boundary are no longer strongly coupled. However, the domains in the lower part of Fig. 7c appear to be pinned between the grain boundary and the defects that have impeded the movement of domains away from the thin edge of the sample. The new domain configuration indicates that the previous state with 90° domain walls was metastable, as only 180° domains are expected in this highly anisotropic material, and this emphasizes the need for caution in extrapolating the magnetization state of thin films to bulk materials.

4.4.2 Fields in Vacuum

Complications in hologram interpretation owing to specimen thickness variations do not arise when the technique is used to image magnetic fields that fringe out into the vacuum. This geometry, for example, has enabled Tonomura and colleagues to image the spatial distribution of fluxons from a Nb superconductor [173]. Magnetic films with pronounced out-of-plane magnetization are of much interest to the magnetic recording industry. Figure 8 shows a hologram and the reconstructed phase image for an epitaxial layer of an $Fe_{0.5}Pt_{0.5}$ ordered alloy that was deposited by molecular beam epitaxy onto an MgO (001) single-crystal substrate [177].

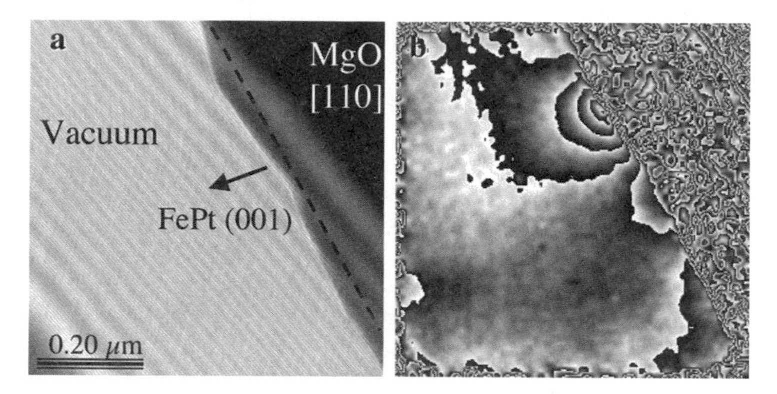

FIG. 8. (a) Off-axis electron hologram of FePt/MgO with (001) easy axis parallel to the film normal; (b) contoured phase image showing magnetic flux extending into vacuum.

In this growth direction, the $L1_0$ ordered alloy phase has high magnetocrystalline anisotropy with the (001) easy axis parallel to the film normal. The reconstructed phase image clearly shows magnetic fringing fields outside the material, and it indicates the presence of domains of opposite polarity.

Significant difficulties arise whenever quantitative information about the distribution of flux in vacuum is desired. One problem is that off-axis electron holography is insensitive to any component of the magnetic field that is parallel to the direction of the electron beam. Quantitative imaging of the magnetic field from the tip of a magnetic force microscope (MFM), for example, requires extensive calculations to model the two-dimensional projection of the three-dimensional fringing field [106]. Figure 9 shows a multipole expansion fitted to an experimental phase map of a rare-earth magnet MFM tip. The fitted contour lines are visible near the tip as noise-free contours. In addition, limited tomographic experiments were needed to confirm the cylindrical symmetry assumed in the calculation [190]. Finally, it should be noted that full three-dimensional simulations are always required for correct quantitative interpretation when the fringing fields extend relatively large distances into vacuum [191, 192].

4.4.3 Small Magnetic Particles

In small magnetic particles, the energy associated with formation of any domain walls becomes increasingly important relative to the total magnetostatic energy of the entire particle. It is then less likely that domain walls will be observed within the small particles in their remanent state (with no

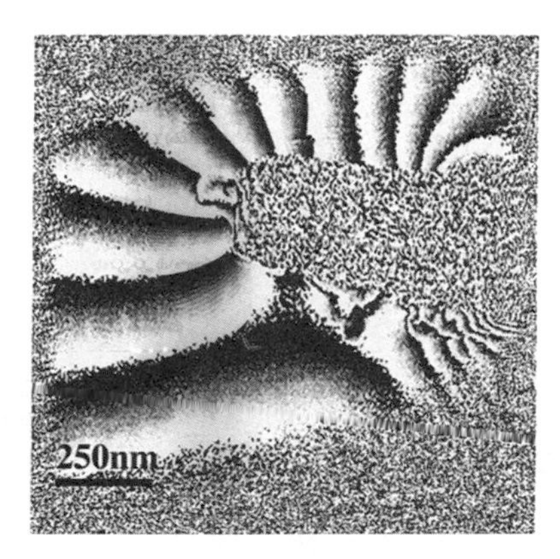

FIG. 9. Multipole expansion simulation of phase contours surrounding MFM tip. (Reprinted with permission from Streblechenko [190].)

externally applied field). Moreover, micromagnetic structure may become difficult to characterize because of the limited resolution of other magnetic imaging techniques. Isolated polyhedral particles of barium ferrite, with sizes ranging from about 0.1 to 1.0 μm, were reported by Hirayama *et al.* [193] to exist as single magnetic domains on the basis of the magnetic flux line geometry surrounding the particles. Flux lines within the ferrite particles, and also in smaller-sized iron particles, were not identified presumably because of the dominant influence of the mean inner potential contribution to the phase and phase gradients [see Eqs. (5) and (6)]. Prior information about the thickness profile of small particles is usually essential before their internal magnetization can be reliably determined [194]. In some special cases such as the CrO_2 needles observed by Mankos *et al.* [195], cross sections may be constant so that thickness variations can be neglected.

For some samples, large variations in the projected thickness of the sample will be unavoidable. For example, holographic analysis of a chain of carbon-coated Co particles, shown in Fig. 10, illustrates the dominant contribution of the mean inner potential to the phase profile [59]. The magnetic contribution to the phase shift of a Co particle is visible as a difference in the value of the phase in the vacuum on either side of the chain of spheres. Contributions to the phase shifts in vacuum can also be expected from the fringing fields of the other particles. In this case, quantitative analysis of the magnetization of crystalline nanoparticles required extensive modeling (see Chapter 2).

For certain sample geometries, that lend themselves to unidirectional remanent states, separation of phase shifts due to thickness and electrostatic effects from those due to magnetostatic effects can be accomplished by *in*

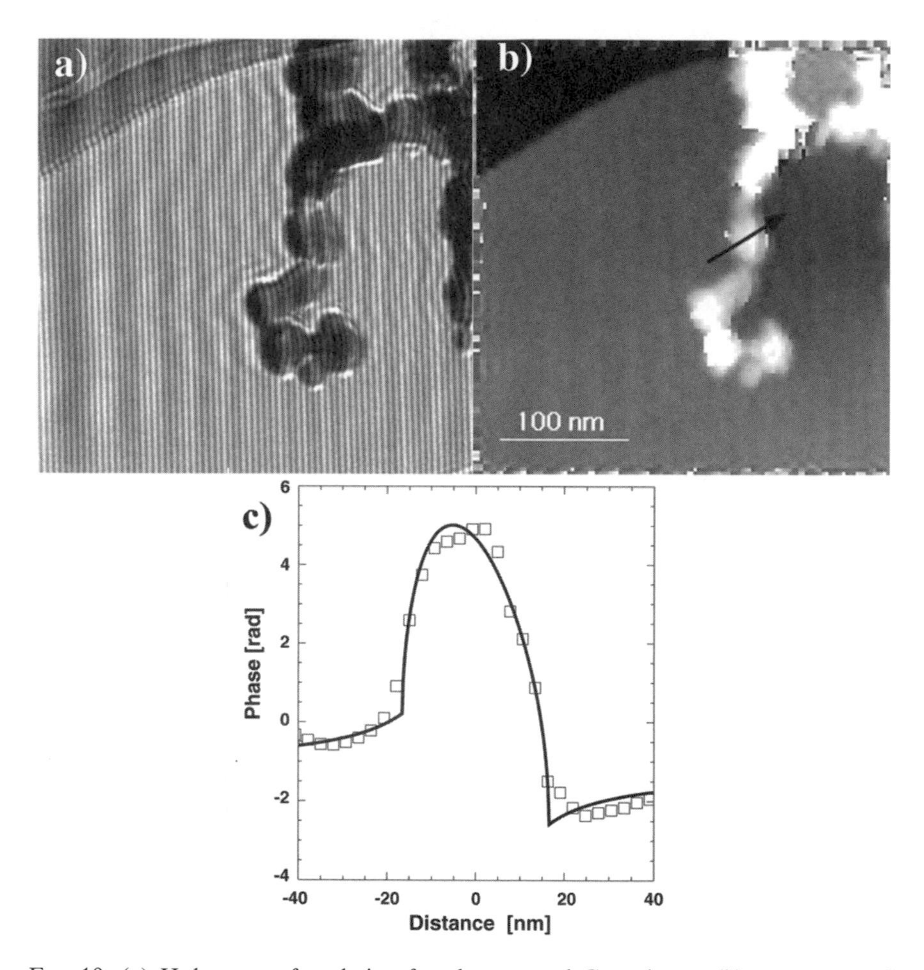

FIG. 10. (a) Hologram of a chain of carbon-coated Co spheres; (b) reconstructed phase image; (b) phase profile along arrow in phase image in (b) (open squares). Solid line in (c) shows the expected phase shift calculated from the isolated sphere.

situ magnetization reversal. This procedure involves acquisition of two holograms with the direction of magnetization reversed in the sample by tilting the sample in the field of the weakly excited objective lens [196]. The sum of the phases of these two holograms then represents twice the mean inner potential contribution to the phase if the magnetization has exactly reversed (see Eq. (5)), whereas the difference of the phases gives twice the magnetic contribution. As an example, Fig. 11a shows a hologram of a chain of magnetite crystals in an aquatic magnetotactic bacterium, *Magnetospiril-*

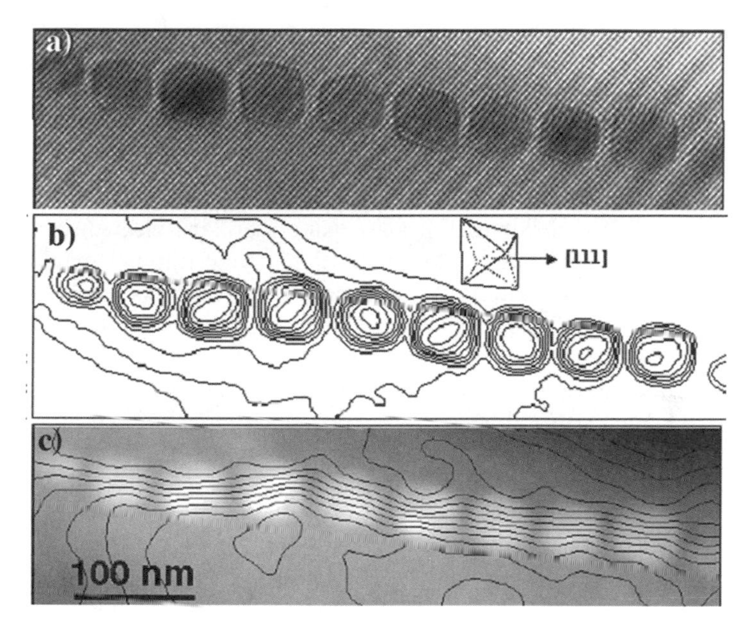

FIG. 11. (a) Hologram of a chain of magnetite crystals in *Magnetospirillum magnetotacticum*; (b) electrostatic contribution to the phase of the reconstructed hologram, where thickness contours indicate an octahedral shape; (c) magnetic contribution to the phase.

lum magnetotacticum. Following magnetization reversal and addition of the phases, Fig. 11b shows the electrostatic contribution in which the thickness contours reveal the crystallites to be octahedral in shape [194]. The magnetic contribution to the phase is shown in Fig. 11c. The contours are parallel to the magnetic induction and have been overlaid onto the mean inner potential contribution so that the positions of the crystals can be correlated with the magnetic contours. This procedure makes the flux visible both within and between the crystallites and also allows the total dipole moment of the chain to be measured.

4.4.4 Nanostructured Elements

The magnetization reversal of individual magnetic elements is of much interest because of their potential utilization in information storage applications. Owing to their small size, magnetostatic energy contributions have a major influence in determining the overall magnetic response. Individual particle geometry as well as proximity to other particles are further important factors that need to be considered. Off-axis electron holography has already played a valuable role during investigations of micromagnetic behavior by enabling the magnetization state of nanostructured elements to be visualized during hysteresis cycling [197–199].

In our initial studies we have observed elements of different shapes and sizes that were prepared by electron-beam evaporation onto self-supporting 55-nm-thick silicon nitride membranes using standard electron-beam lithography and lift-off processes. Magnetic fields were applied by tilting the sample *in situ*, thus enabling determination of the hysteresis loops of individual elements. The extent of any interactions between closely spaced elements could also be assessed. A montage of Lorentz micrographs from an array of rectangular Co elements of nominal thickness 30 nm is shown in Fig. 12. The external field was applied along the line joining the rectangles, and the strengths of the in-plane applied fields are indicated. Domain wall configurations within the particles are visible as narrow bright or wider dark lines. Careful analysis showed that the coercive field increased with decreasing element size. It was also significant that domain structures were not reproduced exactly during successive hysteresis cycles.

Electron holography is required in order to visualize the flux lines within and outside the elements. Figure 13 shows a typical hologram of the two smallest Co rectangles from the array in Fig. 12. Despite the substantial loss of contrast owing to the underlying silicon nitride, holographic reconstruction showing the magnetization was still possible after removal of the mean potential contributions, as described above. The results for a complete

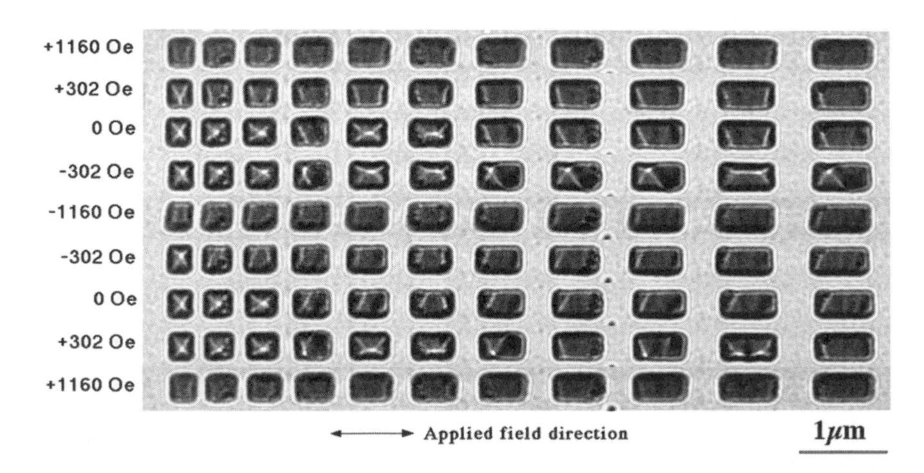

Fig. 12. Montage of Lorentz micrographs from patterned Co nanostructures, recorded during an entire hysteresis cycle with in-plane applied fields as indicated.

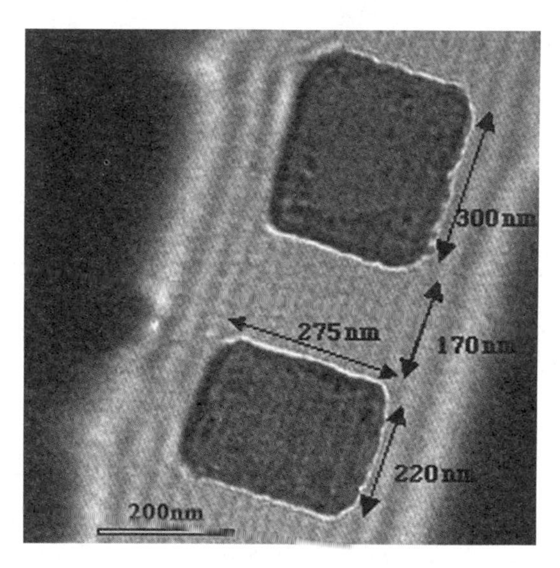

FIG. 13. Representative off-axis electron hologram showing two patterned Co nanostructures from a linear chain of elements.

hysteresis cycle are shown in Fig. 14a (see color insert), with the corresponding in-plane components as indicated. Note that the holograms of this series were recorded with an average out-of-plane field of 3600 Oe, and a color wheel has again been used to represent the different directions of the in-plane magnetization [pure directions are red (right), yellow (down), green (left), and blue (up), with intensity reflecting the magnitude]. The phase contours follow lines of constant magnetization, and their separation is proportional to the magnetic induction integrated in the incident beam direction. It is significant that the fringing fields between the elements are only minimized when the field lines within both of the elements show a solenoidal shape, which is indicative of flux closure.

These phase measurements were compared with simulations based on the Landau–Lifshitz–Gordon equations, as shown in Fig. 14b [198]. Qualitatively, there was reasonable agreement, but subtle differences became apparent after further analysis. For example, simulated vortices matched the experimental results except that they were formed at higher fields in the simulations, presumably because of local defects or inhomogeneities in the real elements. The squareness of the corners also affected the simulated results. Slight changes in the initial state had a strong influence on the subsequent domain formation, and it was demonstrated in the simulations that the close proximity of the larger neighboring cell led to a single, centered vortex state in the smaller cell rather than a paired vortex state that occurred when it was isolated. Simulations also showed that the strength and direction of the applied field had a marked impact on the observed domain structure, thereby emphasizing the value of correlating experimental measurements with micromagnetic simulations whenever possible. Overall, it was concluded that intercell coupling should not be overlooked in the design of high density magnetic storage devices.

The formation of remanent states from different stages in a magnetization reversal loop is an important consideration in the design of magnetic elements for device applications. In normal operation, the element must first be magnetized, and then the external field would be removed, with the object being to retain a nonsolenoidal distribution. As shown in Fig. 15 (see color insert), none of the different states of the two Co elements, which were produced by full magnetization followed by cycling to the value indicated and then final reduction of the applied field to zero, show the nonsolenoidal domain structures visible at the extreme ends of the hysteresis loop. Indeed, new and unexpected domain configurations are observed, such as the double-vortex structure seen in the larger rectangle. It may well be that some of these states are only stable because of the presence of the external vertical field. Nevertheless, the need for reproducibility in the formation of remanent states, and the value of experimental observations, is emphasized.

Following these initial observations of Co rectangles, further studies have focused on patterned, submicron Co (10 nm)/Au (5 nm)/Ni (10 nm) nanostructures, patterned as diamonds, ellipses, and rectangles [199]. Both off-axis electron holography and micromagnetic simulations have been used to investigate magnetization reversal processes as well as interlayer coupling. Figure 16 (see color insert) shows representative results over a complete hysteresis cycle for a selection of the elements, tabulated in three sets of four columns for one element of each shape. The left column of each set shows the magnetic contributions to the experimental holographic phases, where the phase contours (spacing of 0.64π radians) follow lines of constant field strength. The remaining columns show corresponding simulations for 3.5-nm-thick magnetic films of each element shape, with the columns labelled Co and Ni showing the magnetization state for the individual FM layers and those labelled Total showing the computed holographic phase shifts. Several important results emerged from this study. Solenoidal states are observed experimentally for both elliptical and diamond-shaped elements, but they could not be replicated in the simulations despite extensive trial-and-error attempts, suggesting that structural imperfections such as crystal grain size or orientation are likely to have been contributing factors. The Ni layer in each element reverses its magnetization well before the external field reaches 0 Oe, confirming that an antiferromagnetically coupled state would be the normal remanent state that would be obtained after saturation of the element followed by removal of the external field. This coupling is attributed to the strong demagnetization field of the closely adjacent and magnetically more massive Co layer. The occurrence of this flux closure associated with an antiferromagnetic state would contribute to the lack of end domains, as commonly seen in thicker single-layer films or elements of larger lateral dimensions.

4.4.5 Layered Thin Films in Cross Section

Magnetic multilayers in the cross-sectional geometry present a particularly difficult challenge for quantification using electron holography. In such cases, all three significant quantities in Eq. (6), namely, $V(x)$, $B(x)$, and $t(x)$, are likely to vary rapidly as a function of position. Figure 17a shows part of a hologram of a magnetic tunnel junction (MTJ) consisting of 22 nm Co/4 nm HfO_2/36 nm CoFe deposited by sputtering onto a Si substrate. The phase profile shown in Fig. 17b illustrates the difficulty of analyzing the induction in a sample in which the thickness and mean inner potential both vary abruptly. One can now calculate the difference in the gradients (Eq. (6)] of magnetically reversed layers and then divide this difference by the sum of the phases, which contain the thickness variation, to yield an image that is

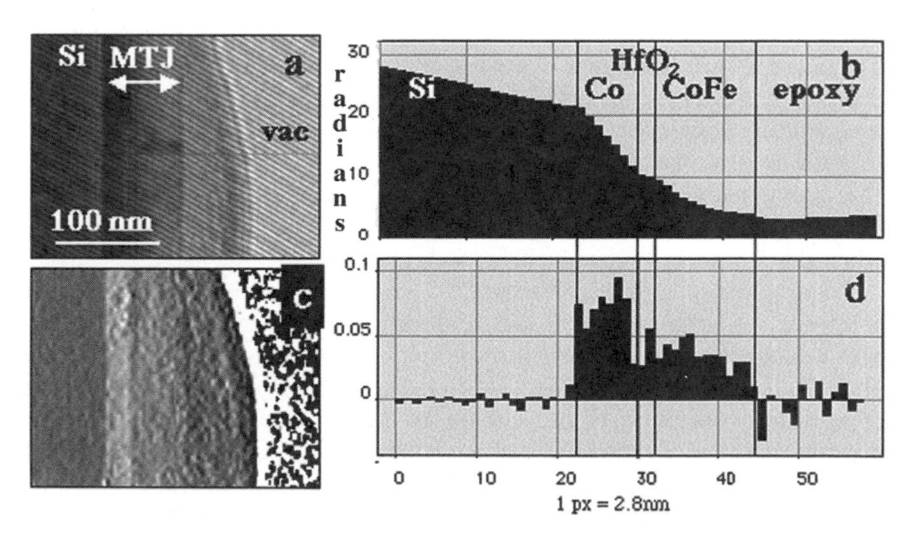

FIG. 17. (a) Hologram of Co/HfO_2 (4 nm)/CoFe magnetic tunnel junction; (b) phase profile; (c) after processing, with contrast proportional to magnetization; (d) profile of (c).

proportional to the magnetization, as shown in Fig. 17c [196]. Although a variety of artefacts may remain in this image, the profile in Fig. 17d shows that it is appropriately nonzero only in the magnetic layers, and analysis yields a value for the magnetization of the Co layer (assuming that Co has a mean inner potential of 25 V) of 1.5 T.

4.4.6 Layered Thin Films in Plan View

Thin magnetic films below some critical thickness will have in-plane magnetization, but domain wall formation will occur in the absence of an applied field in order to reduce stray field energy. Off-axis electron holography can be used to characterize the magnetic behavior of thin films in plan view provided that regions of vacuum can be found that can be overlapped onto the sample. Figure 18 (see color insert) shows Lorentz microscopy and holography results obtained from domain walls in 15-nm-thick Co and $Ni_{40}Fe_{60}$ films sputtered directly onto holey carbon. The Lorentz images at the top show the areas from which the reconstructed phase images were determined. (The black areas were outside the holographic interference region and were scaled to zero in the images, as they contained no phase information.) Prominent domain walls and vortices are visible when the phase images are displayed with contour lines overlaid on them and as color maps. It is particularly interesting to note that the magnitude of the in-plane magnetic induction integrated in the incident beam direction shows dark contrast at some of the domain walls. This darker contrast may be associated with the presence of strong magnetic fringing fields outside Néel walls, and these may result in the magnetization decay of an adjacent magnetically hard reference layer when such a Co or $Ni_{40}Fe_{60}$ film is used to form the sense layer in a spin-dependent tunnel junction [200]. However, care is also required in the interpretation of such data because similar dark contrast at a domain wall may result from the acquisition of a hologram when the sample is imaged far from focus.

An example of a [Co (1.5 nm)/Cu (3 nm)] $\times$ 10 superlattice structure supported on a holey carbon film is shown in Fig. 19 [201]. The Lorentz out-of-focus image in Fig. 19a shows the presence of five domains centered about a vortex, whereas the reconstructed phase image in Fig. 19b enables measurements of the maximum phase gradient to be made in each of the domains. In four of these, the average value differed from that expected for bulk Co (15 nm thick) by less that 1%, implying that these domains were ferromagnetically coupled and penetrated through the entire superlattice. In domain 4, the measured phase gradient was 90% of the expected value. This discrepancy was explained in terms of one of the layers having its magnetization rotated by 90° (but not antiferromagnetically aligned, as this would

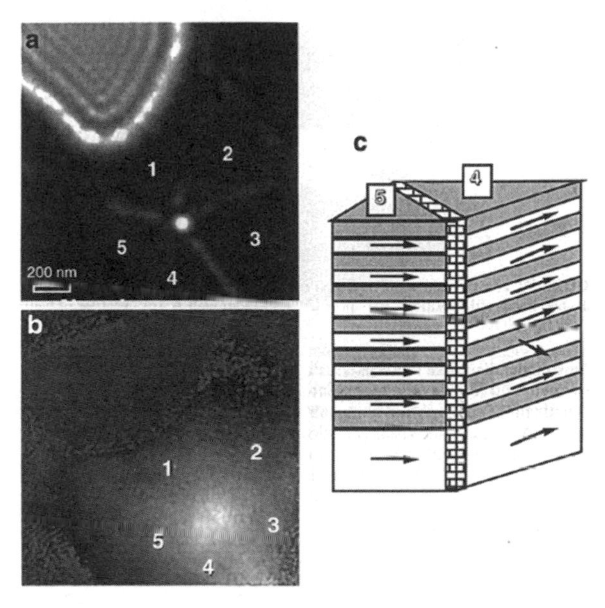

FIG. 19. Co/Cu multilayer structure: (a) Fresnel image recorded with stationary, underfocused probe; (b) partially unwrapped phase image showing domains surrounding the vortex; (c) proposed magnetic structure in the multilayer stack. (Reprinted with permission from Mankos *et al.* [201].)

have caused a 20% reduction in total magnetization), as drawn schematically in Fig. 19c.

4.4.7 Differential Phase Contrast Mode

A major limitation of off-axis holography is the need for a vacuum reference wave. This requirement can be relaxed in the case of nanostructured elements supported on silicon nitride membranes, even though the holographic fringe visibility is reduced substantially. An alternative method involves the use of a biprism located in the illumination system preceding the sample [176, 179], as illustrated in Fig. 20, which results in a form of differential phase contrast (DPC). This mode of operation allows examination of features of interest located far from the specimen edge. This geometry thus removes the concern that the thin edge of the sample has affected the magnetic state of the film.

A preliminary result obtained using this DPC mode of electron holography for a Co film of nominal 30 nm thickness is shown in Fig. 21a. Black and white lines characteristic of domain walls, as well as magnetization

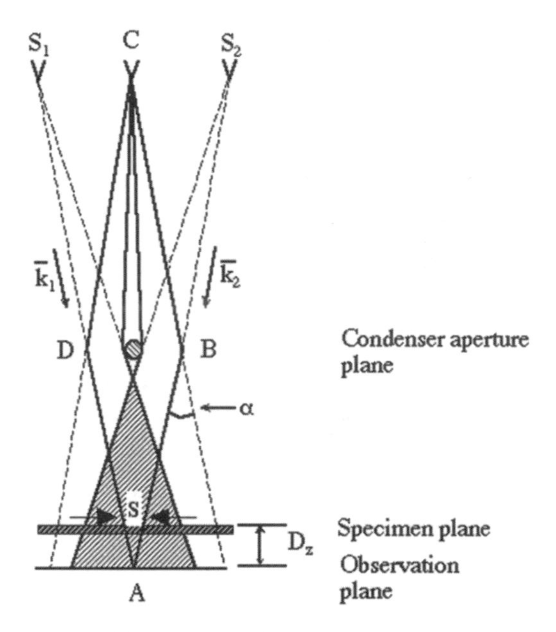

FIG. 20. Schematic ray diagram showing setup used for the differential phase contrast mode of off-axis electron holography in the TEM. The electrostatic biprism located in the condenser aperture plane creates two closely spaced, overlapping plane waves.

ripple, are visible outside the region of interference fringes. Doubling of image features owing to the split incident beam as well as fringe bending owing to locally varying fields are clear in the enlargement shown in Fig. 21b. The reconstructed phase image shown in Fig. 21c was created from a series of eight holograms acquired with the fringe system moved laterally between exposures. The image contrast is proportional to the magnetic field component parallel to the holographic fringes, as indicated. Recall that either a rotating biprism or a specimen holder is required before both components of the in-plane field can be fully characterized.

4.5 Prospects and Concluding Remarks

The direct visualization of dynamic effects caused by an externally varying magnetic field, which is of great importance to a full understanding of micromagnetic behavior, represents a major challenge for the technique of electron holography. Changes in the applied field will affect the electron trajectory and thus also impact the holographic interference process. One

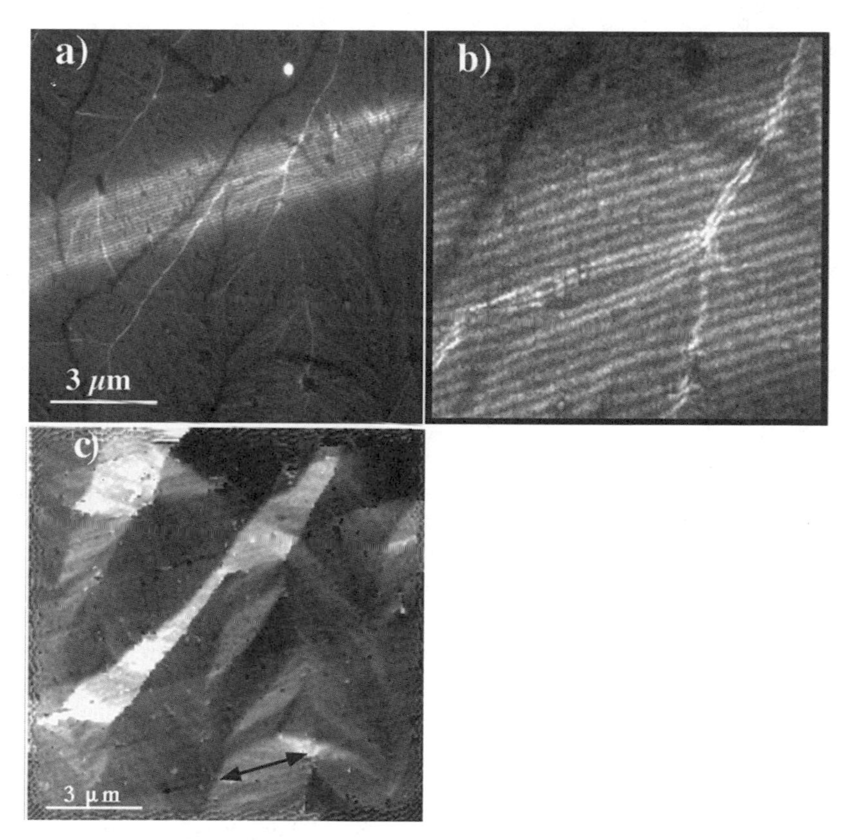

FIG. 21. (a) Differential mode hologram of a 30-nm-thick Co film; (b) enlargement from the area of fringes showing fringe bending and doubling of image features; (c) DPC image corresponding to (a). The field direction is along the direction of the arrow.

solution mentioned above [197] is to tilt the sample *in situ* while it is located in the field of the weakly excited objective lens, and thereby change the in-plane field component. However, the presence of an out-of-plane component of the applied field was shown to result in switching asymmetries during hysteresis cycling of Co nanostructures [198]. A preferable solution, but one that is not yet generally available, is to provide three sets of auxiliary coils: one set is used to apply the field at the level of the sample, and the subsequent sets steer the beam back onto the optic axis after its initial deflection [202].

Because holograms are usually processed off-line, the technique of off-axis electron holography is not well suited to real-time observations. This limitation has been (partially) circumvented by feeding the signal from an intensified TV camera to a liquid crystal (LC) panel, and then using this panel as the input for a light-optical reconstruction [203]. Unlike digital recording and processing, reference holograms are not easily available to correct distortions of the imaging system, and further geometric distortions may result from the TV camera and the LC panel. Nevertheless, by using this type of system, the dynamic behavior of magnetic domains in a thin Permalloy film was observed in real time during (de-) magnetization processes [193]. Lorentz microscopy could certainly be used for similar dynamic observations, but the technique lacks the capability for directly visualizing local magnetization distributions. With the advent of faster computers and the availability of faster CCD cameras, real-time viewing of reconstructed phase images may soon become available.

In conclusion, the results described in this chapter have shown that off-axis electron holography offers the prospect of direct visualization and quantification of magnetic fields with high spatial resolution and sensitivity. The theoretical framework of the technique is well established, and experimental applications to important materials problems are being explored. With the increasing availability of field emission gun TEMs, electron holography should develop into a widely used tool for micromagnetic characterization.

Acknowledgments

Much of the electron holography described here used facilities at the Center for High Resolution Electron Microscopy at Arizona State University. We are grateful to Drs. R. F. C. Farrow, R. Frankel, B. Kardynal, S. S. P. Parkin, M. Postfai, M. R. Scheinfein, and Y. Zhu for provision of samples and for ongoing collaborations.

5. PHASE RETRIEVAL IN LORENTZ MICROSCOPY

Anton Barty, David Paganin, and Keith Nugent
School of Physics
The University of Melbourne
Victoria 3010, Australia

5.1 Introduction

The conservation of energy is a fundamental concept in physics from which it can be shown that the time-averaged flow vector of a statistically stationary wave field is divergence free. Very early in the development of quantum mechanics the concept of flow was developed to produce the so-called hydrodynamic formalism [204], which expressed quantum mechanical ideas in a form describing the flow of probability density. It was claimed, even then, that knowledge of the three-dimensional probability density is sufficient to fully define the field [205]. Although there are some caveats on this claim that were not recognized then, it is this insight that we use in the present chapter to explore phase recovery for an electron wave field.

The flow conservation equation $\nabla \cdot \mathbf{j} = 0$, where $\mathbf{j}$ is the flow vector, is well known to physics undergraduates. However, its use in the recovery of phase was, to our knowledge, first proposed by Teague in 1983 when he wrote down a paraxial form of the flow conservation equation and dubbed it the transport-of-intensity equation (TIE) [71, 206]. Teague pointed out that its solution would permit the recovery of phase from a measurement of the optical intensity distribution in a plane and a measurement of the intensity derivative along the optical axis of the system, and he gave a solution method for the equation based on Green's functions. The adaptive optics community, primarily concerned with the real-time correction of atmospheric distortions of astronomical images, took up these ideas in the method known as "curvature sensing" [207–209]. However, to our knowledge, no formal solution to the equation was described in this work.

In the early 1990s one of the authors (K.N.) explored a number of approaches to the solution of the TIE [67–69, 210]. An analytic, closed form solution was developed for recovery of the phase of a wave in the absence of intensity modulation [211]; however, the computational requirements for the fully general solution were prohibitive for all but the smallest of images [69]. More recent work has led to a new understanding of what we mean

EXPERIMENTAL METHODS IN THE PHYSICAL SCIENCES
Vol. 36
ISBN 0-12-475983-1

ISSN 1079-4042/01 $35.00

by phase, leading to the development of a complete, analytic solution for the phase valid in the presence of intensity modulation. This solution can be implemented extremely efficiently using modern computer hardware [70].

In this chapter we explore the application of this technique of phase measurement to high energy electrons in an electron microscope and apply this work to Lorentz microscopy of magnetic structures at the domain level [212]. In Section 5.2 we review the ideas behind our phase recovery strategy. In Section 5.3 we examine some practical issues that apply to microscopy, and in Section 5.4 we present our results and compare the outcomes with electron holographic work. In Section 5.5 we discuss the implications of our work and possibilities for future developments.

5.2 Phase Retrieval Theory

Here we develop the requisite theory for quantitative noninterferometric phase measurement in Lorentz microscopy using the TIE [71]. Section 5.2.1 gives a derivation of this equation for the case of a monoenergetic electron beam, emphasizing its origins in the hydrodynamic formulation of quantum mechanics. Section 5.2.2 generalizes the formalism to encompass a wide variety of different forms of radiation, both coherent and partially coherent. Finally, Section 5.2.3 discusses a method for the rapid numerical solution of the TIE.

5.2.1 The Transport-of-Intensity Equation

The TIE is central to our later discussions on quantitative phase retrieval in Lorentz microscopy. In the present section we give a derivation of this equation in the context of quantitative noninterferometric phase imaging, emphasizing the historical continuity of this work with the hydrodynamic formulation of quantum mechanics derived by Madelung in the same year that Schroedinger wrote his famous sequence of papers [204, 213]. Note that an alternative derivation of this equation has been presented by De Graef in Chapter 2 in this volume (see Section 2.5), using the standard transfer function formalism of image formation in electron microscopy.

Consider an electron wave that obeys the free-space Schroedinger equation:

$$(\nabla^2 + k^2)\sqrt{\rho(\mathbf{r}_\perp, z)}\; e^{i\phi(\mathbf{r}_\perp, z)} = 0, \tag{1}$$

where ∇^2 denotes the three-dimensional Laplacian operator, $k = 2\pi/\lambda$, λ is the de Broglie wavelength, ρ and ϕ are respectively the probability density and phase of the wave function, and $\mathbf{r}_\perp$ is a position vector in the plane perpendicular to the optic axis z. After a little algebra, we learn that the first term of Eq. (1) may be expanded thus:

$$\nabla^2\sqrt{\rho}\,e^{i\phi} = \sqrt{\rho}\,e^{i\phi}\left(-|\nabla\phi|^2 + i\nabla^2\phi + \frac{i}{\rho}\nabla\rho\cdot\nabla\phi + \frac{1}{\sqrt{\rho}}\nabla^2\sqrt{\rho}\right), \quad (2)$$

where for the sake of clarity we have dropped the explicit functional dependence of the probability density and phase on position. If we now substitute Eq. (2) into Eq. (1), we obtain

$$-\sqrt{\rho}|\nabla\phi|^2 + i\sqrt{\rho}\nabla^2\phi + \frac{i}{\sqrt{\rho}}\nabla\phi\cdot\nabla\rho + \nabla^2\sqrt{\rho} + k^2\sqrt{\rho} = 0. \quad (3)$$

Both the real and imaginary parts of the left-hand side must be equal to zero, leading, respectively, to the equations

$$\nabla^2\sqrt{\rho} + \sqrt{\rho}(k^2 - |\nabla\phi|^2) = 0, \quad (4a)$$

$$\nabla\cdot(\rho\nabla\phi) = 0. \quad (4b)$$

In the case of a coherent field, the flow vector is the probability current (or Poynting vector) and is given by, for a quantum mechanical field, $\mathbf{j} = (1/m)\rho\nabla\phi$ [214]. Thus, Eq. (4b) is the conservation equation discussed in the introduction. Equations (4a) and (4b) form the basis of the hydrodynamic formulation of quantum mechanics that was mentioned earlier. The term "hydrodynamic" is appropriate in view of the fact that the quantity under the divergence sign of the conservation equation [Eq. (4b)] is proportional to the expression for probability current for stationary-state solutions to the Schroedinger equation. We will return in more detail to this point in the following section.

The TIE is simply a paraxial form of Eq. (4b). Paraxial waves may be viewed as weak perturbations of the elementary z-directed plane wave e^{ikz}. More precisely, paraxial waves are those for which all current vectors lie in a cone of small angle about the z-axis. A paraxial wave $\psi(\mathbf{r}_\perp, z)$ satisfies a modified form of Eq. (1), namely [67],

$$\left(2ik\frac{\partial}{\partial z} + \nabla_\perp^2 + 2k^2\right)\psi(\mathbf{r}_\perp, z) = 0, \quad (5)$$

where $\nabla_\perp^2$ denotes the Laplacian operator acting in the plane perpendicular to the optic axis z. If we now explicitly write the wave function as a perturbed plane wave nominally traveling in the z-direction:

$$\psi(\mathbf{r}_\perp, z) \equiv e^{ikz}u(\mathbf{r}_\perp, z), \quad (6)$$

and substitute this into Eq. (5), we arrive at the paraxial wave equation:

$$\left(2ik\frac{\partial}{\partial z} + \nabla_{\perp}^{2}\right) u(\mathbf{r}_{\perp}, z) = 0. \tag{7}$$

To obtain the transport-of-intensity equation, we perform manipulations exactly analogous those that led to the transport equation Eq. (4b). First, we write the wave function in Eq. (7) explicitly in terms of its modulus and phase:

$$\left(2ik\frac{\partial}{\partial z} + \nabla_{\perp}^{2}\right) \sqrt{\rho(\mathbf{r}_{\perp}, z)}\, e^{i\phi(\mathbf{r}_{\perp}, z)} e^{ikz} = 0. \tag{8}$$

Then we expand the result and take the imaginary part. This leads directly to the TIE [17]:

$$\nabla_{\perp} \cdot [\rho(\mathbf{r}_{\perp}, z)\nabla_{\perp}\phi(\mathbf{r}_{\perp}, z)] = -k\frac{\partial\rho(\mathbf{r}_{\perp}, z)}{\partial z}. \tag{9}$$

This is a second-order elliptic partial differential equation that may be solved for the phase of the radiation over the plane of measurement, given the probability density and its longitudinal derivative over this plane.

Importantly, it has been shown that, in the absence of intensity zeroes, Eq. (9) has a unique and stable solution for the phase (up to an arbitrary additive constant) [68, 205, 215, 216]. Thus we have all the prerequisites for a deterministic and robust method of quantitative noninterferometric phase extraction using a focal series of three images.

Before continuing, let us make some interpretive remarks on the TIE. As was pointed out in Chapter 2 [see Eq. (12) of Section 2.3], the phase gradient appearing in Eq. (9) is proportional to the Lorentz deflection angle. Indeed, since we have already identified $\rho\nabla\phi$ as being proportional to the probability current, we see that Eq. (9) may be interpreted in intuitive hydrodynamic terms as stating that *the divergence of the transverse component of the probability current is proportional to the rate at which probability density is focused on propagation.* The negative sign is a simple consequence of the fact that a positively diverging wave must have a negative intensity derivative, and conversely. These remarks are nothing more than a restatement of the principles which underlie qualitative Lorentz microscopy (in the small-defocus regime).

5.2.2 A Generalized Definition of Phase

The concept of "phase" is intimately connected with the complex scalar representation of wave fields. Rather than viewing fields (or their components) in the amplitude/phase terms of "complex scalar functions of space–

time," we might choose instead to work with the local energy flow vector $\mathbf{j}(\mathbf{r}, t)$ and the energy density $P(\mathbf{r}, t)$. Here, $\mathbf{r}$ denotes a vector in three-dimensional space and t denotes time. This picture of energy flow will lead to a rather general equation that forms the starting point for TIE-based phase-retrieval programs using a variety of coherent and partially coherent radiation forms such as electromagnetic waves, electrons, and atoms [70].

Appealing to the general principle of energy conservation and ignoring the particular form of a wide class of wave equations, we may write down the continuity equation that the appropriately defined current vector $\mathbf{j}$ and energy density P satisfy

$$\nabla \cdot \mathbf{j}(\mathbf{r},\ t) + \frac{\partial P(\mathbf{r},\ t)}{\partial t} = 0. \tag{10}$$

To proceed further, assume that the process of interest has statistical properties that do not depend on the origin of time, for a finite-time experiment which is much longer in duration than the characteristic time scale of wave field fluctuations. Take the time average (denoted by angular brackets) of the continuity equation [Eq. (10)] and use our assumption to neglect the influence of the second term, yielding an equation that includes Eq. (4b) as a special case:

$$\nabla \cdot \langle \mathbf{j} \rangle = 0. \tag{11}$$

To emphasize the general (indeed, topological) nature of Eq. (11), recall that the divergence of a vector field $\mathbf{F}(\mathbf{x})$ may be defined as its limiting flux per unit volume (flux density at a point) [217, 218]:

$$\nabla \cdot \mathbf{F}(\mathbf{x}) \equiv \lim_{V \to 0} \frac{1}{V} \oiint \mathbf{F}(\mathbf{x}) \cdot d\mathbf{a}, \tag{12}$$

where V is the volume of the closed surface over which the integral is performed, and $\mathbf{a}$ is the outward pointing unit normal. Equation (11) may therefore be interpreted in terms of a topological tautology: in the absence of sources or sinks, the time-average flow lines may neither begin nor end at any point in the space. Therefore, all streamlines of a field in a source-free region must either form closed loops (vortices) or extend to infinity [219]. This physical description indicates that closed loops of flow do not appear in the definition of divergence. Therefore, any phase recovery method based on this quantity may face difficulties in detecting such loops. This is indeed the case here, and we shall be led to consider the role of vortices, and the associated phase discontinuities, from time to time in our discussion.

Now, we are interested in the practical question of whether one can perform quantitative noninterferometric TIE-based phase extraction using such partially coherent fields. To achieve this, we need a connection between energy flow and phase. This connection begins with a standard means of picturing the phase of a coherent scalar wave that propagates through free space: the local direction of energy flow is perpendicular to the surfaces of constant phase.

To explore this further, we begin with a definition for the normalized probability current in terms of the time-averaged current:

$$\langle \hat{j}(\mathbf{r}) \rangle \equiv m \lim_{\varepsilon \to 0^+} \frac{\langle \mathbf{j}(\mathbf{r}) \rangle}{\langle \rho(\mathbf{r}) \rangle + \varepsilon}. \tag{13}$$

Here, angular brackets denote the time average or, where appropriate, the ensemble average. Over regions of nonzero time-averaged probability density, Eq. (13) describes a well-defined vector field that may therefore be Helmholtz decomposed into a potential and a rotational component in the usual way [217]. Performing this decomposition, we are able to rewrite the probability current in the following form, which is analogous to the expression for the electromagnetic current vector in the presence of both scalar and vector electromagnetic potentials [214]:

$$\langle \mathbf{j}(\mathbf{r}) \rangle = \frac{1}{m} \langle \rho(\mathbf{r}) \rangle \{ \nabla \varphi_S(\mathbf{r}) + \nabla \times \boldsymbol{\varphi}_V(\mathbf{r}) \}. \tag{14}$$

We regard Eq. (14) as defining the scalar phase, φ_S, which is single valued, and the vector phase, $\boldsymbol{\varphi}_V$, which is divergence free, in terms of the ensemble-averaged probability current $\langle \mathbf{j}(\mathbf{r}) \rangle$ and the ensemble-averaged probability density $\langle \rho(\mathbf{r}) \rangle$.

Equation (14) may be inverted to express the phase components in terms of the probability current, $\langle \mathbf{j}(\mathbf{r}) \rangle$ [70]. This decomposition is unique up to a vectorial constant that may float between the two components; we place this vectorial constant in the gradient term. The phase so defined obeys the following Poisson-type differential equations:

$$\nabla^2 \varphi_S(\mathbf{r}) = \nabla \cdot \langle \hat{j}(\mathbf{r}) \rangle, \tag{15a}$$

$$\nabla^2 \boldsymbol{\varphi}_V(\mathbf{r}) = -\nabla \times \langle \hat{j}(\mathbf{r}) \rangle. \tag{15b}$$

Thus these two functions are related to the current vector and thereby to the phase of the wave via the following integrals [217]:

$$\varphi_S(\mathbf{r}) = -\frac{1}{4\pi} \int \frac{\nabla \cdot \langle \hat{j}(\mathbf{r}') \rangle}{|\mathbf{r} - \mathbf{r}'|} \, d^3\mathbf{r}', \tag{16a}$$

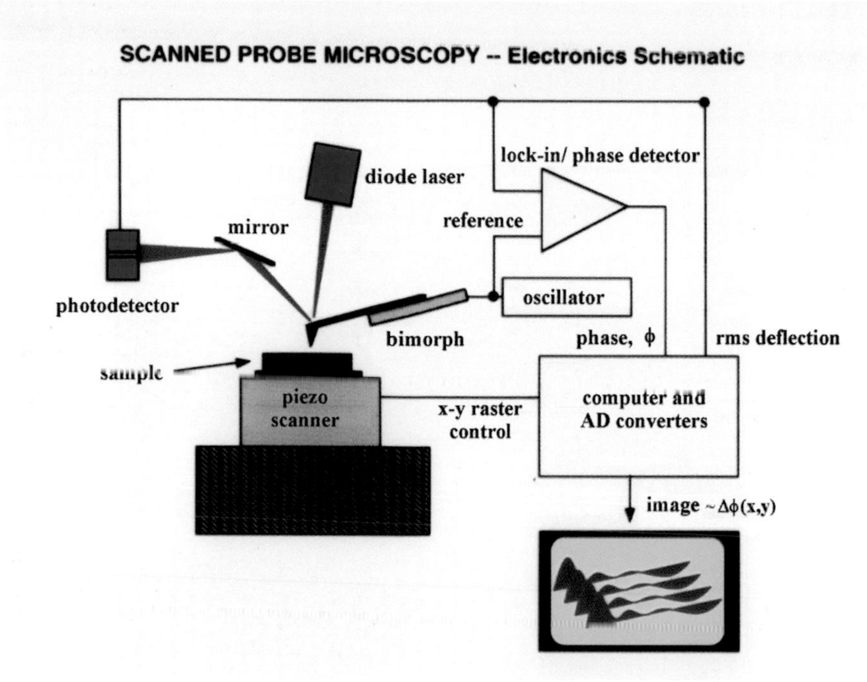

SCANNED PROBE MICROSCOPY -- Electronics Schematic
lock-in/ phase detector
diode laser
reference
mirror
photodetector
oscillator
bimorph
phase, ϕ
rms deflection
sample
piezo scanner
x-y raster control
computer and AD converters
image ~ Δϕ(x,y)

FIG. 2. (Chapter 3; Gomez; p. 73) Electronics schematic for scanned probe microscopy.

+ 0.045 degs
10 μm
- 0.045 degs
10 μm
+0.4 degs.
-0.6 degs.

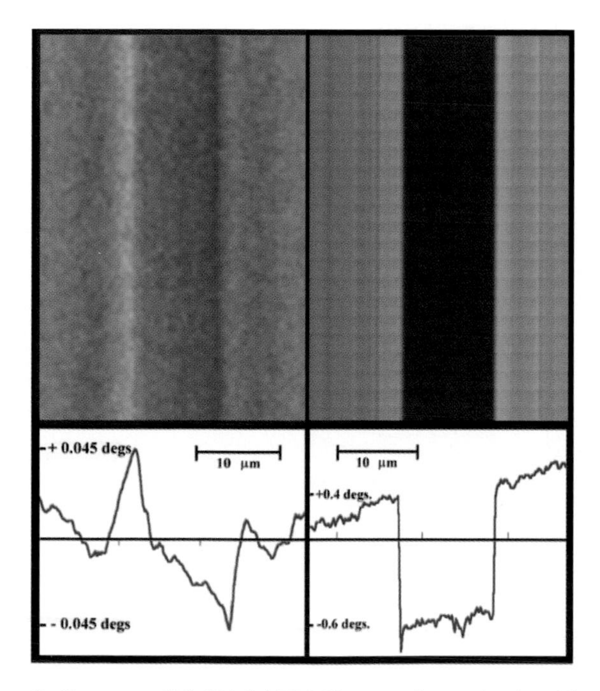

FIG. 4. (Chapter 3; Gomez; p. 87) (Right) EFM image of a conducting 10-μm-wide metal strip at V = 2V. (Left) MFM image of the same strip with 11 mA current. (Bottom) Average line profiles across the strip.

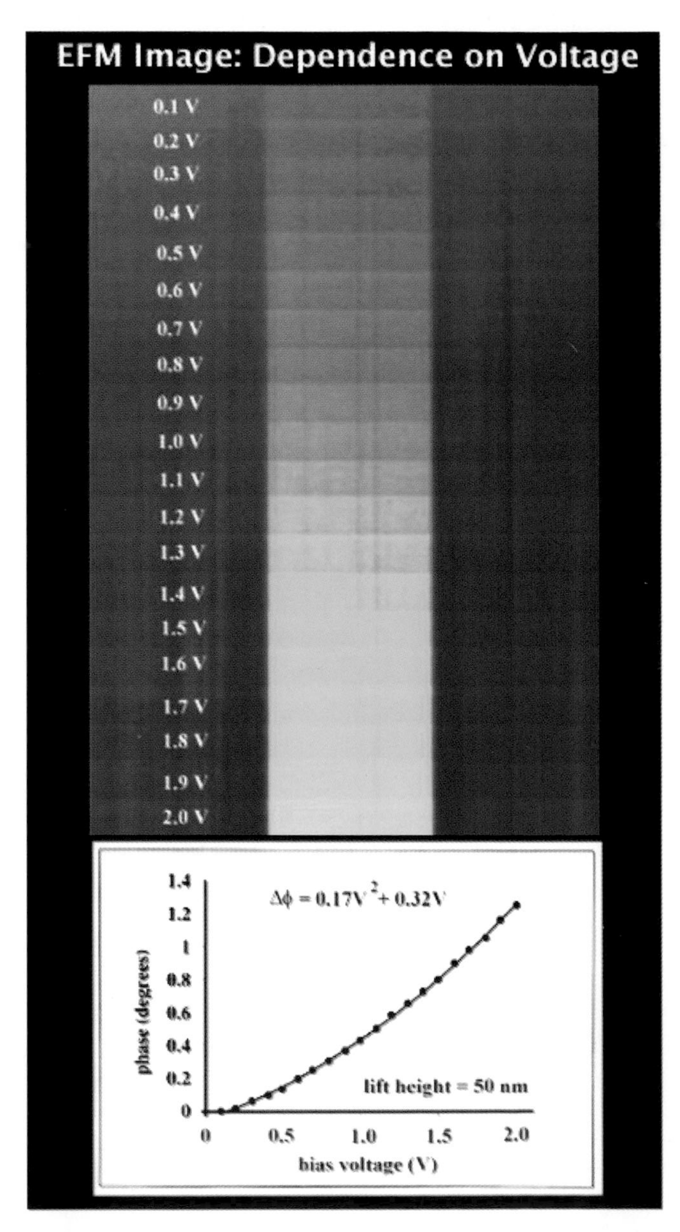

FIG. 5. (Chapter 3; Gomez; p. 88) EFM contrast dependence on bias voltage. (Top) EFM image at a lift height of 50 nm with increasing voltage. (Bottom) Plot of the maximum contrast as a function of voltage, fitted to a quadratic function $\Delta\phi = 0.19V^2$

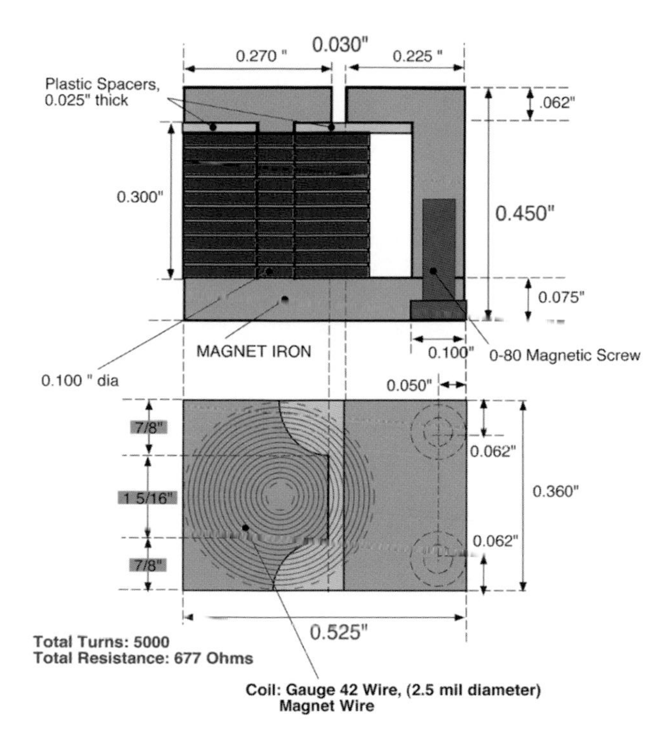

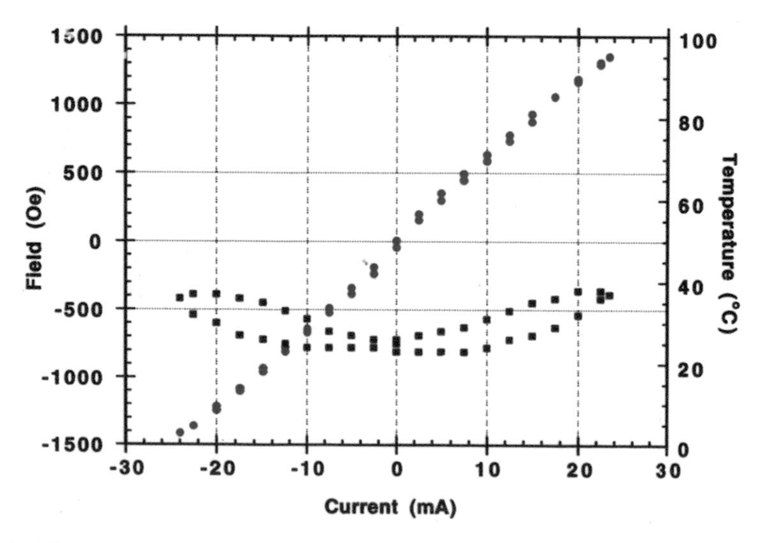

FIG. 6. (Chapter 3; Gomez; p. 91) (a) A magnet design for MFM imaging with an external field. (b) Calibration of the magnet in (a).

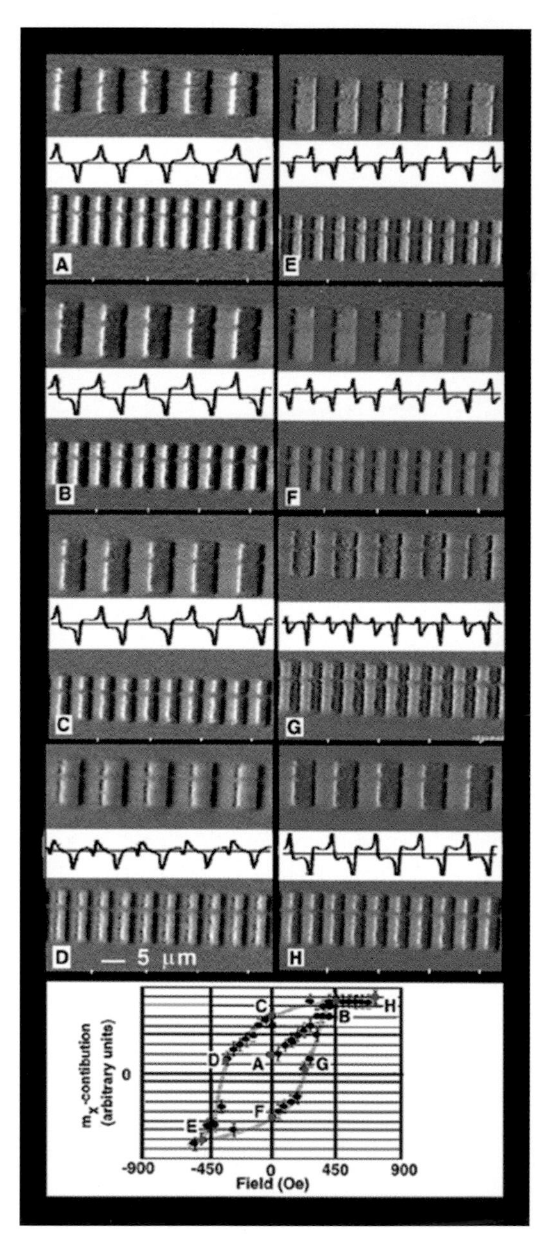

FIG. 7. (Chapter 3; Gomez; p. 94) Contrast change as a function of probe exposure.

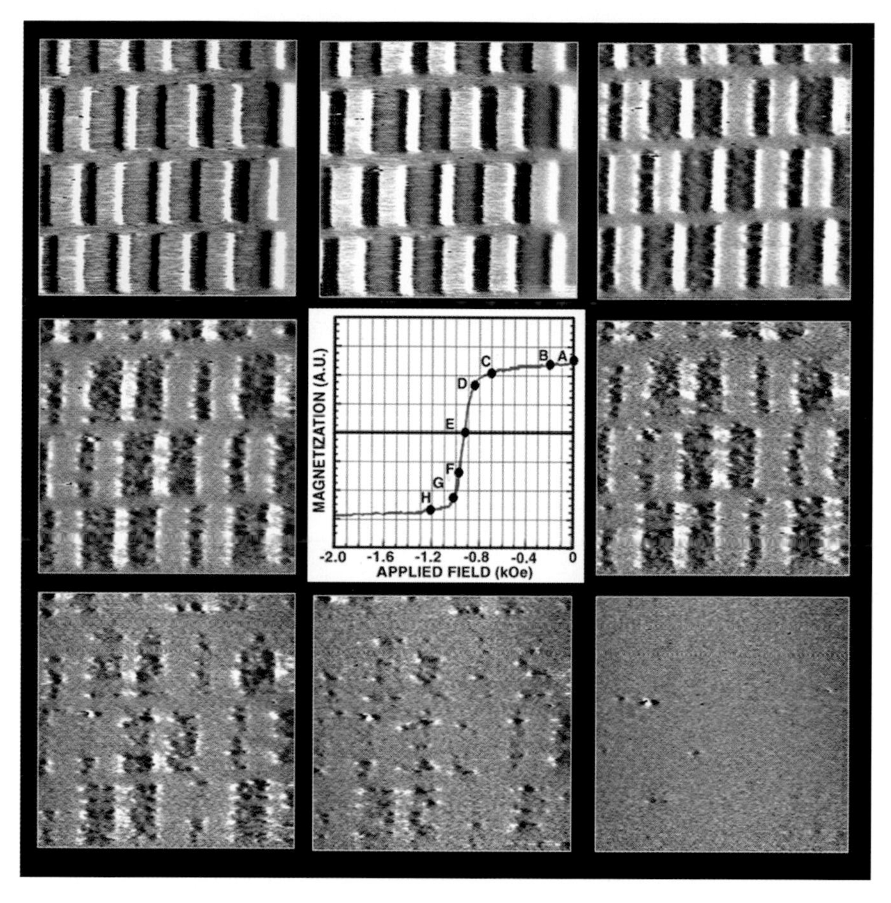

FIG. 8. (Chapter 3; Gomez; p. 96) Erasure process of a thin film recording medium.

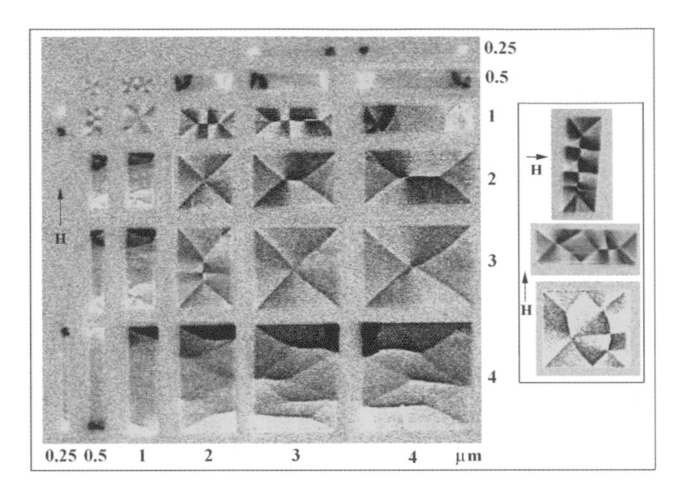

FIG. 9. (Chapter 3; Gomez; p. 99) MFM of an array of 25-nm-thick NiFe islands on silicon oxide.

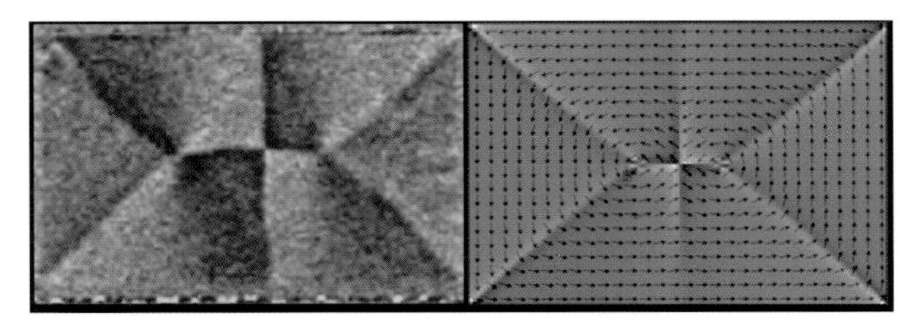

FIG. 11. (Chapter 3; Gomez; p. 101) Type C domain pattern and micromagnetic magnetization vector model.

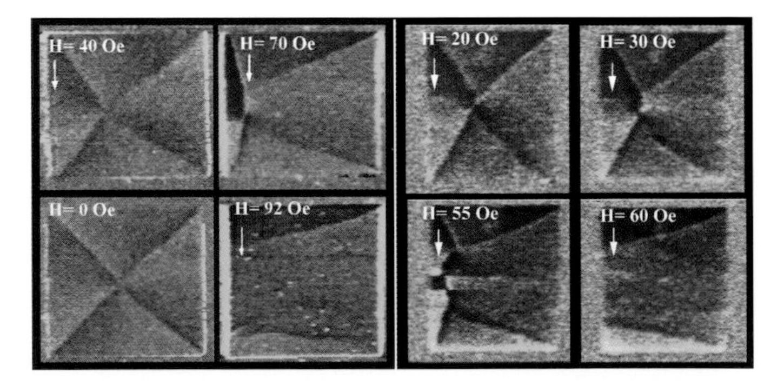

FIG. 12. (Chapter 3; Gomez; p. 103) Evolution of a four-domain closure pattern 3 X 3 μm^2 island as a function of applied field. The field was raised monotonically from zero while imaging. The bottom is a schematic of the charges and the inferred magnetization distribution for the upper left-hand corner of the zero field image.

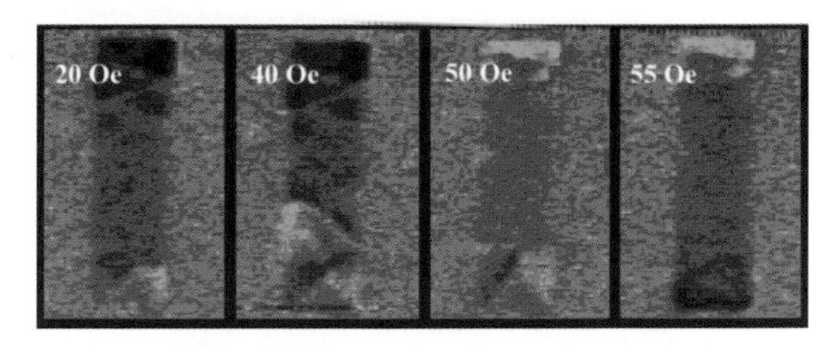

FIG. 13. (Chapter 3; Gomez; p. 104) Reversal of a NiFe element.

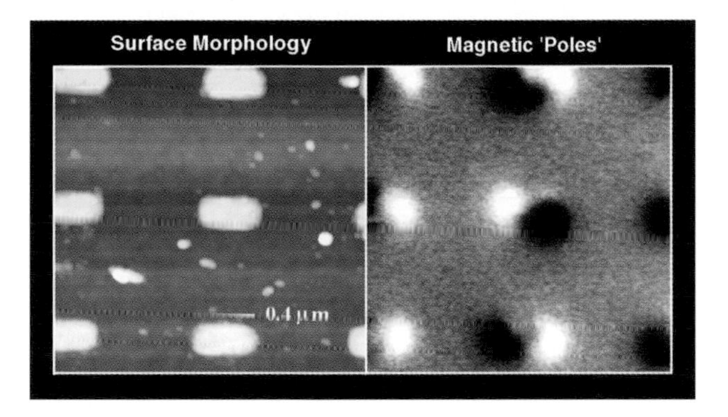

FIG. 15. (Chapter 3; Gomez; p. 107) Topography and magnetic "poles" of 200 nm x 400 nm x 20 nm cobalt islands (courtesy Stanford University).

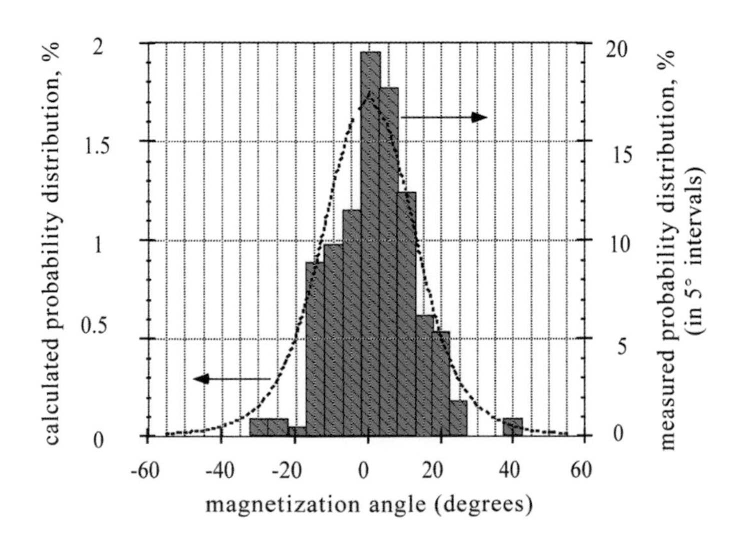

FIG. 16. (Chapter 3; Gomez; p. 107) Model and measured distribution of the easy axis of cobalt islands.

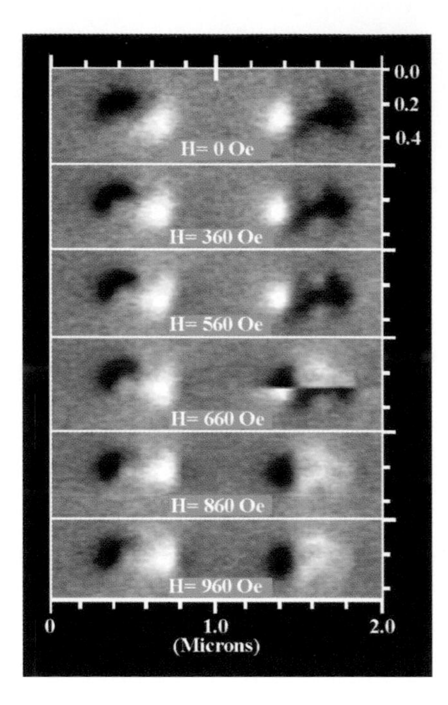

FIG. 17. (Chapter 3; Gomez; p. 108) Moment rotation and switching processes of submicron cobalt islands. The applied is along the horizontal axis from left to right.

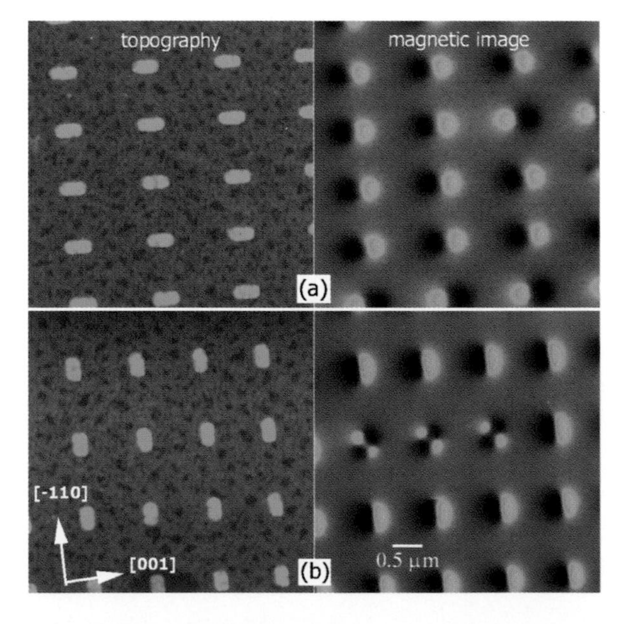

FIG. 18. (Chapter 3; Gomez; p. 109) Epitaxially grown cobalt islands on MgO. The shape and crystalline anisotropy axes are (a) parallel and (b) orthogonal.

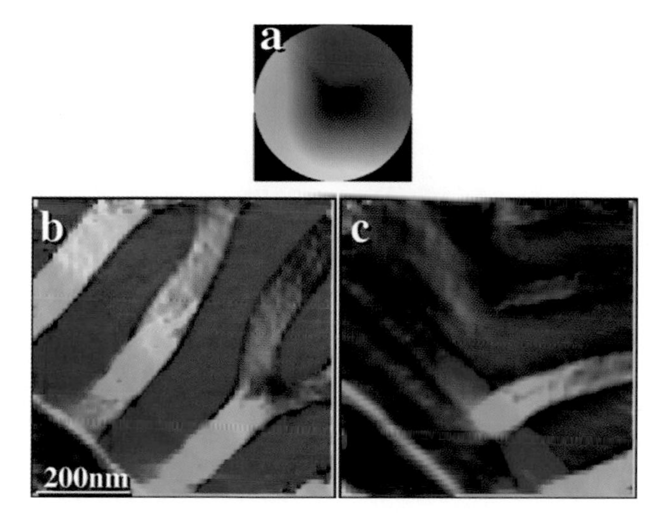

FIG. 7. (Chapter 4; McCartney et al.; p. 123) (a) Hue of color wheel indicates direction of induction, and saturation indicates magnitude; (b) color map of induction of Y-shaped domains in $Nd_2Fe_{14}B$; (c) color map of same area after heating of same area after heating to 300°C shows green and red 180° domains and domain pinning at defects.

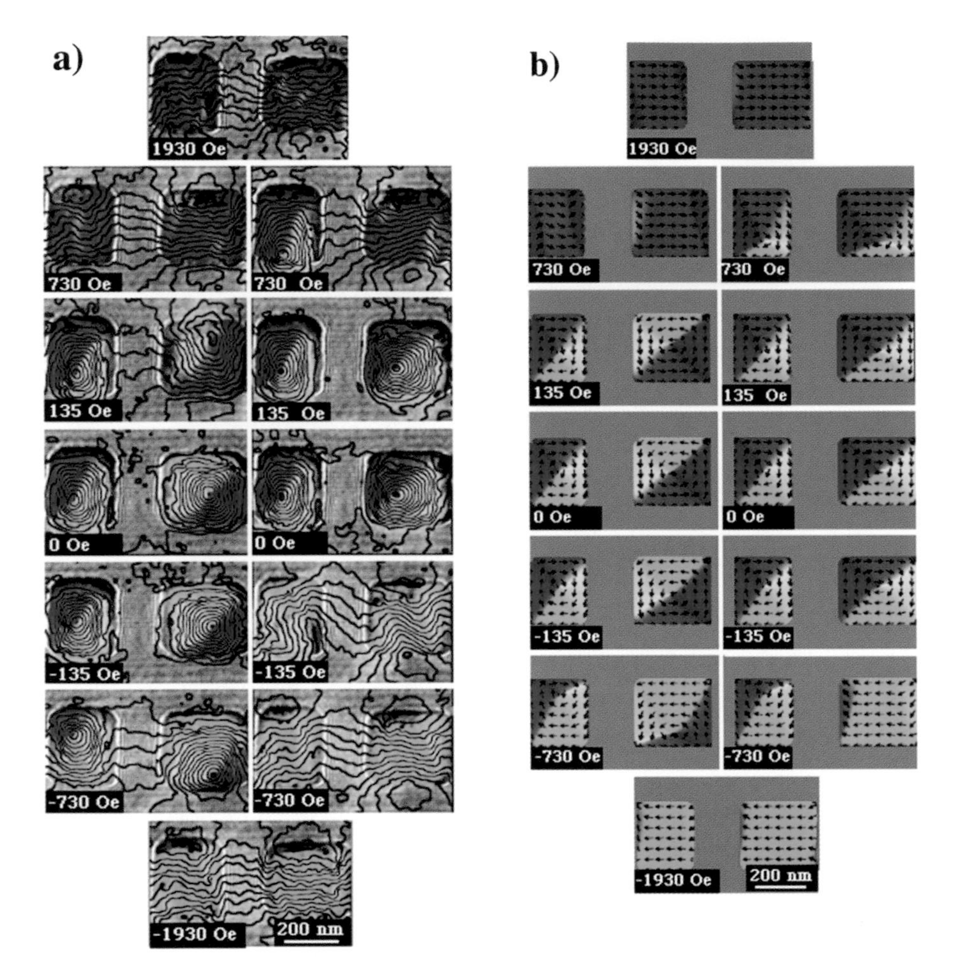

FIG. 14. (Chapter 4; McCartney et al.; p. 130) (a) Magnetic contributions to phases for 30-nm-thick Co elements during a hysteresis cycle. Phase contours are separated by 0.21π radians. The field was applied in the horizontal direction of the figure and should be allowed counterclockwise. (b) Corresponding micromagnetic simulations with square corners on cells and 3600 Oe field into the page.

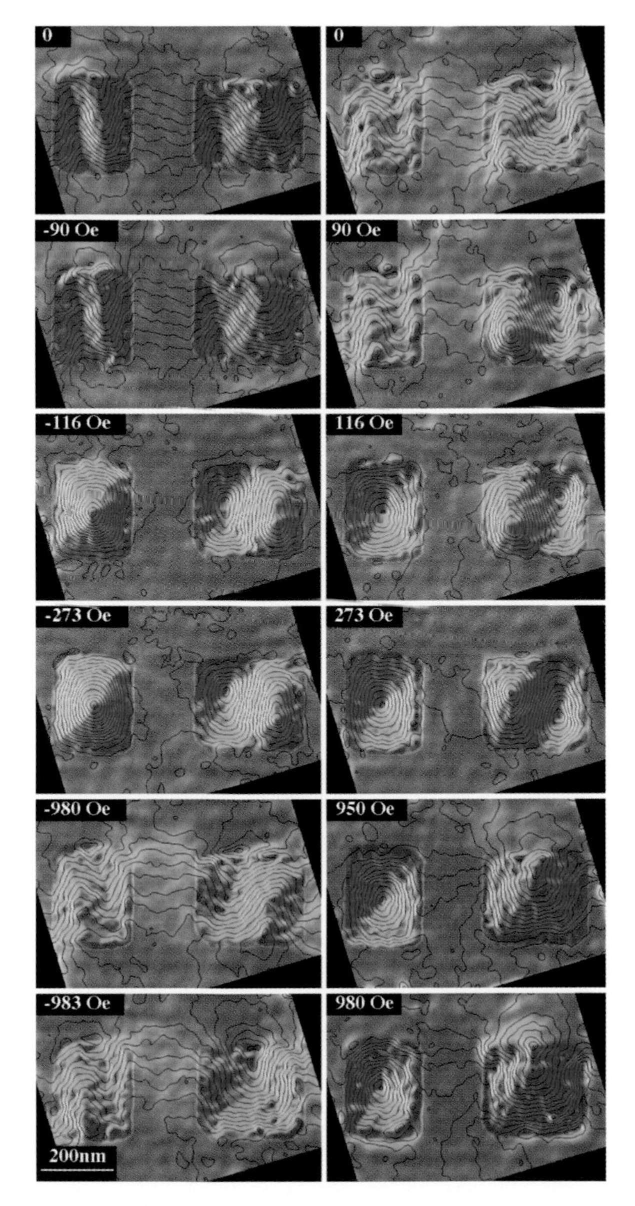

FIG. 15. (Chapter 4; McCartney et al.; p. 130) Remanent states for Co nanostructures shown in Fig. 14, where the in-plane fields indicated were applied directly after a large positive in-plane field for images in the left-hand column or large negative in-plane field for images in the right-hand column.

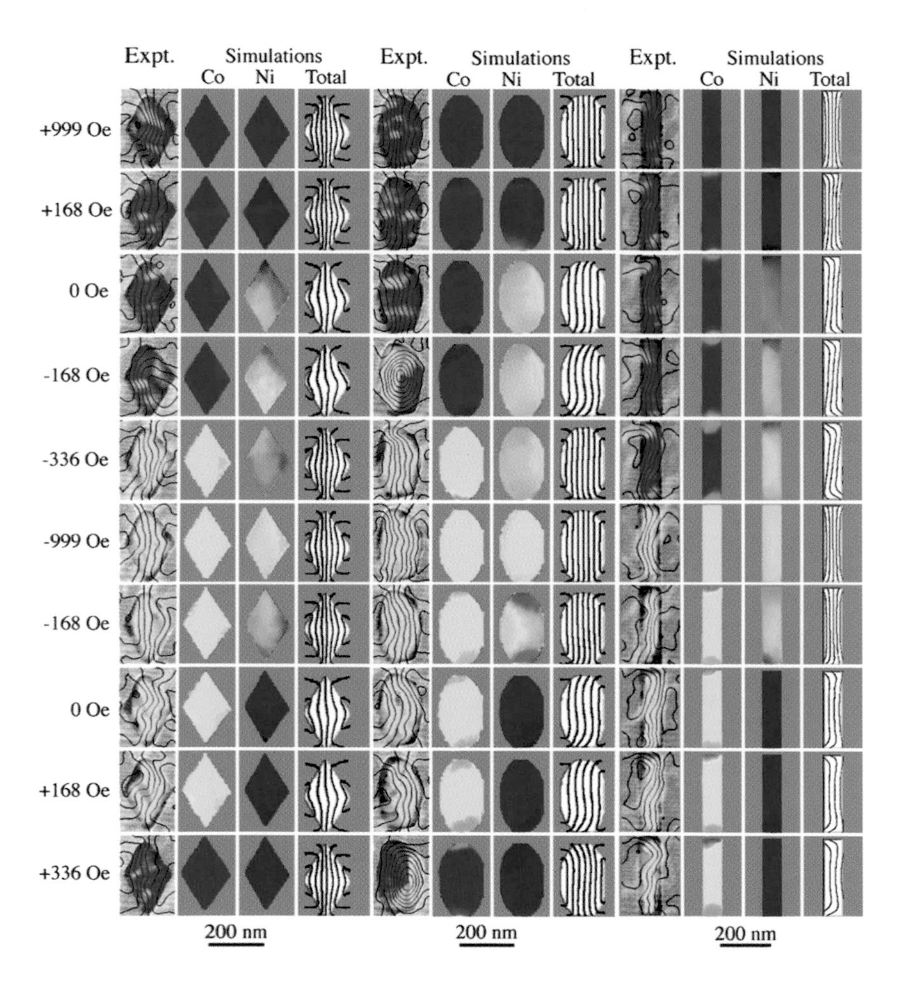

FIG. 16. (Chapter 4; McCartney et al.; p. 130) Comparison of experimental and simulated magnetization states during complete hysteresis cycles for patterned Co/Au/Ni spin-valve elements in the form of diamonds, ellipses, and rectangular bars. Applied in-plane fields are shown at left. Experimental phase contours are separated by 0.64π radians. Columns labeled Co and Ni are simulations for the individual FM layers, and those labeled Total are simulations for the composite Co/Au/Ni structure.

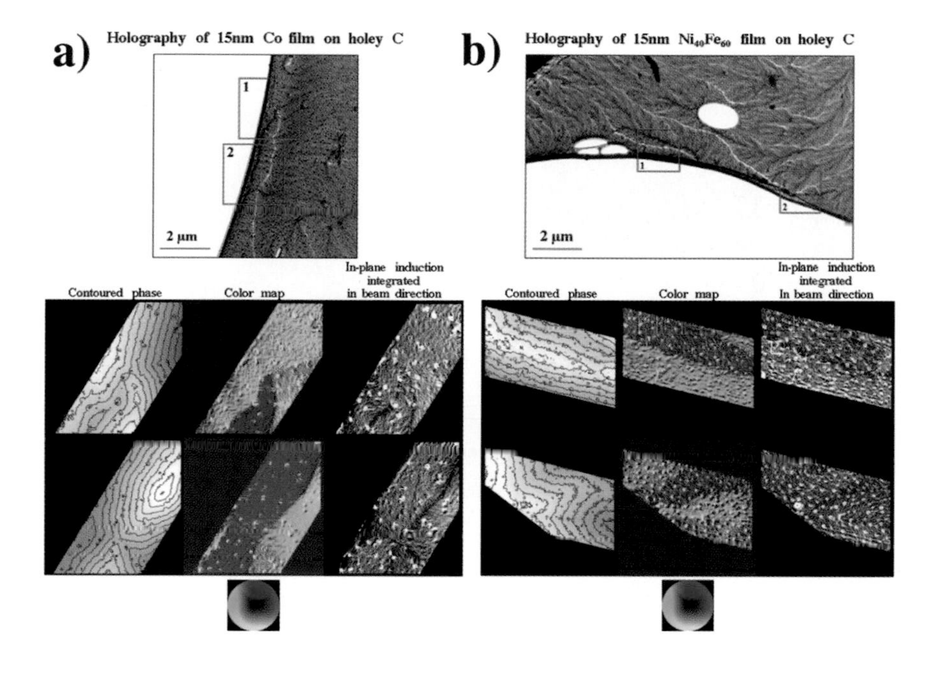

FIG. 18. (Chapter 4; McCartney et al.; p. 132) Lorentz images of domain walls in films of (a) Co (15nm) and (b) $Ni_{40}Fe_{60}$ (15nm), sputtered into holey carbon film. Reconstructed phase images of marked areas are shown in three forms: with contours of spacing 3.3 radians, as color maps, and as the magnitude of the in-plane component of the magnetic induction in the incident beam direction. Features of interest are dark lines at positions of domain walls (small white or black circles are surface oxide particles; their contrast is not magnetic).

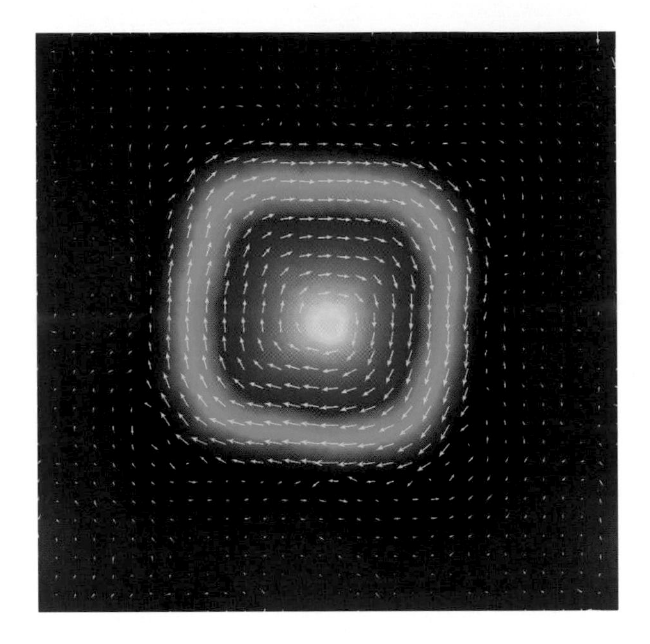

FIG. 9. (Chapter 5; Barty et al.; p. 158) Sample magnetization computed from the recovered phase image. The electrons are deflected perpendicular to the lines of magnetization by an amount proportional to the local magnetic induction; therefore, we can compute both the direction and the magnitude of the magnetization from the sample phase.

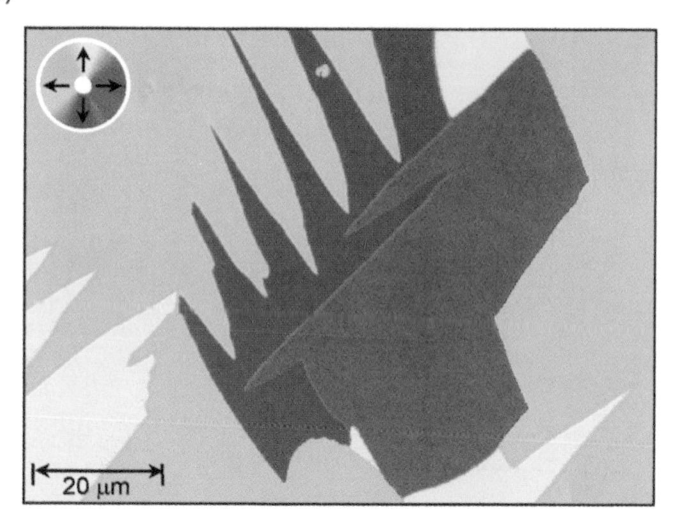

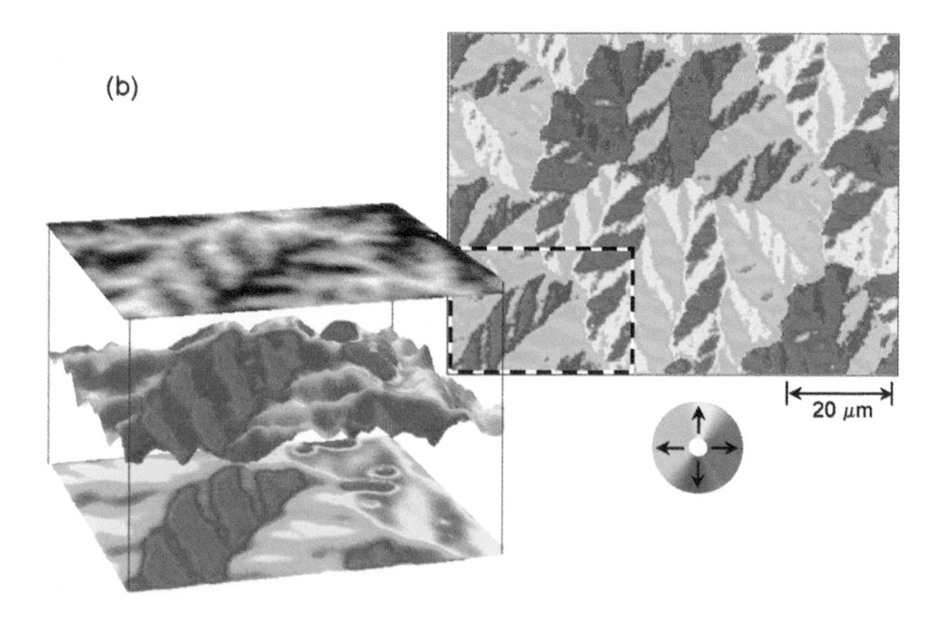

FIG. 16. (Chapter 6; Unguris; pp. 176, 181) (a) SEMPA image of magnetization direction in Fe(100). The relationship between color and direction is given by the color wheel. (b) In-plane Co(0001) magnetization is shown at right. At left a boxed portion of the in-plane image (shown at the bottom) has been combined with the corresponding perpendicular magnetization image (shown at the top) to generate a three-dimensional rendering of the magnetization direction (shown in the middle).

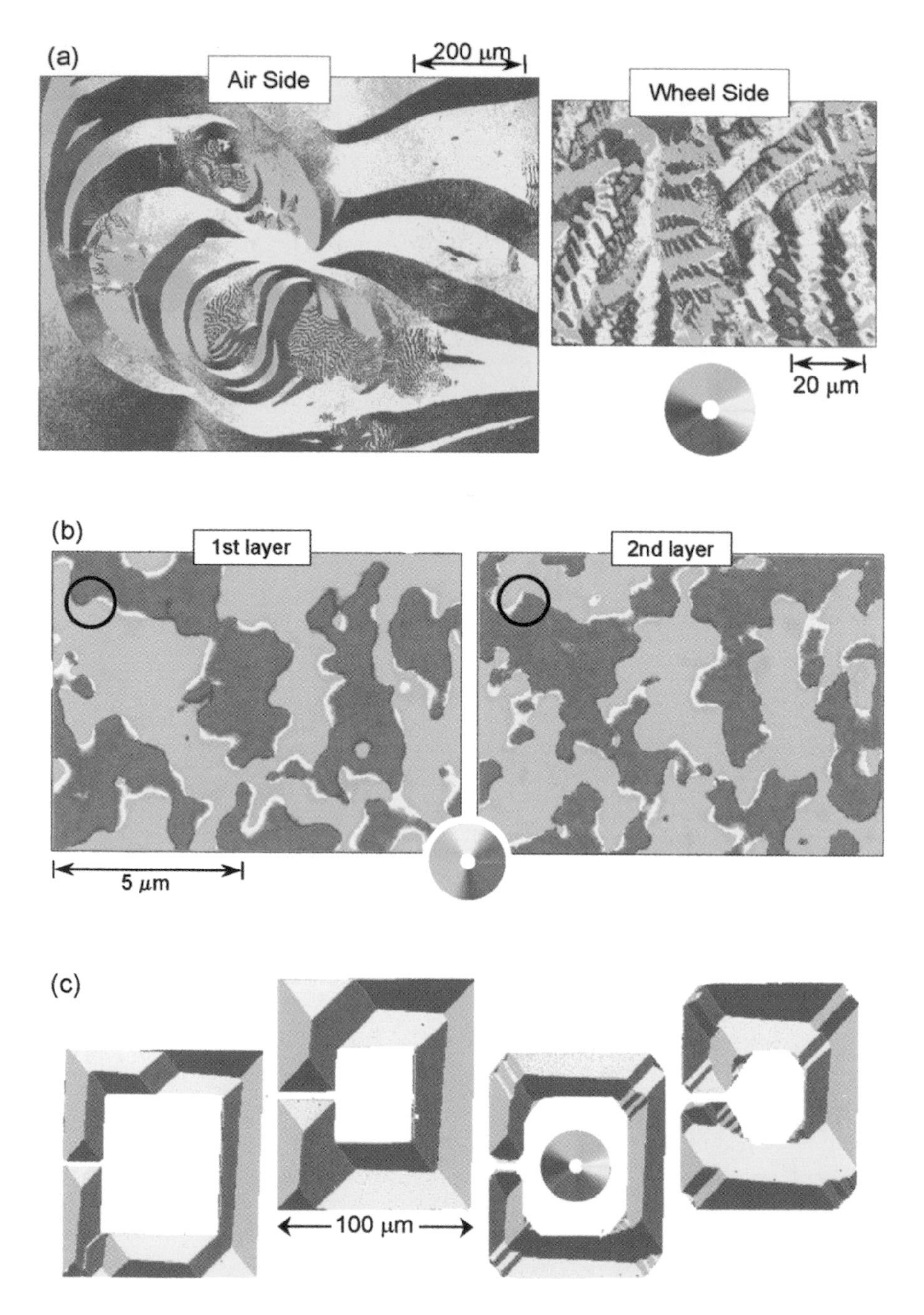

FIG. 17. (Chapter 6; Unguris; pp. 182, 190, 192) SEMPA images of magnetization direction in (a) an amorphous ribbon, (b) a Co/Cu multilayer, and (c) patterned Fe Films. The relationship between color and direction is given by the color wheel.

$$\boldsymbol{\varphi}_V(\mathbf{r}) = \frac{1}{4\pi}\int \frac{\nabla \times \langle \hat{j}(\mathbf{r}')\rangle}{|\mathbf{r}-\mathbf{r}'|}\, \mathrm{d}^3\mathbf{r}'. \tag{16b}$$

These expressions are of considerable utility as they permit the use of the concept of phase, and provide a rigorous definition, even where phase does not ordinarily exist. We have shown elsewhere that the phase so defined has many of the physical properties that are ordinarily associated with phase and so may be used in its place where it is experimentally or theoretically convenient [220].

We conclude this section by showing that the definition of phase given in Eqs. (16a) and (16b) reduces to the usual definition of phase in the coherent limit, provided of course that one is dealing with a scalar wave field. In the coherent limit, where $\mathbf{j}(\mathbf{r}) = [\rho(\mathbf{r})/m]\nabla\phi(\mathbf{r})$ [214], Eqs. (16a) and (16b) reduce to

$$\varphi_S(\mathbf{r}) = -\frac{1}{4\pi}\int \frac{\nabla^2\phi(\mathbf{r}')}{|\mathbf{r}-\mathbf{r}'|}\, \mathrm{d}^3\mathbf{r}', \tag{17a}$$

$$\boldsymbol{\varphi}_V(\mathbf{r}) = \frac{1}{4\pi}\int \frac{\nabla \times \nabla\phi(\mathbf{r}')}{|\mathbf{r}-\mathbf{r}'|}\, \mathrm{d}^3\mathbf{r}'. \tag{17b}$$

Note that the vector phase $\boldsymbol{\varphi}_V(\mathbf{r})$ will vanish if $\nabla \times \nabla\phi(\mathbf{r}) = 0$. Thus, in the coherent limit, the vector phase will vanish if the conventional phase of the wave is single valued and continuous. In the case of a coherent field, then, the vector phase is nonzero only if the phase of the wave field is discontinuous, or multiply valued, and so corresponds to a topological phase [70, 219]. It is also apparent from Eq. (17a) that the scalar phase reduces to the conventional phase when the field is coherent and the phase is continuous [70].

5.2.3 Method of Solution for the Transport-of-Intensity Equation

We have developed a generalized notion of phase that is well defined for a broad class of different types of partially coherent free-space radiation such as visible light, X rays, electrons, and neutrons. In this subsection, we invert a paraxial form of the associated partially coherent TIE in order to develop a rapid and stable deterministic algorithm for unique noninterferometric phase recovery using partially coherent radiation.

As pointed out in Section 2.5, a straightforward solution to the TIE may be initiated by introducing the auxiliary function $\Psi(\mathbf{r}_\perp, z)$ via

$$\nabla_\perp \Psi(\mathbf{r}_\perp, z) \equiv \rho(\mathbf{r}_\perp, z)\nabla_\perp\phi(\mathbf{r}_\perp, z). \tag{18}$$

This approximation will be valid provided the radiation field is vortex free and possesses an intensity that does not vary too strongly over distances of

$\nu \pm$ = the order of the de Broglie wavelength. With this auxiliary function, the TIE becomes the following Poisson-type equation:

$$\nabla_{\perp}^{2}\Psi(\mathbf{r}_{\perp}, z) = -k\,\frac{\partial\rho(\mathbf{r}_{\perp}, z)}{\partial z}, \tag{19}$$

which has the symbolic solution

$$\Psi(\mathbf{r}_{\perp}, z) = -\nabla_{\perp}^{-2}k\,\frac{\partial\rho(\mathbf{r}_{\perp}, z)}{\partial z}. \tag{20}$$

Taking the two-dimensional gradient of this equation and making use of Eq. (18) we obtain the following expression for the phase gradient:

$$\nabla_{\perp}\phi(\mathbf{r}_{\perp}, z) = -\frac{k}{\rho(\mathbf{r}_{\perp}, z)}\,\nabla_{\perp}\nabla_{\perp}^{-2}\,\frac{\partial\rho(\mathbf{r}_{\perp}, z)}{\partial z}. \tag{21}$$

Taking the two-dimensional divergence of this equation immediately yields the following symbolic solution for the phase [70]:

$$\phi(\mathbf{r}_{\perp}, z) = -k\nabla_{\perp}^{-2}\nabla_{\perp}\cdot\frac{1}{\rho(\mathbf{r}_{\perp}, z)}\,\nabla_{\perp}\nabla_{\perp}^{-2}\,\frac{\partial\rho(\mathbf{r}_{\perp}, z)}{\partial z}. \tag{22}$$

We now consider the application of these ideas to microscopy.

5.3 Practical Considerations in Electron Phase Microscopy

Suppose that a z-directed normally incident monoenergetic beam of electrons is incident on a magnetic material of some form. At the exit surface $z = 0$ of this object, the wave function is assumed to have square modulus $\rho(\mathbf{r}_{\perp}, z = 0)$ and phase $\phi(\mathbf{r}_{\perp}, z = 0)$. Let us now assume that a perfect imaging system is used to bring the exit surface wave function of the object into focus over the plane B, as shown in Fig. 1. Suppose furthermore that the intensity (probability density) of the in-focus image is measured over plane B, together with the intensity of slightly defocused images over planes A and C. By "slightly defocused," we mean that planes A and C are sufficiently close to plane B for the difference between images A and C to be proportional to the intensity derivative of the radiation over plane B.

Clearly, the intensities measured over planes A and C, and hence the intensity derivative, will be a function of both the phase and the intensity of the radiation over plane B. This is a consequence of what might be termed propagation-induced phase contrast, a phenomenon well known to electron and visible-light microscopists who use defocus to qualitatively visualize phase information. To quote Zernike: "Every microscopist knows that

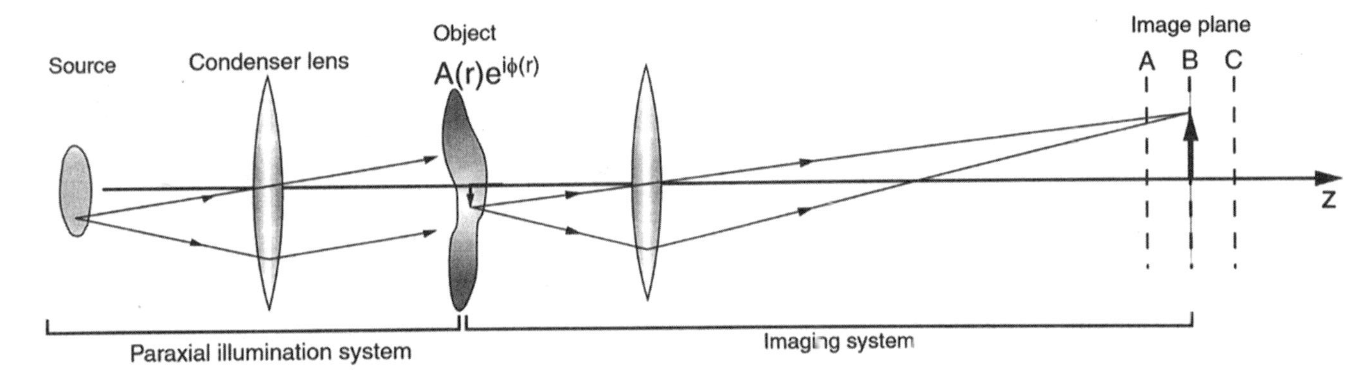

FIG. 1. Generic setup for quantitative phase extraction from Fresnel images.

transparent objects show light or dark contours under the microscope in different ways varying with defocus" [221]. This phenomenon is a simple consequence of the local redirection of energy flow as the radiation passes through the object.

To put these ideas on a concrete footing consider the simulations presented in Fig. 2 in which we have taken two arbitrary, uncorrelated images as the intensity (Fig. 2a) and phase (Fig. 2b) structure of a hypothetical wave field. The intensity data ranges from 0 to 1 arbitrary units of intensity, while the phase structure varies from 0 to π radians. All images are 256×256 pixels $= 1$ cm square, with $\lambda = 632.8$ nm and defocus distances of ± 5mm. When imaged through a microscope such as that described schematically in Fig. 1 the intensity structure (Fig. 2a) will be seen in plane B; in a perfect imaging system, all phase information is lost in the in-focus plane. If we defocus the microscope a very small amount either side of the plane of best focus, corresponding to planes A and C in Fig. 1, we will observe the intensity distributions shown in Fig. 2c, d in which the propagation-induced phase contrast is clearly visible. To recover the phase from these three intensity images (Fig. 2a, c, d), corresponding, respectively, to measurements of the intensity in planes B, A and C in Fig. 1, we use a numerical implementation of Eq. (22) where the in-focus intensity (Fig. 2a) is used for $\rho(\mathbf{r}_\perp, z)$ and the intensity derivative is computed as the difference of Fig. 2c and Fig. 2d. Performing this operation on the data in Fig. 2 produces the recovered phase image shown in Fig. 2e. Comparison of the input and recovered phase images indicate that they are in excellent agreement with the recovered phase, being within a few percent of the original data across the entire image.

Clearly the intensity derivative used as the input data is critical as Eq. (22) seeks the phase distribution that will account for the measured intensity differences between the three planes A, B, and C. In practice, of course, the intensity derivative will contain components that do not correspond to the sample phase, for example alignment errors and noise in the imaging system, which will contribute spurious components to the phase solution. These differences may have a variety of sources, and in the following two subsections we consider the stability of the algorithm in the presence of noisy data and the effects of such spurious difference signals on the recovered phase.

5.3.1 Stability Considerations and the Effects of Noise

To understand the effects of noise on the recovered phase consider, for simplicity, a wave with no intensity modulation but some phase distribution.

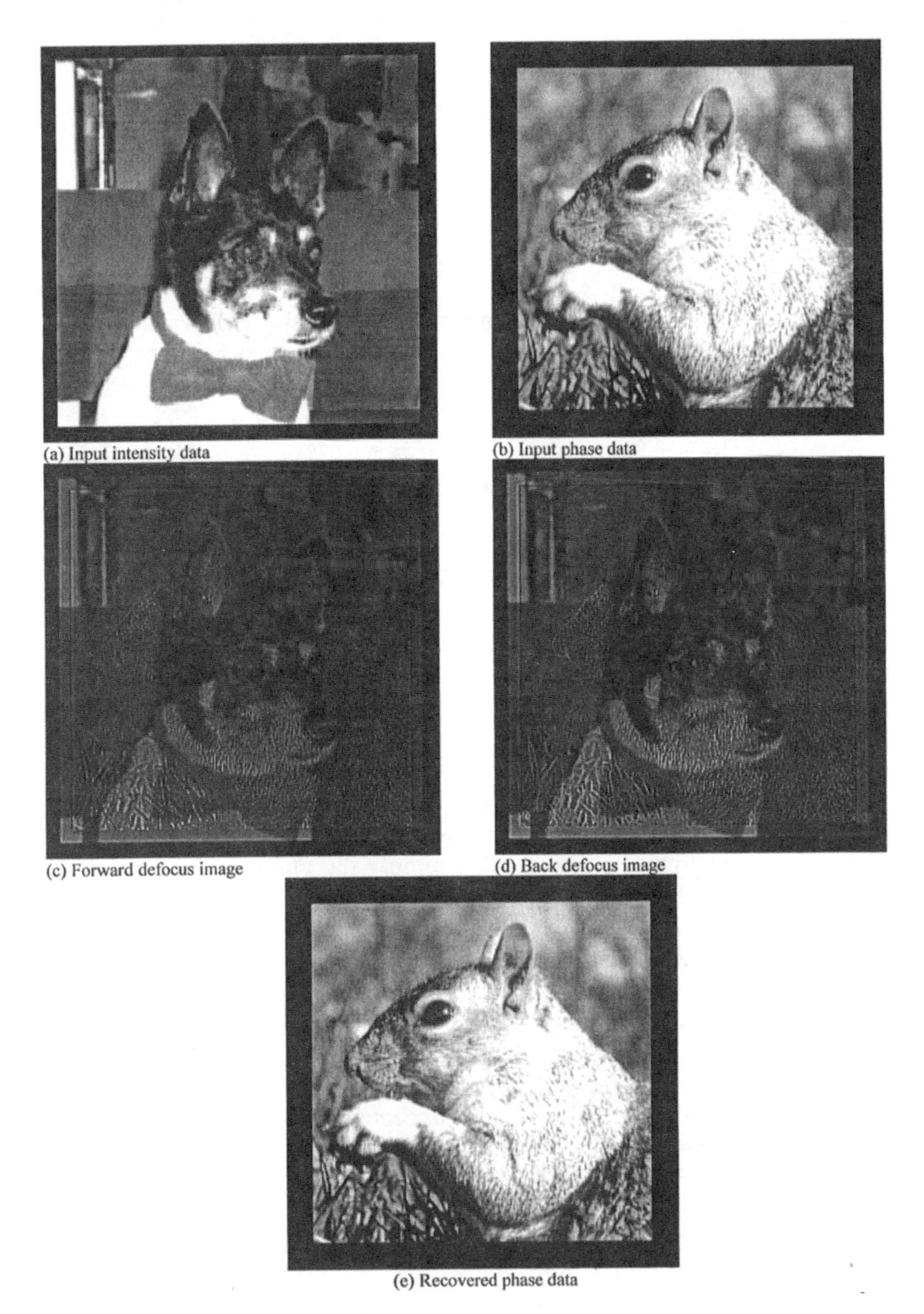

(a) Input intensity data

(b) Input phase data

(c) Forward defocus image

(d) Back defocus image

(e) Recovered phase data

FIG. 2. Demonstration of the phase recovery process: (a) input intensity data; (b) input phase data; (c) forward defocus image; (d) back defocus image; (e) recovered phase data.

In this case, the TIE reduces to

$$k \frac{\partial \rho}{\partial z} = -\nabla^2 \phi. \tag{23}$$

Elementary Fourier theory directly gives

$$\frac{2\pi}{\lambda} \mathscr{F}\left[\frac{\partial \rho}{\partial z}\right] = |\mathbf{q}_\perp|^2 \mathscr{F}[\phi], \tag{24}$$

where $\mathbf{q}_\perp$ is the spatial frequency and $\mathscr{F}$ denotes the Fourier transform, which enables us to write a solution for the phase in the form

$$\phi = \frac{2\pi}{\lambda} \mathscr{F}^{-1}\left[\frac{1}{|\mathbf{q}_\perp|^2} \mathscr{F}\left[\frac{\partial \rho}{\partial z}\right]\right], \qquad |\mathbf{q}_\perp| \neq 0. \tag{25}$$

Thus recovery of the phase from the intensity derivative of data without intensity modulation in the plane of interest is simply obtained via a low-pass filtering operation. This immediately tells us that the phase recovery process, at least in this simple example, is very stable with respect to noise contamination. Put differently, the input data involve the numerical differentiation of experimental data—a notoriously noise-sensitive process—followed by a double numerical integration in the solution to the second-order differential equation.

Examination of Eq. (25) indicates that the effect on the recovery of noise in the image will be dominated by low spatial frequencies. Consider, for example, the addition of white noise to the input intensity images used to recover the phase. Because Fourier transformation is a linear operation, the phase recovered from a noisy signal is simply the phase of the signal plus the "recovered phase" of the noise component of the intensity images; the "recovered phase" of the noise will be dominated by low spatial frequencies.

To model the effect of noise we added some noise to simulated data. Figure 3 shows a matrix of images covering a range of levels of noise and defocus δz. Note how the phase recovery is indeed insensitive to the effects of noise, with a clear image being recovered even in the case of 10% noise on the original data sets. Note also how the noise pollution is dominated by low frequency Fourier harmonics, as expected from the above discussion. We thus see that the phase recovery has remarkable stability with respect to random errors. Moreover, sensitivity to noise is reduced by increasing the amount of defocus, a result of the fact that the magnitude of the intensity derivative signal increases linearly with defocus distance (within the small-defocus regime). Thus it is possible, within reason, to increase the phase signal to compensate for noise effects at the expense of a slight decrease in spatial resolution.

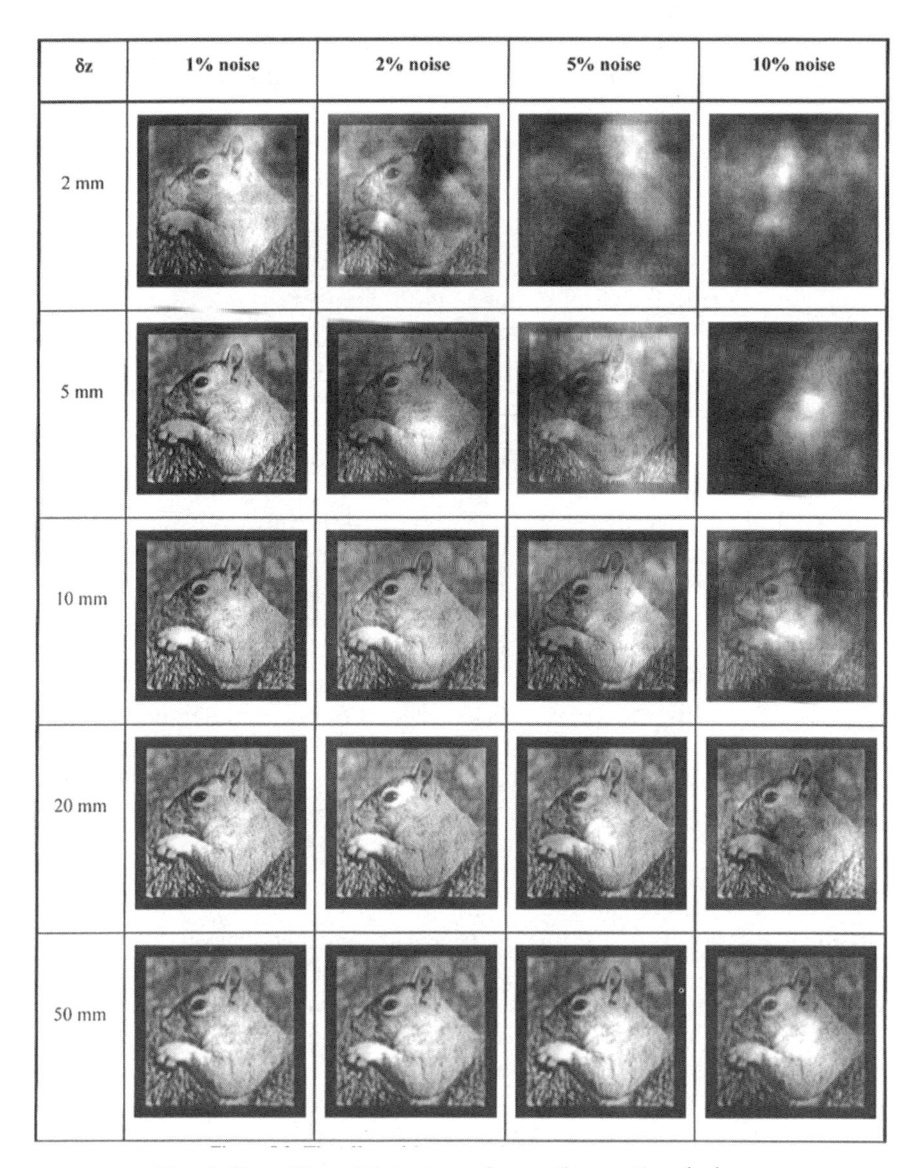

FIG. 3. The effect of detector noise on the recovered phase.

5.3.2 The Effect of Systematic Errors in Data Acquisition

Any movement of the image between the two measurements will, if uncorrected, be included in the intensity derivative and, hence, in the recovered phase, leading to artifacts in the phase image that result from image movement rather than sample structure. We now consider the effect of three classes of alignment that commonly occur in an electron microscope, all of which are easily corrected in practice.

5.3.2.1 Image Shifts. In an electron microscope it is common for the image to drift slightly as the focus is changed, and it is essential to correct for this drift by correct image alignment prior to phase processing. A shift in the image will be generated by off-axis illumination, and thus the effect of a shift on the phase recovery is to generate a phase ramp across the image, thereby mimicking the effect of image misalignment.

A quantitative understanding of this matter is straightforward. Elementary optics teaches us that a plane wave traveling at an angle θ to the optical axis is described by a phase ramp with a linear gradient φ' given by $\varphi' = (2\pi/\lambda)\sin\theta$. Let us suppose that the images are shifted with respect to one another by a distance Δ and that the defocus distance is δz. In this case $\sin\theta \approx \Delta/\delta z$ so that $\varphi' \approx (2\pi/\lambda)\Delta/\delta z$. The overall phase excursion in the recovered phase then depends on the image size. This expression places a limit on the acceptable amount of misalignment that depends on the size of the phase excursions required to be seen in the object.

To confirm these deductions, let us take the data shown in Fig. 2 and deliberately introduce a drift of one pixel in a uniform direction between the in-focus and out-of-focus planes, corresponding to a drift of one-half of a pixel for each step of defocus. The effect of this misalignment on the recovered phase is shown in Fig. 4a, from which it can be seen that the phase recovery is highly sensitive to errors in image alignment, with even a single pixel of misalignment introducing significant errors into the recovered phase. It is thus critical to ensure that all images are properly aligned in order to accurately reconstruct the sample phase.

5.3.2.2 Image Magnification. Just as it is important to ensure correct alignment of the intensity images to ensure correct measurement of the intensity derivative, it is also critical to ensure that all images have the same magnification, as any difference in magnification will introduce artifacts into the recovered phase. We once again take the data shown in Fig. 2 and deliberately introduce a small amount of expansion, 1% between image planes, to simulate the effect of changes in image magnification between measurements; the results are shown in Fig. 4b.

As can be seen, the phase recovery is highly sensitive to changes in magnification, with a magnification change of only 1% introducing significant errors into the recovered phase. To understand why this is the case, consider a wave front with a small amount of spherical curvature passing through an aperture containing our phase object. If we estimate the intensity derivative by subtracting positively and negatively defocused images of the aperture, we will find that one image has expanded with respect to the other

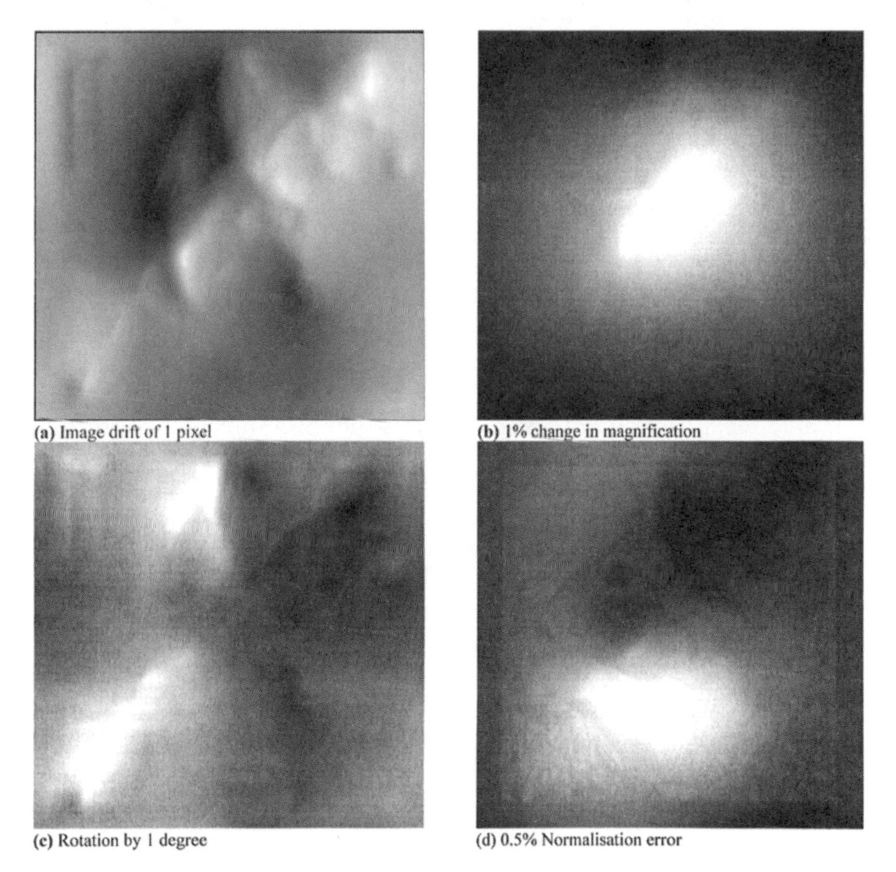

FIG. 4. Common systematic errors and their effect on the recovered phase: (a) image drift of 1 pixel; (b) 1% change in magnification; (c) rotation by 1 degree; (d) 0.5% normalization error.

owing to expansion of the incident spherical wave as the intensity propagates. The effect of a change in magnification is to precisely mimic this effect.

In the case of an astigmatic imaging system, as is common with electron microscopes, the magnification change differs with transverse direction, and thus the defocus-induced magnification artifacts are manifested as an additional cylindrical phase across the image. Note also that a similar quantitative estimate may be made for this case as was done for the previous section, but it will not be repeated here.

5.3.2.3 Image Rotation. The case of rotation is rather more complex. If a phase distribution $\phi = n\theta$, where n is an integer and θ is the azimuthal angle, is placed on an intensity distribution, then that distribution will undergo a differential rotation between planes separated by a differential distance. However, this phase distribution is discontinuous at the origin [68, 219] and is only possible if the intensity distribution itself contains a zero at the origin. Moreover, as has been pointed out at a number of places in this chapter, the phase recovery algorithm we are discussing does not have the capacity to sense zero divergence phase distributions such as that which will create a rotation. The issue of such phase distributions is at the center of our current thoughts on phase recovery, and we are beginning the exploration of appropriately modified algorithms [220].

To obtain some feeling for both the nature of the artifact and the sensitivity to image rotation, we once again take the data shown in Fig. 2 and deliberately introduce a small amount of image rotation. We used a total amount of rotation of one degree in a uniform direction between the in-focus and out-of-focus planes, corresponding to a rotation of half a degree for each step. The results of this calculation are shown in Fig. 4c, from which it can be seen that even this small amount of rotation has seriously distorted the recovered phase map. We see that the additional signal caused by the rotation is converted to local phase gradients, giving rise to the artifacts observed in Fig. 4c.

5.3.2.4 Image Normalization.

The formalism of phase recovery is based on the assumption of conservation of energy. Indeed, this is at the very heart of the approach. It is therefore critical to ensure that the total energy in the two images is identical, or a spurious phase distribution will result. Of course, there are many practical reasons why the two images will differ in integrated signal. It is also, of course, very easy to correct through a simple normalization procedure. Nevertheless, it is worth documenting the effects of a failure to ensure proper normalization.

To get some idea of the sensitivity of this technique to errors in image normalization we once again take the data presented in Fig. 2 and deliberately introduce some normalization error into the data by multiplying the positive defocus image by some constant $(1 + \delta)$ representing multiplicative fluctuations in either source intensity or charge-coupled device (CCD) gain. The results of this simulation are presented in Fig. 4d, from which it can be seen clearly that incorrect image normalization has a dramatic effect on the recovered phase, with variations of less than 1% causing significant image artifacts.

5.4 Application to Lorentz Microscopy

The previous two sections developed a formalism for the recovery of phase information from an electron beam propagating through a transmission electron microscope (TEM), presented in a general form applicable to any object that can be imaged in a TEM. In this section we specialize our application to the imaging of magnetic structures at the domain level in a transmission electron microscope [212]. We directly image the magnetization of microscopic cobalt grains, and the results obtained are shown to agree with structure seen using holographic and Foucault techniques. We also obtain quantitative measurements of the sample phase structure that are in agreement with the existing quantitative phase measurement technique of electron holography. Once again we point out that although the results presented here are demonstrated on magnetic samples, the same technique can be used without modification on any object that may be imaged in a transmission electron microscope.

As is well known, the phase of an electron wave field may be rendered visible using a number of techniques, including simple defocus [222], Zernike phase contrast [223], and Foucault imaging [150], a technique related to optical schlieren imaging. However, quantitative phase measurement of a large scale phase variation has required electron holography [157] using a highly coherent electron source and an extensively modified transmission electron microscope.

A paper by the authors [224] demonstrated quantitative optical imaging of a nonabsorbing phase object using noninterferometric techniques and partially coherent light, work which was subsequently applied to Lorentz microscopy [212]. In the latter paper it was shown that both the phase and the amplitude of a high voltage electron wave may be measured directly using a conventional transmission electron microscope. The technique was tested with a magnetic sample precisely characterized using electron holography, and holographic results were compared with the results of our noninterferometric approach. Apart from a consistent calibration factor, the phase images and the holographic measurements were shown to be in quantitative agreement.

5.4.1 Sample Selection and Preparation

To demonstrate the application of our quantitative phase imaging technique to transmission electron microscopy, the requirement is for a well-characterized sample for which an independent measurement of the phase shift could be obtained. For this purpose we chose to image small magnetized squares of cobalt because cobalt films are a well-characterized sample, with known magnetisation properties, that could easily be imaged in both a conventional TEM and using electron holography. Sample requirements were also determined by the requirements of electron holography because an independent measurement of the magnitude of the phase shift is needed.

It was therefore necessary to manufacture small, isolated objects so that there was sufficient space between samples to pass the electron reference beam required for holography, and to maximize fringe visibility in the holographic experiment the structures had to be thin enough for there to be minimal absorption. Having a thin film also assisted the experiment by ensuring the magnetization was always in the plane perpendicular to the electron wave so that the magnetic field lines would form closed contours within the sample when imaged from above.

In the experiments shown here, $1.65 \times 1.65\ \mu m^2$ cobalt squares with a nominal thickness of 10 nm were fabricated directly on 40-nm-thick silicon nitride membranes supported by silicon frames. First 40-nm-thick silicon nitride was deposited on both sides of Si(100) wafers using low pressure chemical vapor deposition (LPCVD), and then an array of windows was lithographically printed on the rough side of the wafer (we used wafers that had been polished on one side). The size of the window was $120 \times 120\ \mu m^2$ within a frame that fitted into the standard TEM specimen holder. The nitride on the rough side was selectively removed by plasma etch after an extra layer of photoresist was applied to the polished side of the wafer to protect the nitride layer on that side. After removing the photoresist on the polished side, the silicon was etched in potassium hydroxide, and a 0.7 μm-thick layer of JSR photoresist was spun on the membrane side of the wafers. The $1.65 \times 1.65\ \mu m^2$ holes were lithographically printed on the photoresist. Finally, a 10-nm-thick cobalt film was deposited by electron beam evaporation. The photoresist and unwanted cobalt were removed by a lift-off procedure that involved soaking the wafer in acetone.

These silicon nitride membranes have low stress and are strong and robust, making them an ideal substrate for the thin cobalt films. A low magnification electron micrograph of a portion of the wafer showing a number of cobalt dots is shown in Fig. 5. Note the rounded edges of the squares, which are due to the limited resolution of the lithographic projector.

5.4.2 Experimental Results

The cobalt squares just described were imaged using a conventional TEM (Jeol 200CX STEM) operated at 200 keV with a magnification of $5650\times$ that had been modified for the purposes of magnetic imaging. This was achieved by relocating the sample holder to a field-free region of the microscope column by installing a second side entry goniometer into the TEM column above the objective lens. With the objective lens energized the field in the specimen position was measured to be less than 0.5 G, which is sufficiently low to ensure that the sample is not modified by magnetic fields

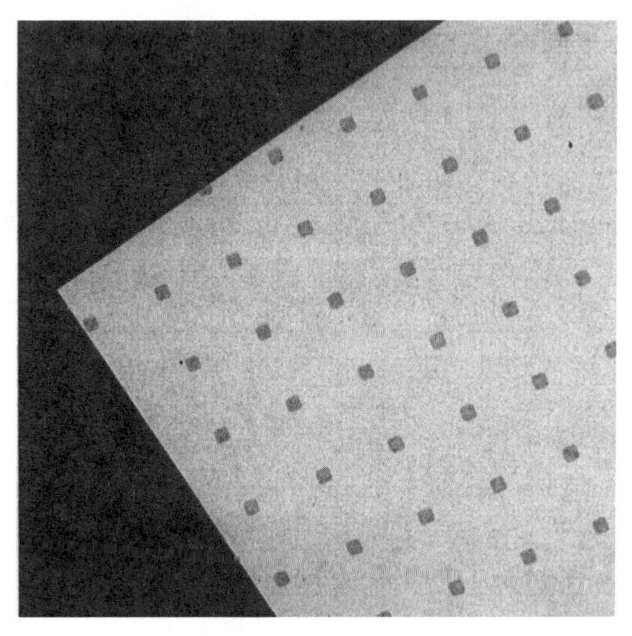

FIG. 5. Low magnification electron micrograph of a portion of the prepared wafer. The dark border is the edge of one of the $120 \times 120\,\mu m^2$ windows in the silicon support frame designed to fit in a standard TEM specimen holder. The dots within the frame are $1.65 \times 1.65\,\mu m^2$ dots of 10-nm-thick cobalt deposited on a 40-nm-thick silicon nitride membrane, which is in turn supported by the silicon wafer frame.

within the microscope. Images were collected in the plane of best focus as well as in five pairs of planes located an equal distance either side of best focus using a 1024×1024 pixel Gatan CCD camera, and one set of images are shown in Fig. 6. The nominal defocus step size used was $0.23\,\mu m$, which was the minimum step possible on the microscope. However, owing to the microscope modifications this figure was not precisely known. Precise calibration of the defocus is essential to obtaining quantitatively correct results using our technique, and thus an independent calibration using Fresnel diffraction around a knife edge was performed. This recalibration indicated that the defocus step size was $23.0 \pm 4.6\,\mu m$, which was the figure used in our phase processing calculations. The uncertainty in this figure is the principal source of uncertainty in our phase measurements.

The images were observed to drift slightly from side to side between defocus distances and were therefore aligned to each other using a Fourier transform based cross-correlation routine prior to image processing. This process consistently aligned the images to within one pixel of each other,

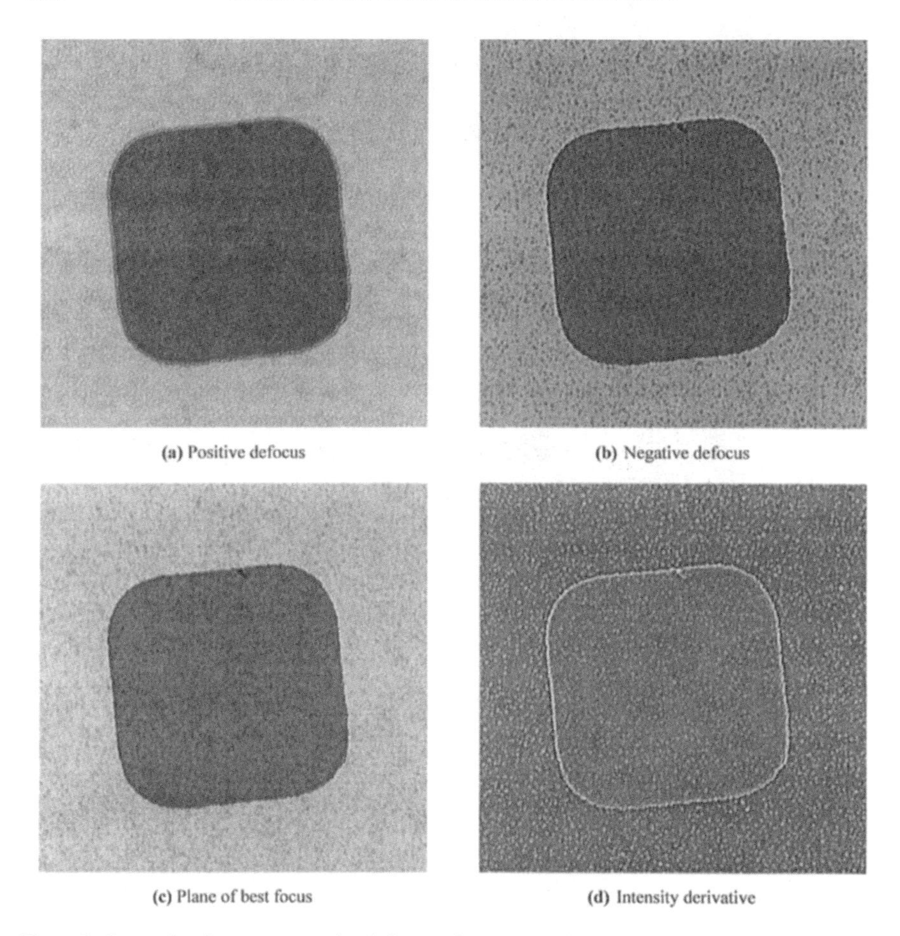

FIG. 6. Intensity images acquired from the TEM. A set of intensity images were acquired using the modified TEM. Image (c) is the intensity distribution in the plane of best focus, whereas the top two images, (a) and (b), are similar intensity images acquired with the microscope defocused 3 units either side of the plane of best focus. The images drifted slightly as the focus was changed and were realigned using a cross-correlation based alignment routine prior to image processing. Image (d) shows the longitudinal intensity derivative formed by subtracting (b) from (a).

which reduced the alignment-induced tilt on the recovered phase to negligible levels. Each pair of planes was independently processed and produced data sets that were indistinguishable apart from slight variations in the magnitude of the recovered phase. This variation in phase magnitude is attributed to slight variations in the defocus distance.

The recovered phase for the pair of images at a defocus distance of ± 3 steps either side of focus is shown in Fig. 7. Note the dimpled appearance of the phase map: these features are due to slight thickness variations in the silicon nitride substrate, as can be seen by the extension of these pits outside the cobalt dot itself onto the surrounding substrate. Also note that it was

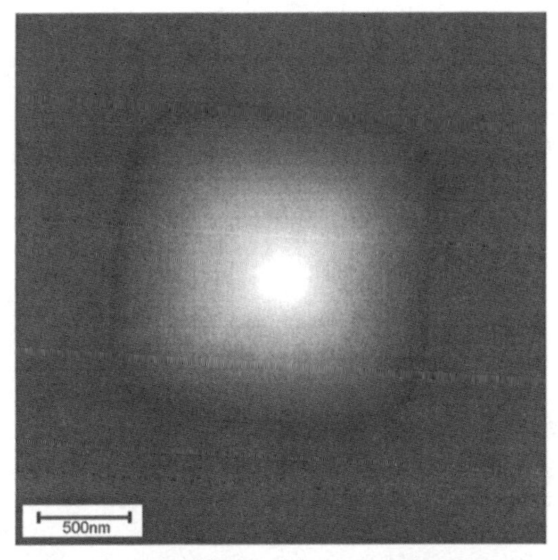

(a) Recovered phase image

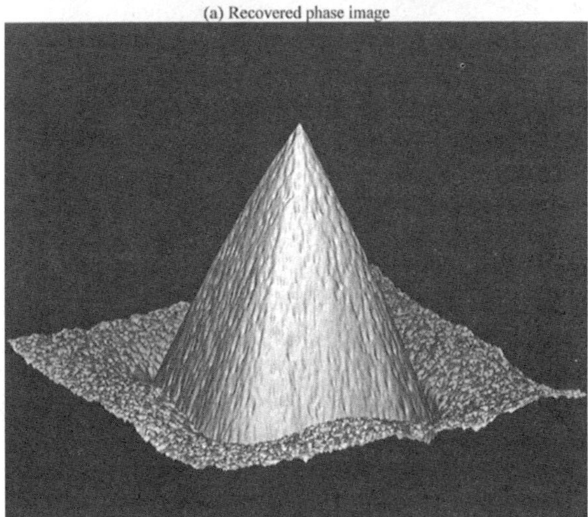

(b) Surface plot of recovered phase

FIG. 7. Recovered phase of the cobalt grain: (a) gray-scale image of the phase recovered from a cobalt grain; (b) surface plot of the same data.

not necessary to do any phase unwrapping on our data even though the data go through many cycles of 2π. This is because our phase processing technique returns the phase directly, unlike interferometric techniques that return the phase modulo 2π. It is also worth noting that the phase recovery algorithm is not especially demanding of computational resources; processing each 1024×1024 pixel data set took about 2 minutes on a 133 MHz Pentium personal computer or 38 seconds on a DEC Alpha 600au workstation including image alignment, phase recovery, and data input/output. Note also that being a deterministic phase recovery algorithm there is no variation in recovery time for different samples; recursive techniques, on the other hand, often have a convergence time that varies between data sets.

5.4.3 Qualitative Evaluation of Recovered Images

To compare the results of our imaging technique with phase structure seen using existing techniques the same cobalt grain was examined using Foucault imaging to reveal the phase gradient of the sample in a particular direction. Given that Foucault imaging shows phase gradients along a particular axis, it is possible to generate a Foucault image from our quantitative phase data by differentiating the recovered phase along a given axis and multiplying by the transmitted intensity, to take account of variations in sample transmittance. The results of this comparison are shown in Fig. 8, and it can be seen that the real Foucault image in Fig. 8a and the "Foucault image" calculated from our phase data shown in Fig. 8b are in excellent qualitative agreement.

It is also possible to compute the magnetic induction directly from the recovered phase because the electrons are deflected perpendicular to the component of magnetization in the image plane by an amount proportional to the magnitude of the magnetization. Reversing this argument enables us to infer the sample magnetization from our phase data: lines of magnetic induction lie perpendicular to local phase gradients, whereas the magnitude of the magnetic induction is proportional to the magnitude of the phase gradient. Because our TIE phase imaging technique recovers the actual phase structure of the sample, it is possible to directly compute the sample magnetization from our recovered phase image. The sample magnetization computed in this way is shown in Fig. 9 (see color insert), and as can be seen this cobalt square consists of the classic vortex magnetization pattern expected of an isolated magnetic domain [189].

5.4.4 Quantitative Comparison with Electron Holography

To independently measure the phase gradients of the cobalt squares, off-axis electron holography was performed using a Philips CM200 TEM

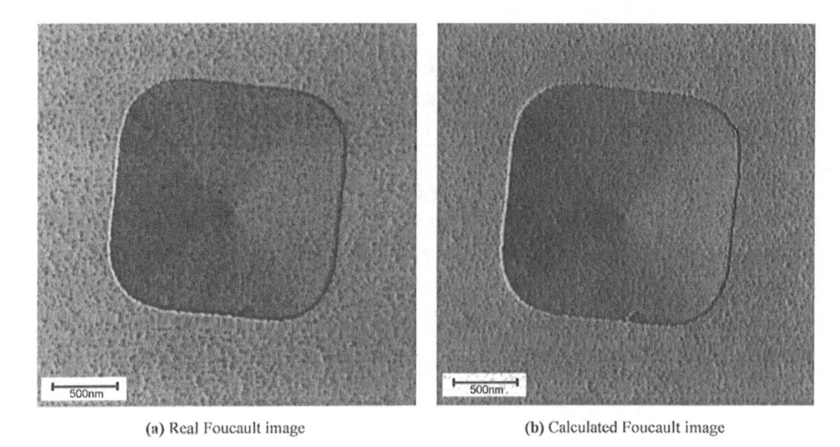

(a) Real Foucault image (b) Calculated Foucault image

FIG. 8. Comparison of real and calculated Foucault images. Image (a) is a real Foucault image taken using an electron microscope operating in the Foucault imaging mode, whereas image (b) is a Foucault image calculated from our quantitative phase data. As can be seen, the image calculated from the recovered phase is in excellent qualitative agreement with the real Foucault image.

modified for electron holography [177, 184, 196, 225]. The microscope was operated at 200 keV with a field emission electron gun, an electrostatic biprism, and a 1024 × 1024 pixel Gatan 794 multiscan CCD camera for collecting the interferograms. The electrostatic biprism consisted of 0.5 μm-thick quartz wire coated with gold, and it carried a voltage of 120 V to cause overlap between the electron wave that has passed through the sample and the silicon nitride membrane and a reference wave that has passed only through the silicon nitride membrane. A Lorentz minilens ($C_s = 8$ m and 1.2 nm line resolution at 200 keV) situated below the lower objective pole piece allowed holograms to be obtained with the objective lens switched off, so that the sample was in almost field-free conditions.

This microscope had a relatively limited field of view compared to the microscope used for obtaining the TIE images; therefore, we had to acquire four holograms per square in order to cover the entire cobalt dot. For each of these holograms, half the field of view was covered with the sample and the other half with the surrounding silicon nitride membrane to provide a reference region with no phase structure; however, the thickness of the silicon nitride membrane is quite uniform, allowing us to assume that the effect of the silicon nitride thickness is the same for both waves. The average contribution of the membrane to the absolute phase of the sample is therefore simply an additional constant phase that is not observable with either the holographic or TIE techniques.

Using electron holography the phase gradient was measured to be 0.028 ± 0.001 rad/nm, whereas our direct phase measurement technique yielded a phase gradient of 0.035 ± 0.007 rad/nm, where the uncertainty in the step size is the principal source of uncertainty. Although both of these values are consistent with the phase gradient expected for a 10-nm-thick layer of cobalt [174], the error bounds of these two measurements only just overlap. We therefore decided to test the repeatability of our phase measurement technique using this sample as a calibration object. Comparing the holographic and TIE phase gradient figures, we deduce that the defocus step size was 29 μm for the modified electron optics system. We then repeated the direct imaging and holographic experiments on a completely different and separately manufactured cobalt sample, and profiles through both our recovered phase image and the holographic phase are shown in Fig. 10.

With the newly recalibrated focal step size of 29 μm, the phase gradient measured using the direct imaging technique was 0.025 ± 0.001 rad/nm, and the phase gradient measured using electron holography was 0.025 ± 0.003 rad/nm. Apparently, correct calibration of step size permits excellent agreement to be obtained. We emphasize, however, that although the step size measurement was based on a different sample, the sample was similarly prepared and of the same element. It cannot therefore be considered completely independent. With this caveat, however, we conclude that the two measurements may be reconciled by an appropriate calibration of the TEM. We therefore see that, once the TEM has been calibrated, we obtain a quantitatively accurate image of both the electron phase and the electron amplitude.

5.5 Critical Evaluation of the Technique

5.5.1 Comparison with Electron Holography

We have presented a technique that permits the direct and quantitative measurement of the phase distribution in an electron microscope. The only competing technique is that of electron holography, and we make a few comments here on the comparative benefits of the techniques.

Primarily, electron holography makes strong demands on the illumination of the sample: there must be room for a reference beam to bypass the sample, and the electron illumination must be highly coherent. The electron microscope must also be fitted with a biprism in order to bring the object and reference beams into superposition. Although the biprism and coherent source are now quite readily available, the object requirement limits the application to objects that are relatively small. By way of example, we point out again that the imaging of the cobalt grain presented here required four holograms in order to obtain measurements over the full face of the grain.

The direct approach discussed here has very limited requirements of the instrument and sample, requiring simply the ability to form a highly linear and high dynamic range image, conditions which are easily fulfilled by the use of a modern CCD camera. One would thus anticipate that the direct approach will find greater application to problems requiring a large field of view and for which a reference beam is not possible. However, the issue of sensitivity is open at this stage. Holographic techniques are able to measure phase excursions of much less than an electron wavelength. This degree of sensitivity is not guaranteed with the noninterferometric approach. As is immediately apparent from an inspection of the simplified expression given in Eq. (23), the signal is proportional to the Laplacian of the phase, that is, to the curvature of the phase. In other words, objects that have a highly structured phase distribution will potentially be imaged very well. However, if the object presents a small phase curvature but nevertheless a large phase excursion, then holographic methods are more likely to be successful. The exploration of these issues will rely on further research and greater experience. We note, however, that the application to high resolution transmission electron microscopy is likely to be successful by the criteria raised in this discussion. In this case, the matter of aberrations is worthy of further consideration.

5.5.2 Application to the Negation of Aberrations

Thus far, our discussions have ignored the effects of any aberrations that may be present in the imaging system. This deficiency does not generate any problems for Lorentz imaging: as was pointed out in Section 2.3.3 spherical and higher order aberrations are irrelevant in this regime. However, as one moves to higher resolution transmission electron microscopy, spherical and higher order aberrations become significant. In the present section we develop a simple formalism for negating such aberrations in contexts where they become important. We give two simple demonstrations of these ideas, namely (1) negating the effects of spherical aberration in transmission electron microscopy and (2) eliminating the twin image of in-line holography.

Let us denote the true exit surface wave function of a given sample by $\psi_{\text{true}}(\mathbf{r}_\perp)$. Then the action of an arbitrary shift invariant linear optical system that attempts to form an image of this wave function may be described using the transfer function formalism. To wit, the Fourier transform of the

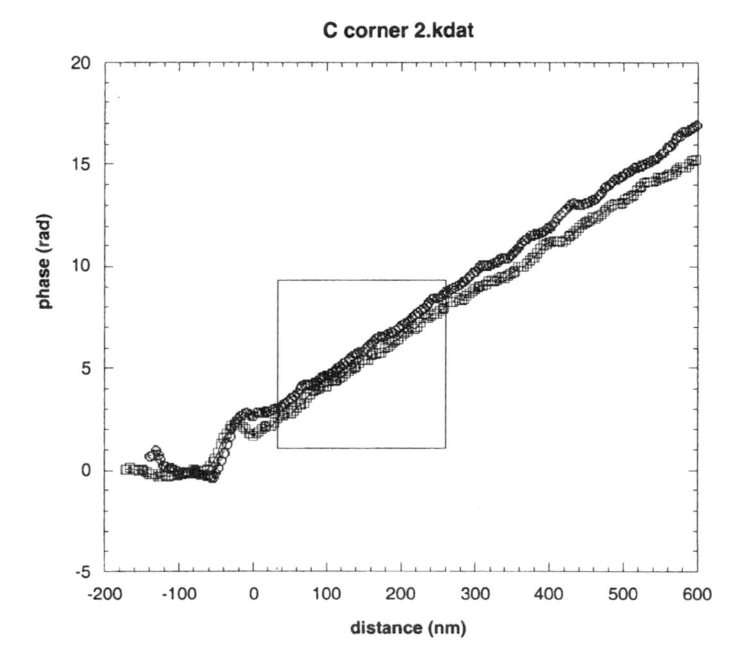

(a) Phase recovered from holography

FIG. 10. Profiles through both the holographic (a) and recovered phase (b). With the newly recalibrated focal step size of 29 μm, the phase gradient measured using the direct imaging technique was 0.025 ± 0.001 rad/nm, and the phase gradient measured using electron holography was 0.0246 ± 0.003 rad/nm.

resulting image is given by

$$\psi_{\text{aberrated}}(\mathbf{q}_{\perp}) = \mathscr{F}[\psi_{\text{true}}(\mathbf{r}_{\perp})]\mathscr{T}(\mathbf{q}_{\perp}), \tag{26}$$

where all symbols are as defined in Section 2.3.3, with the exception of the descriptive subscripts. Taking the inverse Fourier transform $\mathscr{F}^{-1}$, we arrive at an integral transform that relates the aberrated exit surface wave function being imaged to the true wave function:

$$\psi_{\text{true}}(\mathbf{r}_{\perp}) = \tilde{\mathscr{F}}^{-1}[\psi_{\text{aberrated}}(\mathbf{q}_{\perp})\mathscr{T}^{-1}(\mathbf{q}_{\perp})]. \tag{27}$$

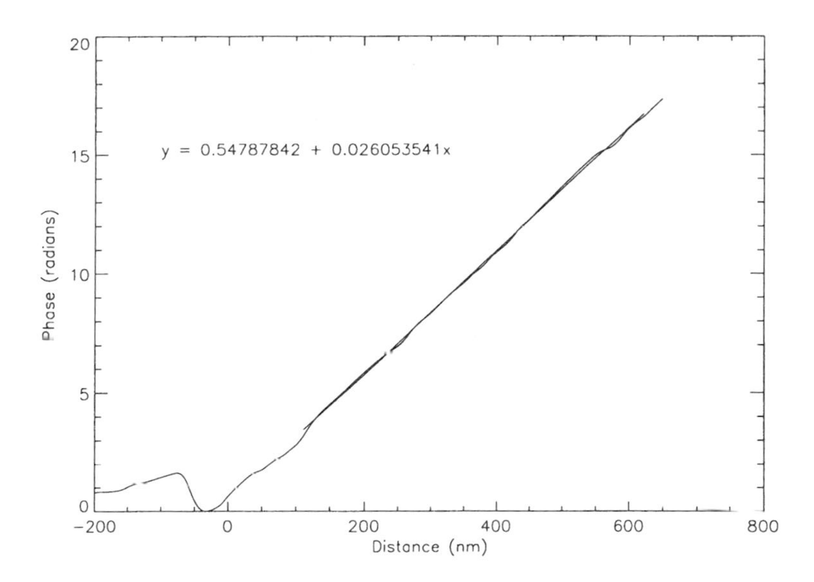

(b) Direct phase measurement

FIG. 10. Continued.

The tilde over the inverse Fourier transform operator indicates that we should restrict the region of integration in Eq. (27) to those values of $\mathbf{q}_\perp$ for which the modulus of the microscope transfer function is sufficiently different from zero. This amounts to admitting ignorance of those spatial frequencies that are blocked by the imaging system.

Now, the image $\psi_{\text{aberrated}}(\mathbf{q}_\perp)$ of the exit surface wave function is readily measured by taking the Fourier transform of the complex wave field that is obtained using the method described in the previous section. Specifically, the modulus of the aberrated wave function may be measured directly and the phase measured using the phase-retrieval procedure. Thus, one now has knowledge of $\psi_{\text{aberrated}}(\mathbf{q}_\perp)$. One may then use the integral in Eq. (27) to negate the effects of the aberrated imaging system, provided the transfer function is known.

As a first example of the application of these ideas, let us consider an out-of-focus electron microscope, operating in the Fresnel imaging regime, which is characterized by the microscope transfer function given by substituting Eqs. (16) and (17) of Section 2.3.3 into Eq. (15) of the same section:

$$\mathscr{T}(\mathbf{q}_\perp) = A(|\mathbf{q}_\perp|)e^{-i\pi\lambda\Delta f|\mathbf{q}_\perp|^2 - i\pi C_s\lambda^3|\mathbf{q}_\perp|^4}e^{-[(\pi\theta_c\Delta f)^2/\ln 2]|\mathbf{q}_\perp|^2}. \tag{28}$$

All symbols are as defined in Chapter 2. The presence of the aperture function A in this expression limits the integral in Eq. (27) to values of $\mathbf{q}_\perp$ for which $A(\mathbf{q}_\perp)$ is nonzero. This, of course, is related to the well-known resolution limit of an imaging system with given numerical aperture. Depending on the strength of the damping envelope, the integral in Eq. (27) may be restricted to an even smaller subset of values for $\mathbf{q}_\perp$, serving to further degrade the resolution available from the imaging system.

As a second application of these ideas, we consider the twin-image problem of in-line holography. In this case, the linear optical system under consideration is simply a slab of free space of thickness Δf. The transfer function for free space (Fresnel propagator) is given by "turning off" the aperture function, spherical aberration, and damping function in Eq. (28), leaving

$$\mathscr{T}(\mathbf{q}_\perp) = e^{-i\pi\lambda\Delta f|\mathbf{q}_\perp|^2}. \tag{29}$$

Thus, one is led to the following solution to the twin-image problem of in-line holography. One first measures the phase, using the method of Section 5.2.3, of the in-line hologram by measuring the intensity of three closely spaced holograms. This gives the exit surface wave function of the linear optical system given by a slab of free space of thickness Δf. One then reconstructs the true exit surface wave function of the object of interest by applying the integral transform [Eq. (27)] to the propagated wave function, using the transfer function given in Eq. (29). Of course, this is equivalent to performing an inverse Fresnel diffraction integral on the propagated wave field [226, 227].

5.6 Summary and Conclusions

Although the discussion in this chapter is firmly placed in the context of magnetic imaging, the same techniques are equally applicable and can be used without modification on any object imaged in a transmission electron microscope. For example, electron phase imaging can be used to image the potentials formed by electrostatic charges [170], and the TIE phase imaging techniques presented here could be directly applied without modification to such an object. In the context of biological microscopy, transmission electron microscopes can be used to study biological samples; however, the low coefficient of absorption of electrons by thin biological specimens makes it necessary to stain samples using heavy metal salts to increase contrast [228, 229]. The use of phase-contrast techniques is not favored in biological electron microscopy because of interpretational problems in the variation of contrast with defocus, and it would be interesting to see whether the use of our phase imaging technique could overcome some of these limitations. In the context of high resolution electron crystallography, algorithms for determining three-dimensional crystal potentials require knowledge of both the amplitude and phase of the electron wave function exiting the sample, and there is much interest in means for reconstructing this so-called exit wave function [230].

In this chapter we have successfully imaged phase structure in microscopic samples using a transmission electron microscope and directly imaged both the direction and magnitude of the magnetization in a microscopic cobalt grain using TIE-based phase retrieval techniques. We have also demonstrated that the TIE technique can be used to measure the phase structure of the electron wave exiting from the sample, once appropriate calibration of the microscope system has been performed, and that the measurements so made agree with the results of electron holography on the same sample. As our technique measures the phase directly and there is no need for phase unwrapping, and because we solve for only the phase component of the transmitted electron wave, we are able to clearly separate the phase and intensity structure of the sample from each other. Furthermore our phase measurement technique can be implemented without the need to insert specialized optics such as knife edges or beam splitters into the electron column, making it easy to implement on existing TEM hardware. As such, the technique presented in this chapter has the potential to be a powerful method for extending and simplifying both phase imaging and phase measurement in transmission electron microscopy.

We would like to stress that the TIE technique for electron phase imaging is not limited to magnetic samples and can, in principle, be used to image any sample, magnetic or otherwise, that introduces a phase shift into an electron beam. The use of our technique for reconstruction of the complex exit wave function and, thereby, the reconstruction of three-dimensional crystal potentials is the subject of investigation by Les Allen and colleagues at the University of Melbourne. However, the use of detector-plane phase information for image restoration and as a means of correcting for spherical aberration in TEM imaging remains uninvestigated, as does the application of quantitative TEM phase imaging to biological specimens. Although we have not yet investigated applications other than the imaging of magnetic materials, we suggest that the phase measurement technique presented here provides an unambiguous means of measuring the phase of electron wave fields and is a potential solution to the phase problem of electron microscopy. As such, the technique presented in this chapter also has the potential for broad application in electron imaging beyond Lorentz microscopy.

Acknowledgments

The experimental work described in this chapter is the result of an extensive collaboration with Dr. Sasa Bajt and Mark Wall of the Advanced Microtechnology Program, Lawrence Livermore National Laboratories, and Molly McCartney of Arizona State University. The image of Jack (the dog) (Fig. 2a) is courtesy of Jane V. Micallef, and the squirrel image (Fig. 2b) is courtesy of Public Domain Images, http://www.PDImages.com/.

6. SCANNING ELECTRON MICROSCOPY WITH POLARIZATION ANALYSIS (SEMPA) AND ITS APPLICATIONS

John Unguris

Electron Physics Group
National Institute of Standards and Technology
Gaithersburg, Maryland

6.1 Introduction

Scanning electron microscopy with polarization analysis (SEMPA) is a technique for directly imaging the magnetic microstructure of surfaces and thin films. SEMPA relies on the fact that secondary electrons emitted from a magnetic sample in a scanning electron microscope (SEM) have a spin polarization which reflects the net spin density in the material. This spin density, in turn, is directly related to the magnetization of the material. By measuring the secondary electron spin polarization, SEMPA can be used for direct, high resolution imaging of the direction and relative magnitude of a sample's magnetization, in the same way that an SEM can be used to image topography by measuring the secondary electron intensity.

The SEMPA technique has evolved from measurements during the 1970s and 1980s in which various conventional electron spectroscopies were combined with electron spin sensitivity to investigate the magnetic properties of surfaces and thin films. Several reviews of this early work are available [231–234]. SEMPA microscopes, also sometimes referred to as spin-polarized electron microscopes (spin-SEMs), made their initial appearance in the mid 1980s [235–237]. Several reviews of the SEMPA technique and instrumentation have appeared since then [238–241].

SEMPA is one of several methods used for imaging magnetic microstructure. Because each method has its own particular strengths and drawbacks, the sample and type of imaging required determine which of these techniques is most useful. The basic features of most of these techniques have been compared in several reviews [21, 25, 242, 243]. SEMPA is particularly well suited for the high resolution imaging of magnetic structures at surfaces and in thin films. For these samples, SEMPA can provide a direct picture of the magnitude and direction of the magnetization (not magnetic field) that is inherently independent of the topography. Moreover, this imaging is done with the high spatial resolution, the large depth of field, and the ease of use of the SEM.

EXPERIMENTAL METHODS IN THE PHYSICAL SCIENCES
Vol. 36
ISBN 0-12-475983-1
ISSN 1079-4042/01 $35.00

The purpose of this chapter is to present a brief overview of the technique and to describe the types of magnetic imaging applications for which SEMPA is best suited. The origin of the magnetic contrast will be described as well as the instrumentation required for its measurement. The review will highlight SEMPA characteristics by presenting a sampling of examples of SEMPA imaging.

6.2 Spin Polarized Secondary Electron Magnetic Contrast

The magnetic contrast in SEMPA is due to the spin polarization of secondary electrons emitted from a magnetic sample. This polarization is related directly to the net spin density and hence the magnetization of the sample. For simple transition metal ferromagnets this polarization can be quite large, and, given the fact that a large number of secondary electrons are produced in an SEM, the raw magnetic signal in a SEMPA measurement can be quite substantial.

The electron spin contribution to the magnetization is

$$M = -\mu_B(n_\uparrow - n_\downarrow), \tag{1}$$

where $n_\uparrow(n_\downarrow)$ is the number of spins per unit volume that are aligned parallel (antiparallel) to the magnetization, and μ_B is a Bohr magneton. The minus sign results from the fact that the magnetic moment of the electron is directed opposite to its spin. For transition metal ferromagnets, such as Fe, Co, or Ni, in which the orbital moment is quenched, this spin contribution to the magnetization is a close approximation to the total magnetization.

The spin polarization of the emitted secondary electrons directly probes the spin part of the magnetization. Since electrons retain their spin orientation during the emission process, the secondary polarization is along the same direction as the magnetization. The spin polarization, like the magnetization, is a vector quantity. For the purposes of SEMPA, it is adequate to consider each component of the polarization separately. For example, the polarization along the z direction is

$$P_z = \frac{(N_\uparrow - N_\downarrow)}{(N_\uparrow + N_\downarrow)}, \tag{2}$$

where $N_\uparrow(N_\downarrow)$ is the number of electrons with spins parallel (antiparallel) to the $\pm z$ direction. Note that the polarization is independent of the total number of electrons and that it may have values $-1 \leqslant P \leqslant 1$.

A sketch of the typical energy distribution of the polarization, $P(E)$, and the number, $N(E)$, of secondary electrons emitted from a ferromagnet is shown in Fig. 1. The secondary electron intensity distribution shows the

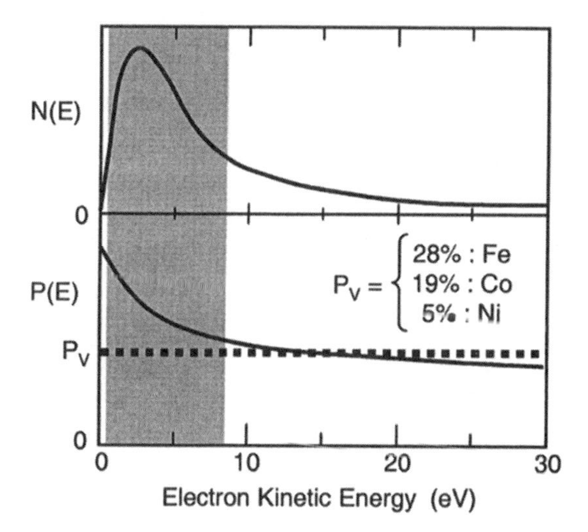

FIG. 1. Schematic energy distribution of the number, $N(E)$, and polarization, $P(E)$, of secondary electrons emitted from a typical transition metal ferromagnet. The shading highlights the approximate energy range accepted by the polarization analyzer used in the SEMPA measurements.

familiar peak at low energies that is due to the secondary electron cascade process. The total secondary electron yield is on the order of 10% of the incident electron beam current and depends sensitively on the sample and the incident beam energy and angle. On the other hand, the polarization is relatively insensitive to the incident beam conditions. The low energy cascade electrons are the result of electron–hole pair excitations in the valence band. To the extent that the electron cascade represents the uniform excitation of electrons from the valence band, the expected polarization is

$$P(E) = \frac{n_B}{n_V}, \tag{3}$$

where n_V is the number of valence electrons per atom and n_B is the number of Bohr magnetons per atom. The number of Bohr magnetons equals the difference between the number of majority spin and minority spin electrons per atom. For Fe, Co, and Ni this simple model predicts polarizations of 28%, 19%, and 5%, respectively. These values agree well with measured polarizations for secondary electron energies between 10 and 20 eV for Fe [244], Co [244], and Ni [245]. At energies below 10 eV, the secondary polarization increases from the predicted values. This enhancement is the result of spin-dependent filtering of the lowest energy electrons [246, 247]. Minority spin electrons are more likely to be lost from the secondary distribution, because in a ferromagnet there are more empty minority states than majority states available for the electrons to decay into before emission.

Although the secondary electron polarization is directly proportional to the magnetization, in practice, the constant of proportionality may be difficult to determine. In general, the energy dependence of the polarization for a particular sample composition may not be known, and the range of secondary electron energies accepted by a polarization analyzer may not be well defined. In general, SEMPA can therefore directly measure the magnetization direction and relative magnitude of the magnetization, but it usually is not possible to determine the absolute value of the magnetization.

A final important feature of the secondary electrons is their short mean free path in the solid. Although the energy of the incident electron beam is deposited several hundred nanometers into the sample, the mean free path of the low energy secondary electrons is only of the order of a nanometer. For spin polarized secondary electrons the $1/e$ sampling depth, or average spin attenuation length, ranges from 0.5 nm for a transition metal such as Cr [248] to about 1.5 nm for a noble metal such as Ag [249]. The effective probing depth of SEMPA is therefore about 1 nm. The combination of shallow probing depth and small beam diameter means that very little magnetic material is required for a SEMPA measurement. SEMPA can sense the magnetization of as few as a thousand ferromagnetic Fe atoms.

6.3 Instrumentation

The essential elements of a SEMPA imaging system are (1) a SEM column to form and raster the incident electron beam, (2) an ultrahigh vacuum chamber with associated instrumentation for surface preparation and surface analysis, (3) spin polarization detectors consisting of electron optics for collecting the secondary electrons and polarization analyzers for measuring their polarization, and (4) a data acquisition system for collecting, storing, and displaying the magnetization images. An overview schematic of our SEMPA apparatus is shown in Fig. 2.

6.3.1 Electron Microscope and Specimen Chamber

The spatial resolution of SEMPA is ultimately determined by the size, intensity, and stability of the focused incident electron beam. Naturally a small spot size is desirable, but beam intensity and stability are just as

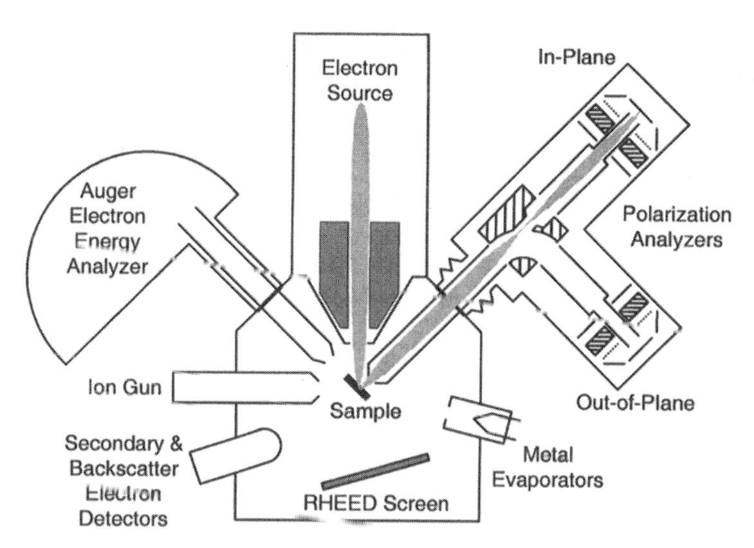

FIG. 2. SEMPA apparatus schematic. The incident electron beam, polarization analyzers, and RHEED screen are positioned as shown. The other components are not shown in their true positions.

important, because the inefficiency of the spin polarization analyzers mandates long exposure times. Furthermore, the working distance between the sample and column must be large enough to allow access to the secondary electrons and minimize the objective lens magnetic field at the sample. Typical operating conditions are the following: a 10 mm working distance, a stray field at the sample of <80 A/m (1 Oe), and a 10 keV incident electron beam energy. The best SEMPA resolution so far is about 20 nm and has been achieved using a high brightness field emission electron source [250]. Most of the images presented here have been obtained using a LaB_6 thermionic emission cathode, which provides 50 nm resolution in a reasonable (<1 hour) acquisition time. With recent improvements to commercial field emission SEMs, it should be possible to obtain $\leqslant 10$ nm SEMPA resolution.

The choice of electron microscope and specimen chamber are further constrained by the surface sensitivity of SEMPA. An ultrahigh vacuum surface analysis environment is required. The vacuum should be better than 10^{-7} Pa to avoid sample contamination that could significantly diminish the magnetic contrast. Conventional surface science preparation techniques, such as ion sputtering and annealing, and surface analysis techniques, such as Auger spectroscopy, to measure the surface chemical composition are desirable. Such requirements are conveniently met by commercial scanning

Auger microprobes such as the one shown in Fig. 2. This apparatus has a hemispherical energy analyzer for Auger analysis, a phosphor screen and electron multiplier for reflection high energy electron diffraction (RHEED) measurements of surface and thin film atomic scale order, and various evaporators for growing thin films. The sample is mounted on a high stability stage that allows heating to 800°C and optional liquid nitrogen cooling. A great advantage of such a system is that sample preparation and analysis can all be done completely *in situ*. In fact, SEMPA imaging can be used to continuously monitor the magnetic structure during the entire sample preparation process.

6.3.2 Spin Polarization Analysis

The spin polarization detectors consist of an electron optical system for collecting the emitted secondary electrons and one or more spin polarization analyzers for measuring the various components of the polarization vector. In our SEMPA apparatus a +1500 V bias is applied to the front of the transport optics and is brought to within 12 mm of the sample. This bias voltage helps collect most of the secondary electrons and accelerates them so that they spend less time in the potentially depolarizing stray magnetic fields of the sample and objective lens. The transport optics are designed to transmit nearly all secondary electrons with an initial kinetic energy between 0 and 8 eV to the spin analyzer. The transport optics also contain electrostatic deflection plates that center the beam in the polarization analyzer and keep the beam stationary at the analyzer while the incident beam is scanned across the sample. Without these descan optics the beam motion during large area scans could introduce instrumental artifacts in the SEMPA images. As seen in Fig. 2, the transport optics also contain a 90° spherical deflector to switch the beam between two orthogonal spin analyzers. Two analyzers are required to completely resolve all of the components of the magnetization vector, as a single detector can only measure two transverse polarization components. In our geometry, one detector measures the two in-plane magnetization components, while the second measures the out-of-plane component along with a redundant in-plane component that is useful for cross-calibration of the analyzers. Alternatively, all three magnetization components can be measured using a single detector along with a polarization rotator [251, 252].

The key element of the SEMPA apparatus is the spin polarization analyzer. The characteristics of various spin analyzers have been compared [253, 254], and the application of some of them to SEMPA has been discussed [240, 254]. Unfortunately, although several different types of

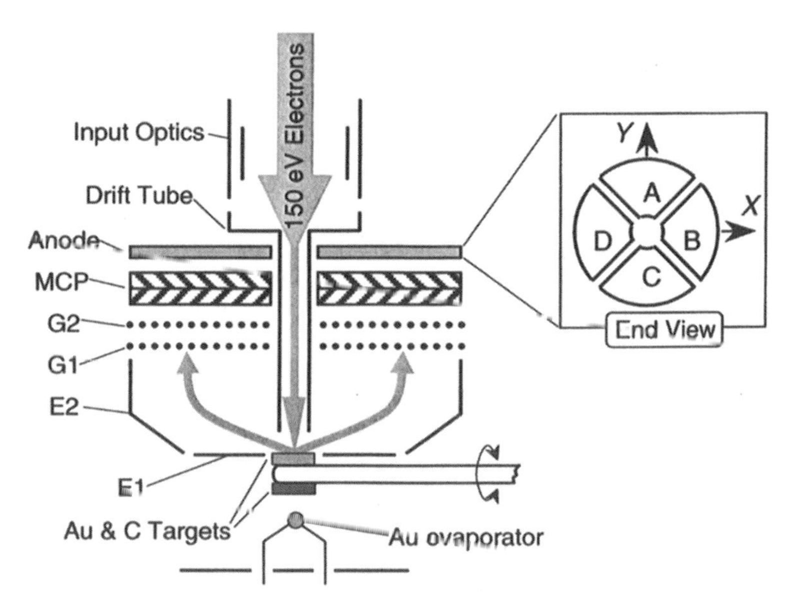

FIG. 3. Cross-sectional schematic of a low energy diffuse scattering polarization analyzer. The inset shows the split quadrant anode arrangement.

analyzers are available, they all share the common characteristic of very low efficiency. The most efficient spin analyzers have a figure of merit of about 10^{-4}. This means that a polarization measurement will take 10^4 times longer to achieve the same signal-to-noise ratio as an intensity measurement that has nearly 100% efficiency. The data acquisition rate in SEMPA imaging is therefore limited much more by the detector than by the secondary electron signal. In other words, if more efficient spin polarization analyzers could be found, SEMPA could image magnetic structure nearly as fast as a SEM images topography.

Most commonly used spin analyzers rely on the asymmetry of spin–orbit scattering as the basis of their spin sensitivity. Figure 3 shows a schematic of our spin analyzer. It is a low energy diffuse scattering analyzer in which the polarized electrons with 150 eV of kinetic energy are scattered from an amorphous Au film [253, 256]. The polarized electrons are scattered diffusely by the Au target and are then deflected by the electrodes $E1$ and $E2$ so that their trajectories are approximately normal to grids $G1$ and $G2$, which filter out the low energy inelastically scattered electrons from the Au target. The remaining electrons are amplified by the microchannel plates and then collected by the four-quadrant anode shown in the inset of Fig. 3. Two orthogonal components of the transverse polarization are measured simultaneously:

$$P_x = \frac{1}{S}\frac{(N_C - N_A)}{(N_C + N_A)} \tag{4}$$

and

$$P_y = \frac{1}{S}\frac{(N_B - N_D)}{(N_B + N_D)}, \tag{5}$$

where N_i is the number of electrons counted in quadrant i, and S is the instrumental Sherman function. The Sherman function is a measure of the spin sensitivity of the detector and is equal to the measured, normalized asymmetry for 100% polarized incident electrons. The overall efficiency of the analyzer is given by the figure of merit,

$$F = S^2\left(\frac{I}{I_0}\right), \tag{6}$$

where I_0 is the incident beam current and I is the current reflected from the Au target. For this analyzer, $S = 0.10$ and $I/I_0 = 2 \times 10^{-2}$, so that the efficiency is 2×10^{-4}.

The low energy diffuse scattering detector has several characteristics that are desirable for SEMPA imaging. First, the detector is sufficiently compact and robust so that it can be added to the electron microscope without adversely affecting the operation of either the detector or the SEM. Second, the relatively high efficiency is almost constant over the 8 eV energy spread of the incident electrons. Third, the electron optical phase space acceptance of the detector is large enough to collect most of the secondary electrons. Note that other spin analyzers have also been successfully used for SEMPA imaging and can offer other features that may be desirable in particular applications [240].

6.3.3 SEMPA Imaging Example

An example of a SEMPA measurement is shown in Fig. 4. The sample is nominally the (100) surface of a Fe crystal, but the surface has been roughened by momentarily heating the sample above the body-centered cubic (*bcc*) to face-centered cubic (*fcc*) phase transition at 910°C. The simultaneously measured in-plane magnetization components, M_x and M_y, and topography, I, are shown in Fig. 4a, b, c, respectively. The detector axes are approximately aligned with the in-plane $\langle 100 \rangle$ easy axis directions. Figure 4a is therefore sensitive to magnetic domains that point to the right (white contrast) and left (black), whereas Fig. 4b is sensitive to domains that either point up (white) or down (black) relative to and in the plane of the

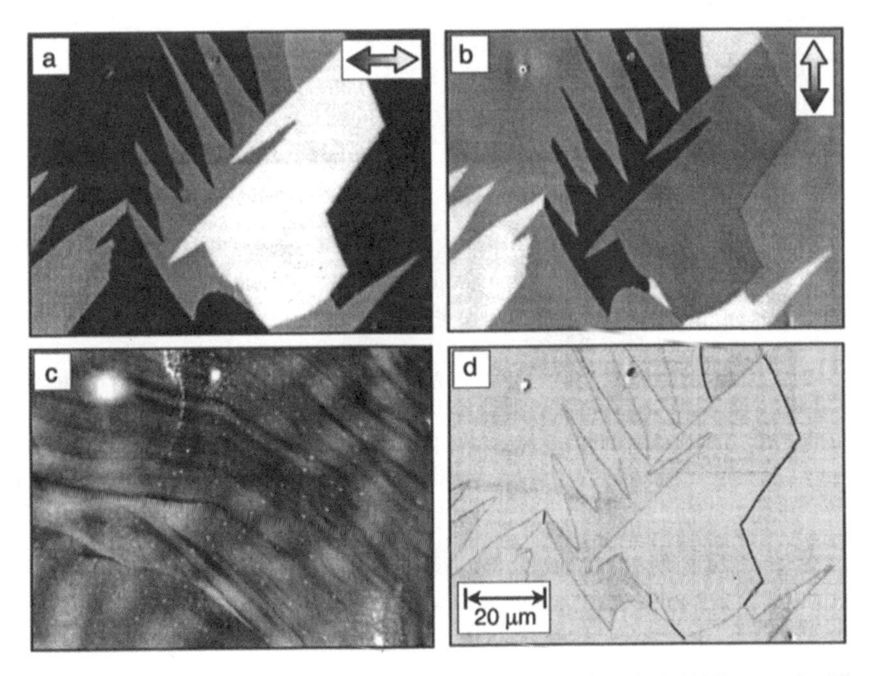

FIG. 4. SEMPA measurement example from a recrystallized Fe(100) sample. The in-plane (a) horizontal and (b) vertical magnetizations, as well as (c) the topography, were measured simultaneously. (d) Derived magnitude of the magnetization. The derived magnetization angle is shown in Fig. 16a (see color insert).

page. Note that, aside from a few nonmagnetic defects, the topography is essentially absent from the magnetization images. From the two components of the magnetization, the direction, θ, and magnitude, $|M|$, of the resultant magnetization vector can be simply derived:

$$|M| = \sqrt{M_x^2 + M_y^2} \qquad \text{and} \qquad \theta = \tan^{-1}\left(\frac{M_y}{M_x}\right). \tag{7}$$

The magnitude is displayed in Fig. 4d. As expected $|M|$ is constant except for the nonmagnetic defects and some missing magnetization at the domain walls. The missing magnetization at the domain walls is due to averaging the magnetization from adjacent domains with an electron probe diameter that is comparable to the domain wall width. This also explains why the 180° walls appear darker than the 90° walls. Smaller probe diameters can be used to resolve the internal structure of the domain walls and, in particular, the unique structure of the walls at the sample surface [257–259].

The direction of the magnetization derived from the images in Fig. 4 is shown in Fig. 16a (see color insert). In this image color is used to represent the direction of the magnetization. A color wheel inset in the image provides the key for mapping color into direction.

6.3.4 Instrumental Asymmetries

A common problem with spin analyzers is the elimination of false polarization signals owing to instrumental asymmetries. Constant polarization offsets, such as those due to mechanical misalignment or unbalanced electronic gains in the different channels, can be measured and accounted for in a reasonably straightforward way. More troubling are asymmetries resulting from changes in the position or the angle of incidence of electrons at the spin analyzer target. For example, the electron beam at the analyzer target may move as the incident electron beam is scanned over the sample. This false asymmetry can be minimized by descan deflection plates in the transport optics.

The sample may also introduce instrumental asymmetries. The application of an extraction voltage makes the sample part of the electron optics, and therefore the electron trajectories may be influenced by the geometry of the sample and sample holder. Variations in the work function, stray magnetic fields, and topography of a sample may also introduce false asymmetries. These instrumental effects may be reduced by designing the input electron optics such that beam displacements cause compensating changes in the incident angle at the analyzer target [256]. Instrumental effects can also be minimized by using high voltage Mott spin analyzers, which have a greater electron optical phase space acceptance.

In cases where the sample geometry deviates significantly from planar, such as spheroidal or wire samples, the secondary electron trajectories can be sufficiently disturbed that topographic features are visible in the polarization image. In such cases, a nonmagnetic reference image can be used to remove the topographic contrast. In the analyzer shown in Fig. 3 this is accomplished by replacing the Au target with a low atomic number graphite target. A reference image is then acquired that only shows the nonmagnetic signal and not the spin-dependent contribution. The graphite target image is then subtracted from the Au target image to obtain the magnetization image [238].

6.3.5 Acquisition Time

The time required to acquire a SEMPA image with a given signal-to-noise ratio and the spatial resolution of the image are closely coupled, as the incident electron beam current of the SEM decreases rapidly as the spot size is reduced. The signal-to-noise ratio, SNR, of a measurement of the polarization, P, after counting N electrons and using a detector with a figure of merit F is

$$\mathrm{SNR} \equiv \frac{P}{\Delta P} = P\sqrt{NF}. \tag{8}$$

The number of electrons that arrive at the detector during a time τ is

$$n = \delta_{\mathrm{se}} \eta I_p \frac{\tau}{e}, \tag{9}$$

where δ_{se} is the secondary electron yield, η is the collection and transport efficiency of the polarization detector input optics, I_p is the beam current incident on the sample, and e is the electron charge. Assuming perfectly efficient electron counting, the total time required to acquire a SEMPA image consisting of n_{pix} pixels is then

$$T = \frac{n_{\mathrm{pix}}(\mathrm{SNR})^2 e}{P^2 F \delta_{\mathrm{se}} \eta I_p} = C \frac{n_{\mathrm{pix}}(\mathrm{SNR})^2}{I_p}, \tag{10}$$

where C is a number that combines all of the instrumental and sample-dependent quantities. For our spin polarization optics and detector, we measure $C = 2.0 \pm 0.5 \times 10^{-13}$ A·s using a 10 keV incident beam and a clean Fe sample.

The spatial resolution is ultimately determined by the incident electron beam diameter, which decreases with decreasing beam current, I_p, in a manner that depends on the specific electron source and probe forming optics. Figure 5 shows the relationship between image acquisition time and resolution for two electron microscope columns: one using a thermal LaB_6 source and the other using a thermally assisted field emission source. These calculations use the measured dependence between I_p and beam diameter for 10 keV incident beam energies and a 10 mm working distance. These acquisition times are for a 128×128 pixel image of Fe with a SNR of 5.

A series of simulated SEMPA images across the top of Fig. 5 demonstrates how the SNR affects the image quality. The magnetic structure in the simulated image consists of two oppositely magnetized vertical domains that contain a small, $\pm 5°$, periodic "ripple." Although the primary signal, the 180° domain wall, is clearly visible with a SNR of 1, a SNR of at least 5 is required before the ripple fine structure becomes clearly visible.

From Fig. 5 it can be seen that acquisition times can range from less than minutes to hours. With a LaB_6 source, a low resolution probe is therefore used to quickly survey a sample and find specific regions of interest. Long scans with high resolution probes are then used for higher quality images of

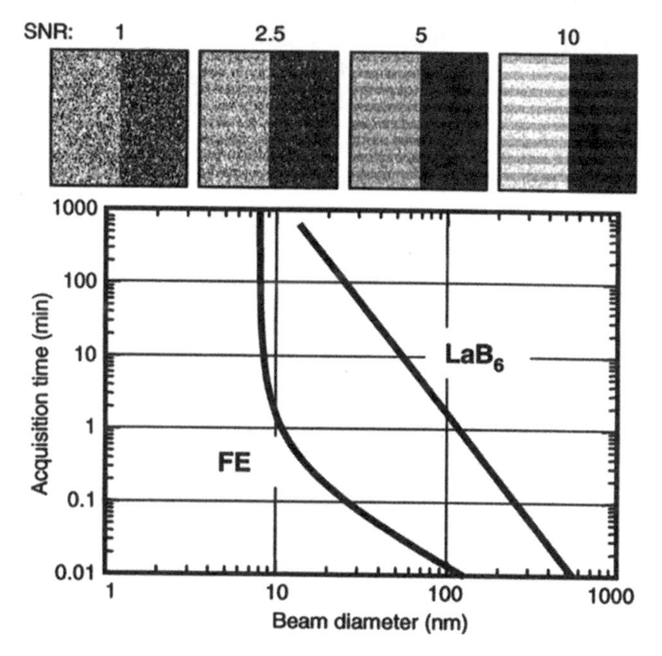

FIG. 5. Simulated SEMPA images with different signal-to-noise ratios (SNR) are shown across the top. The simulated magnetic structure in the images is a 180° Fe domain wall with superimposed $\pm 5°$ "ripple." The acquisition time for a 128×128 pixel SEMPA image of Fe with SNR = 5 is plotted below as a function of incident electron beam diameter for LaB_6 and thermally assisted field emission (FE) electron sources.

small magnetic structures. If the samples have a low spin polarization or if more pixels with higher signal-to-noise ratios are required, high resolution images may take hours to accumulate. In these cases, special care must be taken to avoid or to compensate for slow drifts in the position of the sample or electron source.

Finally, it is worth noting that although the above discussion considered only the image acquisition time for a single polarization component, one other orthogonal polarization component and the topography are measured simultaneously. No additional time is required to acquire these additional images.

6.3.6 Applied Magnetic Fields

Because SEMPA is an electron beam based technique, imaging cannot easily be carried out in the presence of a large applied magnetic field. A magnetic field can not only deflect the incident beam and the emitted secondary electron trajectories, but it can also cause the secondary electron spins to precess, thereby distorting and degrading the polarization signal. Of course these effects depend sensitively on the specific orientation of the magnetic field relative to the trajectories and polarization of the electrons, but the magnitude of the effects is roughly the same; that is, a field which causes a 10° deflection will also cause about a 10° rotation of the polarization.

Usually ambient magnetic fields less than 80 A/m (1 Oe) are not a problem. Larger applied magnetic fields, up to 8 kA/m (100 Oe), have been accommodated using various methods that depend on the specific sample being analyzed. For example, the sample can be made part of a closed loop magnetic circuit, reducing the stray magnetic field [260]. In perpendicularly magnetized samples the applied field effects can be minimized, since the field, polarization, and some of the secondary electron trajectories are all aligned [241]. Finally, for very small samples, it should be possible to apply very localized magnetic fields, using recording heads, for example, that minimize the exposure of the electrons to the applied field.

6.4 Examples of SEMPA Applications

6.4.1 Iron Coatings for Polarization Enhancement

Ultrathin ferromagnetic coatings can occasionally be used to improve the magnetic contrast of samples with inherently low secondary electron spin polarization. For example, Ni or Ni based alloys such as Permalloy have relatively weak polarization contrast. Similarly, magnetic materials such as garnets not only have very low secondary electron polarization, but are also insulators. In these cases, one can take advantage of the surface sensitivity of SEMPA and coat the samples with ultrathin films of higher spin polarization materials, usually Fe or Co. In most cases, the magnetic structure of the sample is not appreciably altered, since only a few monolayers of Fe are needed to significantly boost the secondary electron polarization. And, in the case of certain insulators, a few monolayers may also be sufficient to reduce sample charging.

Figure 6 shows a series of SEMPA images from the same area of a Permalloy sample that is coated with 0, 3, and 6 atomic layers of Fe (yielding films 0, 0.43, and 0.86 nm thick). The Permalloy sample is a patterned 80-nm-thick film with Néel walls that contain cross-ties that are visible in the uncoated sample. The images show a dramatic improvement in contrast with just a thin coating of Fe. Details, such as magnetization ripple,

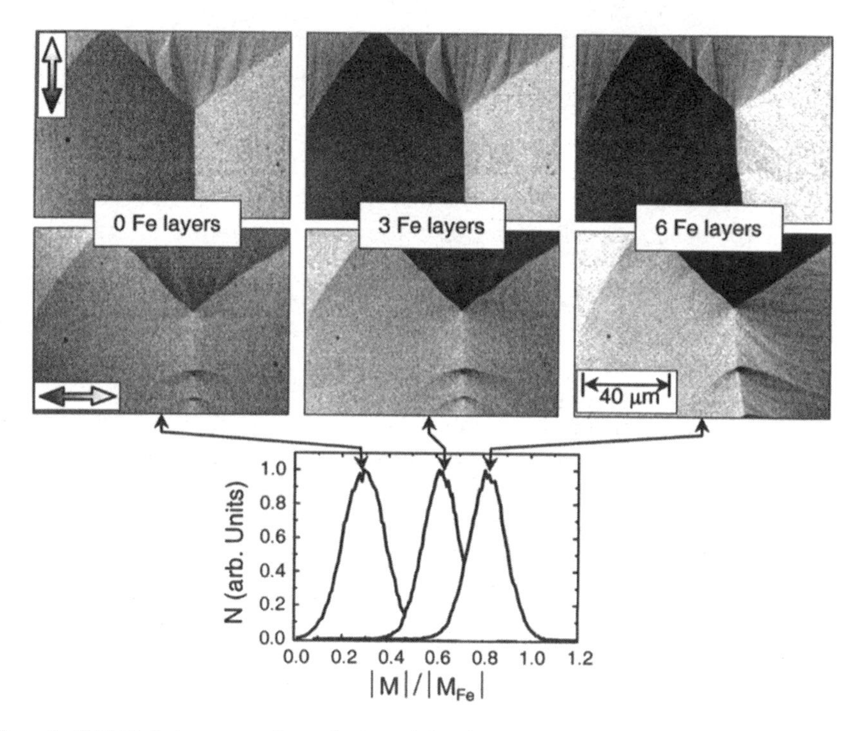

FIG. 6. SEMPA images of vertical and horizontal in-plane magnetizations in a Permalloy thin film sample, uncoated and coated with 3 and 6 atomic layers of Fe. Histograms of the corresponding magnetization magnitudes normalized to pure Fe are plotted.

that are barely seen in the uncoated sample are clearly visible in the coated Permalloy sample. The improvement in contrast can be made more quantitative by plotting the histograms of the magnetization magnitudes from the three images, as shown at the bottom of Fig. 6.

6.4.2 Three-dimensional Magnetization Imaging

Complete three-dimensional imaging of the magnetization vector is relatively straightforward in SEMPA imaging. An example of a three-dimensional magnetic structure imaged using SEMPA is shown in Fig. 7. These SEMPA images show the surface magnetic microstructure of a Co crystal. The hexagonal close-packed (*hcp*) Co crystal has uniaxial crystalline anisotropy, with the easy magnetization axis along the *c*-axis. SEMPA has been used to investigate the surface domain structure of Co with the *c*-axis lying

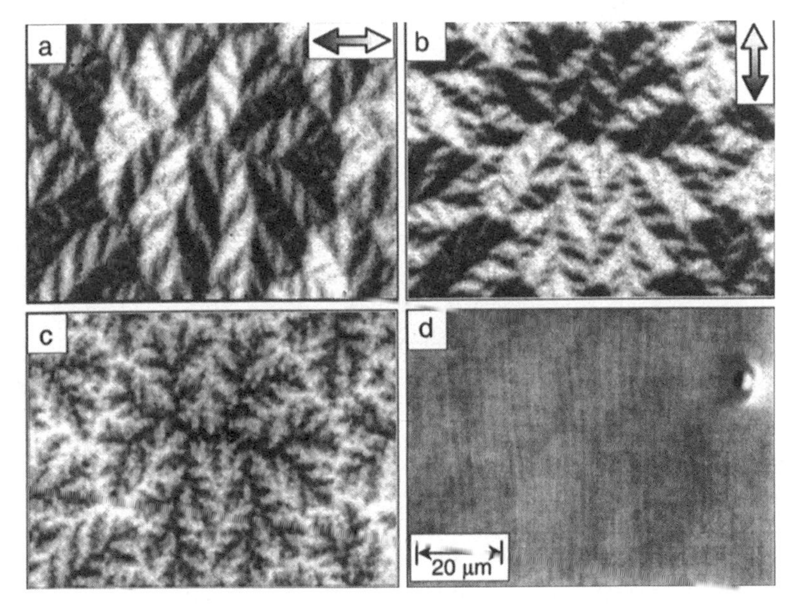

FIG. 7. SEMPA images of the in-plane (a) horizontal, (b) vertical, and (c) the out-of-plane magnetization components of a Co(0001) surface domain structure. (d) The simultaneously measured topography. The magnetization direction is shown in Fig. 16b (see color insert).

in-plane [261] and normal to the surface plane [262]. In this case the c-axis is perpendicular to the surface. Although the domains inside the sample bulk are relatively large and aligned along the c-axis, at the surface the large amount of magnetostatic energy associated with a perpendicular component of surface magnetization favors the formation of a complex closure domain structure with a substantial in-plane magnetization component. Figure 7 shows all three magnetization components and the corresponding intensity topograph. The two in-plane components were measured simultaneously, whereas the out-of-plane component was measured in a separate image. The simultaneously measured topography images were used to align the two polarization measurements. The derived in-plane magnetization direction is shown in Fig. 16b (see color insert).

Although determining the magnetization direction at any point on the sample's surface is a relatively straightforward procedure, displaying this complex three-dimensional structure is difficult. In Fig. 16b (see color insert) a portion of the in-plane magnetization image has been combined with the out-of-plane component to generate a three-dimensional rendering of the magnetization direction with color corresponding to the in-plane magnetization direction and height corresponding to the polar component. This image gives a visual impression of the complex closure domain structure of the Co, where the magnetization points into (out of) the surface in the valleys (hills).

6.4.3 Imaging Rough Surfaces

One major advantage of using an SEM for imaging is the relatively large depth of focus. SEMPA can therefore be used to image the magnetic structure of rough samples with surfaces that have considerable three-dimensional topography [240, 241]. Although samples with very nonplanar geometries can disturb the secondary electron extraction field and introduce instrumental artifacts in the images, surfaces that deviate a millimeter or less from a planar geometry are generally tolerated.

An example of SEMPA imaging from a rough surface is presented in Fig. 8, which shows images from both sides of a melt spun amorphous ferromagnetic glass ribbon. Color images of the magnetization direction are shown in Fig. 17a (see color insert). The as-cast ribbon has a smooth side (the air side) and an optically matte, rough side (the wheel side). The wheel side reflects the roughness of the cooling wheel and includes many voids owing to trapped air bubbles. The domain structures of the two sides are very different. The air side reveals mostly large domains, usually attributed to the bulk domain structure. These domains, which are usually long, parallel, and aligned with the ribbon axis, are shown in Fig. 8 (top) interacting with a dimple in the magnetic ribbon. The presence of such defects can affect the domain wall motion. The ability of SEMPA to acquire simultaneous yet independent images of the magnetic and topographic structure has been useful for understanding the magnetization dynamics in such a sample [260].

The magnetic structure of the wheel side of the ribbon is much more complex. The magnetization on this side is dominated by fine scale, mazelike domains that are the result of the magnetoelastic response to the strains in this surface. In fact, SEMPA images of similar domain structures have been used to analyze the strain fields in other ferromagnetic glass samples [263]. The large depth of focus allows SEMPA to image the finer domain structure inside the bubble-induced pocket as shown at the bottom of Fig. 8.

6.4.4 Recording Media

SEMPA has been successfully used to investigate the magnetic structure of hard disk [264, 265] and magneto-optic [266, 267] media. It is interesting to compare SEMPA with magnetic force microscopy (MFM) which is the more common imaging method for magnetic recording media, to a large

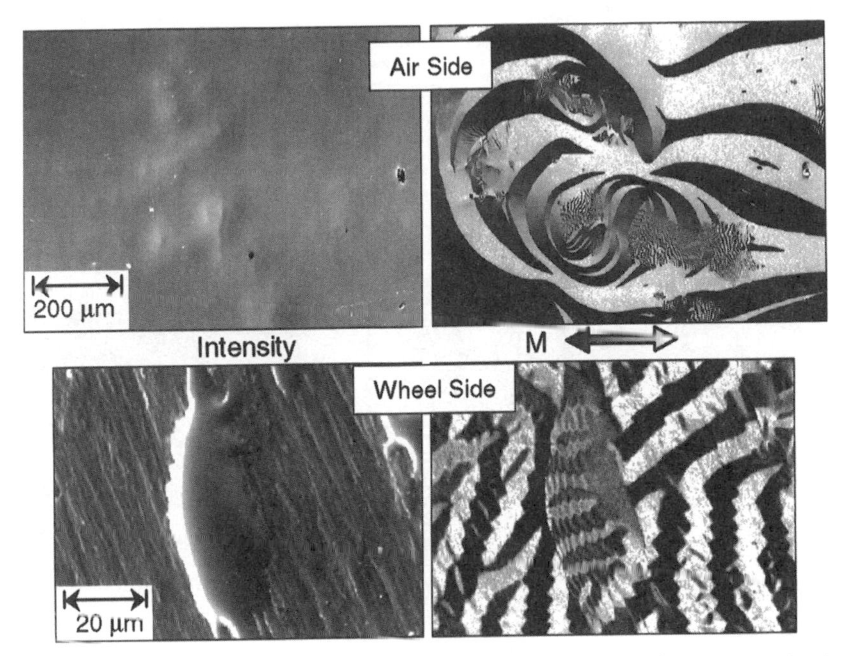

FIG. 8. SEMPA images of the topography and horizontal in-plane magnetization component from the smooth, air side and the rough, wheel side of a melt spun ferromagnetic metallic glass ribbon. The magnetization direction images are shown in Fig. 17a (see color insert).

extent because it can be performed in air with little or no sample preparation [268]. SEMPA can provide valuable additional information, however, as the two techniques probe different yet complementary aspects of the magnetic structure. SEMPA images the magnetization directly, whereas the MFM is sensitive to the magnetic fields arising from the magnetization.

SEMPA and MFM images of test patterns recorded in a special thin film medium sample [269] are compared in Fig. 9. A patterned Au film was deposited on top of the otherwise uncoated magnetic thin film, so that the images could be exactly aligned and the same recorded bits compared. The difference between the magnetic contrast mechanisms of the two techniques is highlighted by two significant differences between the SEMPA and MFM images. First, the magnetic structure underneath the Au film pattern is not visible in SEMPA but is visible with the MFM. SEMPA is more sensitive to the local, surface magnetic structure. Second, since the MFM is sensitive to the gradient of the magnetic field, the strongest MFM contrast occurs at the transitions between the recorded bits. SEMPA, on the other hand, is

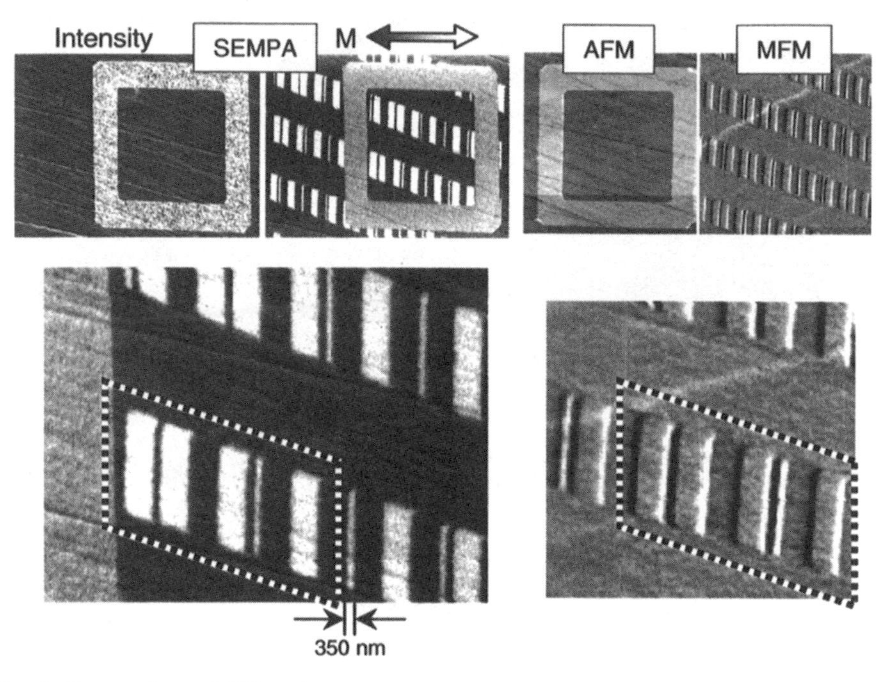

FIG. 9. Comparison of SEMPA and MFM images from a thin film hard disk medium test sample. The images at bottom are magnified views of magnetic structure inside the Au box pattern. Dashed boxes outline the same bits in the SEMPA and MFM images. The smallest written bit is 350 nm.

only sensitive to the magnetization direction. Comparisons such as this have been useful to help understand the contrast mechanisms in MFM imaging, since the magnetization measured using SEMPA can be used to derive the magnetic fields sensed by the MFM.

6.4.5 Spin Reorientation Transitions

A fruitful area of research exploiting the surface sensitivity of SEMPA has involved investigations of spin reorientation transitions in ultrathin magnetic films [270, 271]. In these films, which are only a few atomic layers thick, the magnetization orientation is determined by the balance between surface anisotropy, which favors perpendicular magnetization, and shape anisotropy, which favors in-plane magnetization. Relatively small changes in parameters such as the film thickness, temperature, applied field, or chemical composition can alter this balance and change the magnetic orientation

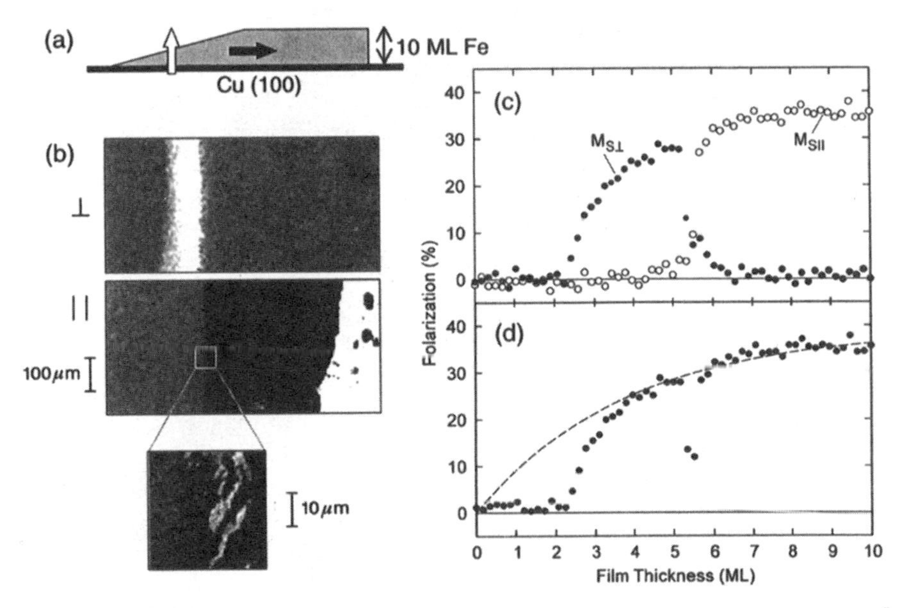

FIG. 10. Spin reorientation in Fe/Cu(100) epitaxial films. SEMPA measurements of in-plane and perpendicular magnetization components of a 0 to 10 atomic layer Fe wedge are shown in (b) images and (c) line scans. The inset shows a high resolution image of the domain structure near the reorientation thickness. (d) Total magnetization. [Reprinted with permission from R. Allenspach and A. Bischoff, *Phys. Rev. Lett.* **69**, 3385–3388 (1992). Copyright (1992) by the American Physical Society.]

between in-plane and perpendicular. SEMPA not only has the surface sensitivity to easily measure the magnetization direction in these films, but SEMPA can also image the domain microstructures, which can be quite complex.

An example of such an investigation involves SEMPA measurements of the spin reorientation in epitaxial Fe/Cu(100) films by Allenspach and Bischof [272]. They imaged the magnetic structure at several temperatures of a 0 to 10 atomic layer thick wedge of Fe grown on a Cu(100) single-crystal substrate. SEMPA images and line scans from a wedge measured at 175 K are shown in Fig. 10b,c, respectively. The magnetic orientation clearly changes from perpendicular to in-plane between a thickness of 5 to 6 atomic layers of Fe. During this transition, very small micrometer sized domains appear, as shown in the inset in Fig. 10b. The presence of these domains could lead a nonimaging magnetization measurement to mistakenly find that the magnetic moment decreases during the transition. In fact, earlier measurements showing reduced magnetizations had led to speculations that the balanced anisotropies at the transition might produce a reduced or paramagnetic Fe moment [273]. The derived magnitude of the magnetization measured by SEMPA plotted in Fig. 10d, however, does not show any reduction in the Fe magnetic moment. Any reduction in the observed magnetization can be accounted for by the finite SEMPA probing depth and a reduced Curie temperature for Fe less than a couple of atomic layers thick. SEMPA images also showed that the domain structures are very sensitive to temperature, varying not only in orientation, but also size and shape as the temperature is varied [272].

6.4.6 Iron/Chromium/Iron Exchange Coupling

Advances in thin film growth techniques have lead to an interesting and useful new class of magnetic materials that consist of magnetic multilayers separated by ultrathin nonmagnetic spacer layers [274]. Technologically, these multilayers are significant because they exhibit giant magnetoresistance (GMR) effects that occur when the layers are switched from ferromagnetic alignment to antiferromagnetic alignment. This GMR is the basis for various new magnetic sensors and memory devices [275]. Scientifically, these multilayers are interesting because they can exhibit unusual long range oscillatory exchange coupling between the magnetic layers. This coupling is mediated by the nonmagnetic spacer layers, and the direction of the coupling, whether the layers are ferromagnetically or antiferromagnetically aligned, varies with the thickness of the spacer. The surface sensitivity of SEMPA, along with the ability to prepare and study films *in situ*, have made SEMPA a very useful technique for understanding the origins of this oscillatory exchange coupling. In particular, SEMPA has been very useful for measuring how the alignment of the magnetic layers depends on the thickness of the nonmagnetic spacer layer [276, 277].

Several combinations of magnetic and nonmagnetic layers have been investigated using SEMPA, but Fe/Cr multilayers are particularly interesting because the lattices are well matched, allowing good epitaxial growth, and the Cr can either be in a paramagnetic or antiferromagnetic state [278]. Conceptually, these measurements are relatively simple. The magnetization alignment of the top and bottom ferromagnetic layers of a magnetic sandwich are imaged with SEMPA for various thicknesses of the nonmagnetic spacer. In practice, a great deal of effort goes into finding the optimal conditions for atomically smooth layer-by-layer growth, since roughness as small as a fraction of an atomic layer can significantly affect the coupling. Furthermore, rather than growing and measuring one spacer thickness at a time, the entire thickness dependence of the coupling is measured in a single SEMPA image by preparing samples with variable thickness, wedge-shaped spacer layers.

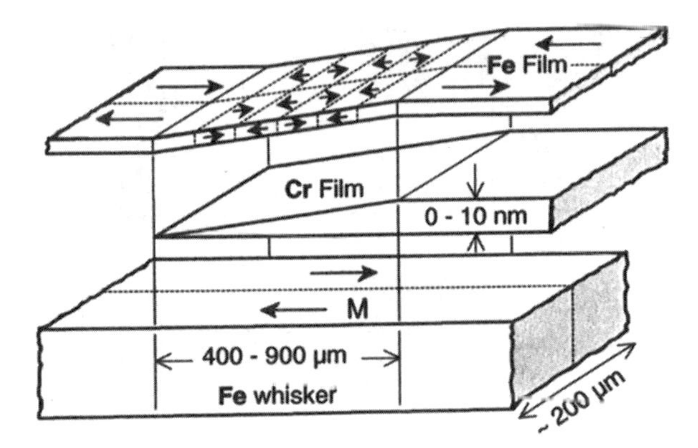

FIG. 11. Schematic expanded view of Fe/Cr/Fe exchange coupling sample showing the Fe whisker substrate, the Cr wedge, and the Fe overlayer. Arrows in the Fe correspond to magnetization directions.

A schematic of the multilayer sandwich structure used in these measurements is shown in Fig. 11. A single-crystal Fe whisker is used as the substrate. These whiskers, grown by thermal decomposition of $FeCl_2$ in a H_2 atmosphere, are among the most perfect metal crystals known [279]. More important for thin film growth, nearly perfect (100) surfaces may be obtained by *in situ* ion sputtering of the whisker surface and thermal annealing. Scanning tunneling microscopy (STM) measurements of these surfaces reveal step densities as low as a single atomic step per micrometer [280]. The long whiskers are usually divided into two opposite domains, a useful feature for establishing the zero of the polarization measurement. The Cr wedge is grown by molecular beam epitaxy (MBE) while moving a shutter in front of the whisker. The first two Cr layers are grown at 100°C to minimize interdiffusion, whereas the rest of the Cr is grown at 300°C to encourage layer-by-layer growth. The thickness of the wedge is measured using spatially resolved RHEED [281]. Figure 12c shows a line scan from a RHEED image showing oscillations of the specular beam intensity, a sign of nearly layer-by-layer growth. Using the measured RHEED oscillations the Cr thickness at any point along the wedge can be measured to within one-tenth of a Cr layer (± 0.014 nm). The sample is then coated with a thin epitaxial Fe film, and the coupling is determined from the direction of the magnetization in this top layer.

The magnetization of the top Fe film is shown in the SEMPA image in Fig. 12a. The whisker is split into two opposite domains, so that in the lower half of this SEMPA image white (black) contrast corresponds to ferromagnetic (antiferromagnetic) alignment of the overlayer magnetization with

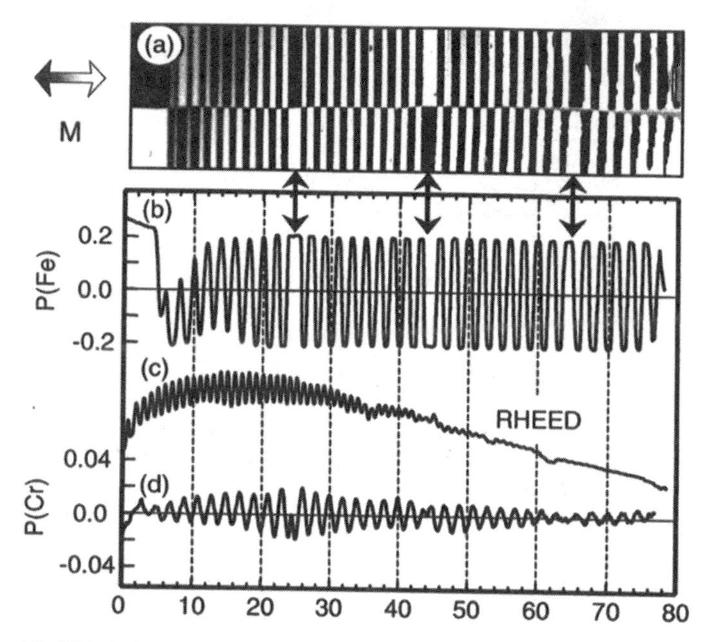

FIG. 12. (a) SEMPA image of the Fe overlayer magnetization for the Fe/Cr/Fe wedge structure shown in Fig. 11. (b) Line scan of the Fe overlayer magnetization. Arrows point to phase slips in the coupling oscillations. (c) RHEED oscillations from the uncovered Cr wedge used for thickness calibration. (d) Spin polarization of electrons emitted from the uncovered Cr wedge.

respect to the whisker substrate, whereas the opposite is true for the upper half of the whisker. A line scan of the magnetization from the lower half of the whisker is plotted in Fig. 12b. From this measurement one can see that the coupling oscillates between ferromagnetic and antiferromagnetic with a period of nearly two atomic Cr layers. The period is not exactly two layers, but instead it is 2.105 ± 0.005 atomic Cr layers. This small incommensurability between the lattice and the magnetic coupling is responsible for the phase slips observed at approximately 24, 44, and 64 layers.

SEMPA also provides information about the antiferromagnetic order of the Cr moments in the uncovered Cr film [248]. Because of the shallow probing depth, the polarization measured by SEMPA from the antiferromagnet does not average to zero, but instead it weighs the polarization of the outermost Cr layer more than the rest of the Cr. A line scan of this bare Cr surface polarization, with the background Fe polarization removed, is shown in Fig. 12d. The uncovered Cr has the same oscillatory periodicity as the exchange coupled Fe film. This is not surprising, as the incommensurate spin density wave nature of Cr is important in both the Cr antiferromagnetism and the Fe/Cr/Fe exchange coupling [278]. Note, however, that the sign of the polarization is reversed, indicating that antiferromagnetic coupling is favored at the Fe–Cr interface. The ability to grow atomically well-ordered magnetic multilayers and examine them *in situ* makes SEMPA a valuable tool for understanding the exchange coupling in epitaxially grown magnetic multilayers. Although short period oscillatory effects are present in other measurements, they are usually difficult to observe, since average interlayer thickness fluctuations of only a few tenths of an atomic layer can average these fine structures away [282]. SEMPA measurements of these nearly ideal systems have therefore provided meaningful information about the origins of the exchange coupling as well as quantitative tests of theoretcal models [283].

6.4.7 Depth Profiling the Magnetization of Cobalt/Copper Multilayers

Not only can SEMPA be used to investigate interlayer coupling in magnetic multilayers as they are grown, in certain cases SEMPA can be used with ion milling to take apart and depth profile magnetic structures in the same way that ion milling and Auger spectroscopy are used to depth profile chemical structure. Magnetic depth profiling works best in weakly coupled magnetic multilayers in which the magnetic structure is pinned in place by some structure in the film. Sputter deposited Co (6 nm)/Cu (6 nm) multilayers meet these requirements [284]. The 6 nm spacing between magnetic layers ensures weak coupling, whereas the inherent fine scale granularity of a sputter deposited film provides the defect structure for pinning the magnetic structures in place. The outermost magnetic Co layers may therefore be removed by ion milling without significantly disturbing the remaining magnetic structure of the multilayer [285].

The main goal of depth profiling the magnetic structure of the Co/Cu multilayers was to understand how correlations between the domain structure of adjacent layers affect the GMR of these potentially useful sensor materials. Figure 13 shows a test crater ion milled using 2 keV Ar^+ ions into the $[Co\ (6\ nm)/Cu\ (6\ nm)]_{20}$ multilayer. The SEMPA topography images in Fig. 13 show light and dark bands corresponding to the Cu and Co layers, respectively. An Auger line scan taken from one edge of the crater is plotted in Fig. 13. Although the Auger depth profile shows some interfacial chemical mixing owing to the ion milling, the Co and Cu layers are still clearly differentiated. The SEMPA magnetization images in Fig. 13 show that the magnetic domain structure of the adjacent Co layers is also well separated. For this multilayer system, depth profiling could clearly resolve at least the outermost 10 Co layers.

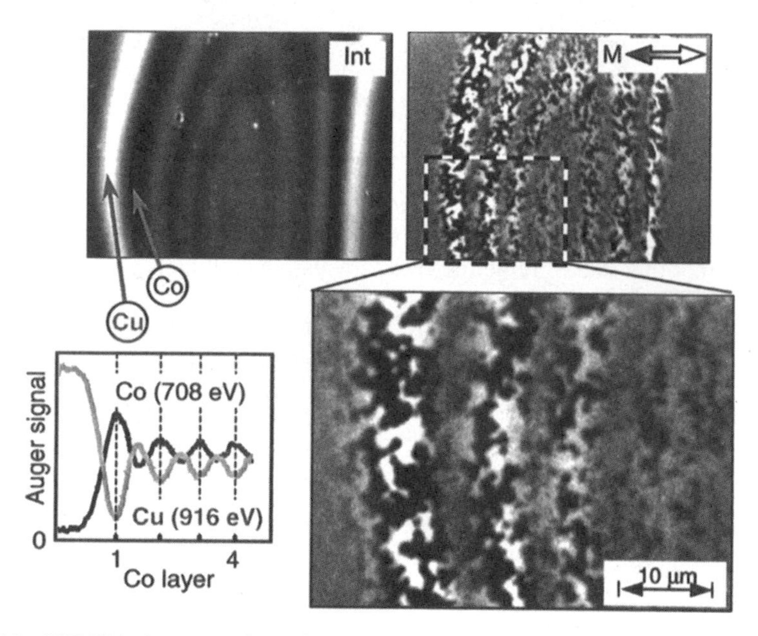

FIG. 13. SEMPA images of an ion-milled crater in a Co/Cu multilayer. The topography is shown at left, while the magnetization images at right show the magnetic domains in the separate Co layers. The plot shows Auger line scans from one edge of the crater.

To correlate the magnetic structure in adjacent layers from the same area, the sample was ion milled to a uniform depth over a large part of the sample. Auger spectroscopy was used to determine the depth of the ion milling. SEMPA images from the outermost two Co layers are shown in Figs. 14 and 17b (see color insert). The magnetization images reveal that within the layers the domains have random shapes that are nominally 1 μm in size, with uniaxial alignment of the magnetization. As can be seen from the images the magnetization in adjacent layers is strongly anticorrelated. This observation can be made more quantitative by taking the pixel-by-pixel difference of the magnetization direction, $\Delta\phi$, between the two layers. A histogram showing the resulting distribution of alignments is plotted in Fig. 14. The plot shows that antiferromagnetic alignment between the adjacent layers is clearly preferred. The antiferromagnetic alignment even extends to structures as small as the domain walls. Within the layers the domain walls are Néel-like with in-plane magnetization and random chirality. Domain walls in adjacent layers, such as the ones highlighted by the circles in Fig. 17b, (see color insert), are oppositely magnetized so that the chirality of the

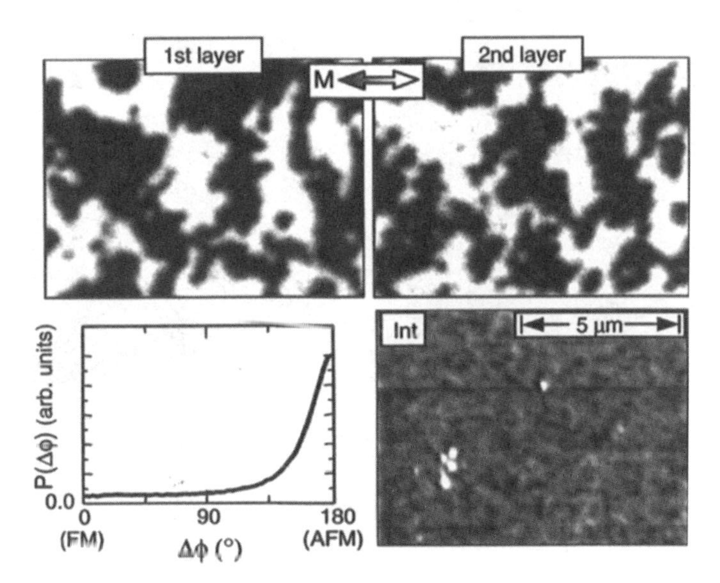

FIG. 14. SEMPA images of the outermost two layers of a Co/Cu multilayer exposed by ion milling. The strong anticorrelation of the domain structures is shown in the histogram distribution of the difference in magnetization directions between the two layers. The magnetization directions are shown in Fig. 17b (see color insert).

walls is the same in the adjacent Co layers. The observed anticorrelations are found to persist for at least the first 10 Co layers [285].

The SEMPA measurements of the interlayer magnetic correlations have been used to quantitatively explain the observed GMR in these multilayers [285]. Relative field-dependent changes in the GMR are found to be in good agreement with the GMR derived from the SEMPA measured correlations, assuming that the GMR is simply proportional to $-\cos\Delta\phi$. In addition, combining the SEMPA measurements with polarized neutron reflectivity measurements that measure the average magnetic structure has proved to be especially useful for understanding the magnetic structure of these multilayers [286].

6.4.8 Patterned Magnetic Structures

The use of patterned magnetic structures in magnetic technology is ubiquitous. Patterned high density recording media, GMR sensors, and magnetic random access memories (MRAM) are all current examples of magnetic technology where small magnetic structures are important. SEMPA has been used to investigate how the magnetic structure in these

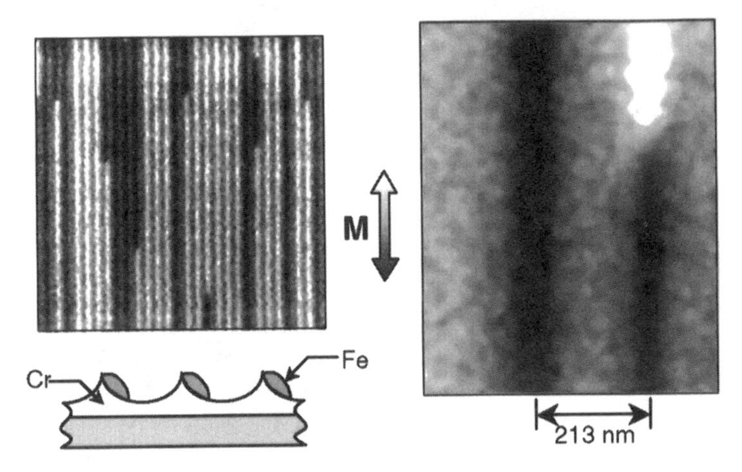

FIG. 15. SEMPA images of Fe wires formed by grazing incidence deposition on a Cr grating.

systems, or in idealized versions of these systems, is influenced by the shape of the device and the presence of physical or chemical defects. Figures 15 and 17c (see color insert) show examples of SEMPA images of patterned magnetic structures with very different length scales. Figure 15 shows the domain structure of very fine Fe wires deposited *in situ* by grazing incidence evaporation onto a Cr grating [287]. In this case the size and shape of the wires restrict the magnetic structure to single domains with magnetization directions along the length of the wires. Figure 15 also shows a high magnification SEMPA image of the domain transition, which is at roughly 45° to the wire.

SEMPA images of much larger lithographically patterned Fe structures are shown in Fig. 17c (see color insert). These samples, provided by G. Prinz at the Naval Research Laboratory, consist of Fe films grown epitaxially on GaAs substrates. The patterns are all variations of a rectangular basic shape with sides that are aligned along the (001) easy magnetization axis of Fe. The simplest domain structure for this basic shape would therefore be a "picture frame" structure with a single domain along each side. Instead of this simple domain pattern, however, the structures show what happens to the magnetization when a small gap is introduced in one side of the frame. A more complex structure is formed, allowing the formation of closure domains in the gap that reduce the magnetostatic energy associated with free magnetic poles in the gap. Figure 17c also shows the sensitivity of the domain structures to the size of the structures and to any shaping of corners.

In addition to shape, film thickness can also strongly affect domain structure. This is especially true as the magnetic patterns become nearly two-dimensional. The surface sensitivity of SEMPA imaging has made it a useful tool for investigating these systems, which can be only a few layers thick. So far the results are somewhat conflicting. For example, for small Co structures grown *in situ* on Cu(001) substrates, some groups have reported that the magnetic structure is independent of lateral size and shape of structures [288], whereas others have seen significant dependencies [241]. Clearly, the issue of domain formation in these very small and very thin structures is not as simple as one might anticipate.

6.5 Conclusion

The preceding examples show that SEMPA can be used to image the magnetic microstructure in a wide variety of structures and materials. The surface sensitivity of SEMPA makes it especially well suited for the direct, quantitative mapping of the magnetization direction in thin films and at the surface of magnetic materials. Comparisons between magnetic and physical structure in these systems are facilitated by the natural ability of SEMPA to easily separate the magnetic and topographic contrast. When combined with other compatible surface analytical techniques such as Auger, RHEED, and STM, SEMPA can also provide information about the relationship between the magnetic structure and the chemical structure and atomic scale order.

In the future, advances in electron microscope design should allow the resolution of SEMPA imaging to improve to less than 10 nm. In addition, the development of higher efficiency polarization analyzers could dramatically improve the resolution as well as the speed of SEMPA. More complex sample holders will also allow the application of localized magnetic fields as well as make simultaneous *in situ* electrical measurements of devices possible. As the lateral size of magnetic structures used in magnetic technologies, such as magnetoelectronics and magnetic storage, continues to shrink from the microscale to the nanoscale, and as the thickness of the magnetic films used in these structures decreases to only a few atomic layers, SEMPA will be ready to provide valuable information about the magnetic microstructure and novel magnetic properties of these systems.

Acknowledgments

I wish to thank colleagues who have worked with me in developing SEMPA, especially R. J. Celotta, D. T. Pierce, M. H. Kelley, and M. R. Scheinfein. This work was supported in part by the Office of Naval Research.

7. HIGH RESOLUTION LORENTZ SCANNING TRANSMISSION ELECTRON MICROSCOPY AND ITS APPLICATIONS

Yusuke Yajima
Hitachi Research Laboratory
Hitachi, Ltd.
Kokubunji, Tokyo, Japan

7.1 Introduction

Lorentz scanning transmission electron microscopy (Lorentz STEM) is a relatively new version of Lorentz microscopy that used to be performed mostly on conventional (i.e., projection-type) transmission electron microscopes (TEM). The Lorentz STEM utilizes a focused fine probe scanned across the observation area, giving microscopic distributions of magnetic induction as raster images. From the principle of magnetic induction detection that shall be discussed later, the Lorentz STEM is also referred to as differential phase contrast (DPC) Lorentz microscopy. Soon after DPC microscopy was first proposed as a simple means to observe amplitude and phase contrasts separately but simultaneously [289], it was successfully applied to magnetic induction mapping [63]. Some earlier works on DPC Lorentz microscopy as well as other techniques of Lorentz transmission electron microscopy have been reviewed quite thoroughly [150, 290, 291].

When compared with conventional TEM-based Lorentz microscopy, Lorentz STEM has the following notable features. First, the image contrast is quantitatively related to magnetic induction. This stems from the nature of the detected signal, which is, with sufficient accuracy, proportional to the Lorentz deflection of the probe arising after transmitting through a particular position on the specimen. Second, easy access to digital data acquisition and further image processing is ensured by the sequential electrical signals constituting the immediate output of the detector. Third, conventional bright-field and/or dark-field images could be obtained simultaneously with induction maps with perfect spatial registration. Furthermore, such microanalytical methods using fine electron beams as electron energy loss spectroscopy and energy dispersive X-ray spectroscopy could be combined easily. These features make the Lorentz STEM particularly well suited not only for obtaining highly resolved magnetic induction maps but for unraveling relations among microscopic magnetic, structural, and compositional properties. The rapid growth of current STEM technology should also

EXPERIMENTAL METHODS IN THE PHYSICAL SCIENCES
Vol. 36
ISBN 0-12-475983-1

ISSN 1079-4042/01 $35.00

prompt further development of the Lorentz STEM itself and the area of investigation where it applies.

In this chapter, we first describe the principle of Lorentz STEM from the viewpoints of both what we are looking at and how we can look at it. Then some of its applications are presented and discussed.

7.2 Principle

7.2.1 The Nature of Small Lorentz Deflections

In this section, we show an appropriate way to understand the behavior of the electron beam under typical Lorentz STEM conditions. The "small" Lorentz deflection in the section title infers that the momentum change caused by the Lorentz force is sufficiently smaller than the original momentum of the electrons. This is always the case in Lorentz STEM performed with the combination of transmission electron microscopes operating with $\sim 10^5$ V accelerating voltages and magnetic objects with dimensions of 10^{-6}–10^{-7} m. We discuss here some properties of small Lorentz deflection by relating it to the wave nature of electrons.

In the presence of a vector potential $\mathbf{A}$, the electron wave function Ψ satisfies the Schrödinger equation ($e \equiv -|e|$)

$$\frac{1}{2m}(-i\hbar\nabla - e\mathbf{A})^2\Psi + V\Psi = E\Psi, \tag{1}$$

and it is written as

$$\Psi = \Psi^{(0)}e^{i(e/\hbar)\int_s d\mathbf{s}\cdot\mathbf{A}}, \tag{2}$$

where $\Psi^{(0)}$ is the electron wave function for the Schrödinger equation without the vector potential but with the same scalar potential V and energy E as those in Eq. (1):

$$\frac{-\hbar^2}{2m}\nabla^2\Psi^{(0)} + V\Psi^{(0)} = E\Psi^{(0)}. \tag{3}$$

By putting

$$\Psi = |\Psi|e^{i\phi}, \tag{4}$$

the continuity condition of $|\Psi|^2$ leads to the current density vector expressed as

$$\mathbf{J} = \frac{e\hbar}{m}|\Psi|^2\nabla\phi - \frac{e^2}{m}|\Psi|^2\mathbf{A}. \tag{5}$$

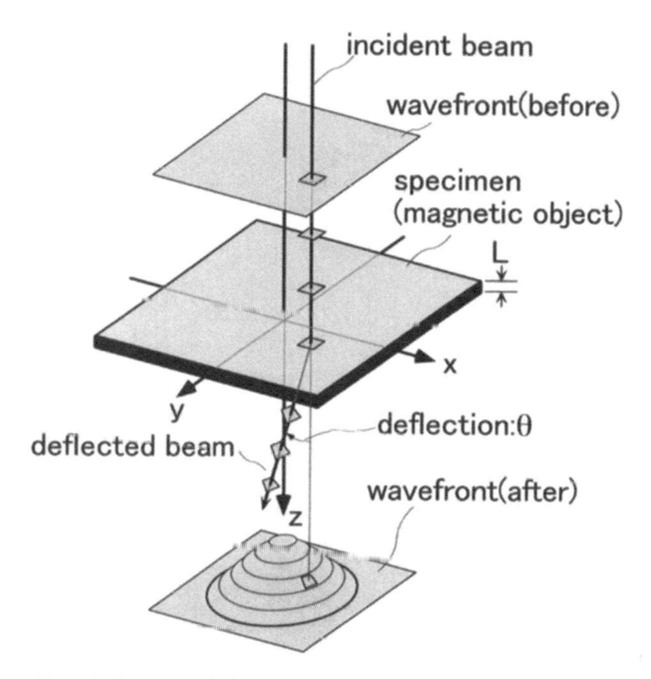

FIG. 1. Lorentz deflection and wave front deformation.

Note that any irrotational component in the vector potential, that is, the component giving rise to no magnetic induction, does not affect the above current density vector, since it appears in both terms of the right-hand side of Eq. (5) with opposite signs. Equation (5) therefore indicates that, in the presence of a vector potential with a finite solenoidal component, $\nabla \times \mathbf{A} \neq 0$, electrons do not flow along the surface normal of the wave front, ϕ = constant; this is the origin of the Lorentz deflection.

Next we consider the small Lorentz deflection of an electron beam with definite energy flowing initially toward the z-direction (Fig. 1). The initial wave function $\Psi^{(0)}$ is then written as

$$\Psi^{(0)} = \sqrt{\frac{J^{(0)}m}{|e|\hbar k}}\, e^{ikz}, \tag{6}$$

where the normalization constant is conveniently taken so as to describe the flow of electrons with current density $J^{(0)}$ $(=|J^{(0)}|)$. These electrons are deflected when they traverse the specimen, a magnetic object with solenoidal vector potential. In the case of small deflection, the line integral in Eq. (2) could be carried out in parallel with the z-axis throughout. This leads to, for the magnetic object extending in the xy-plane between $z = -L$ and $z = 0$,

$$\phi = \begin{cases} kz & (z \leqslant -L), \\ kz + \dfrac{e}{\hbar}\displaystyle\int_{-L}^{z} \mathrm{d}z A_z & (-L < z < 0), \\ kz + \dfrac{e}{\hbar}\displaystyle\int_{-L}^{0} \mathrm{d}z A_z & (0 \leqslant z), \end{cases} \tag{7}$$

and, accordingly,

$$\mathbf{J} = \begin{cases} -J^{(0)}(0,\ 0,\ 1) & (z \leqslant -L), \\ -J^{(0)}\left(\sqrt{\dfrac{|e|}{2mV_{\mathrm{ac}}}}\displaystyle\int_{-L}^{z} \mathrm{d}z\, B_y,\ -\sqrt{\dfrac{|e|}{2mV_{\mathrm{ac}}}}\displaystyle\int_{-L}^{z} \mathrm{d}z\, B_x, 1\right) & (-L<z<0), \\ -J^{(0)}\left(\sqrt{\dfrac{|e|}{2mV_{\mathrm{ac}}}}\displaystyle\int_{-L}^{0} \mathrm{d}z\, B_y,\ -\sqrt{\dfrac{|e|}{2mV_{\mathrm{ac}}}}\displaystyle\int_{-L}^{0} \mathrm{d}z\, B_x, 1\right) & (0 \leqslant z), \end{cases} \tag{8}$$

where we have used, for instance,

$$\int_{-L}^{z} \mathrm{d}z\left(\frac{\partial A_z}{\partial x} - \frac{\partial A_x}{\partial z}\right) = -\int_{-L}^{z} \mathrm{d}z\, B_y, \tag{9}$$

and the electron accelerating voltage is denoted by V_{ac}:

$$V_{\mathrm{ac}} \equiv \frac{(\hbar k)^2}{2m|e|}. \tag{10}$$

Under typical Lorentz STEM conditions (e.g., $V_{\mathrm{ac}} \approx 100$ kV, $L \approx 10^{-7}$ m, and $B_{xy} \approx 1$ T), the passage through the object causes the electron flow to deviate from the original z-direction by about 10^{-4} radian only; this allows the small deflection description as stated earlier.

At $z \geqslant 0$ where $\mathbf{A} = 0$ again, the deflection components toward x- and y-directions,

$$\theta_{x,y}(x,\ y) \equiv \pm\sqrt{\frac{|e|}{2mV_{\mathrm{ac}}}}\int_{0}^{L} \mathrm{d}z\, B_{y,x}(x,\ y,\ z), \tag{11}$$

satisfy

$$\nabla\phi = \frac{\sqrt{2m|e|V_{\mathrm{ac}}}}{\hbar}(\theta_x,\ \theta_y,\ 1). \tag{12}$$

This infers that the Lorentz deflection is proportional to the wave front deformation, or gradient, arising as a result of the passage through the magnetic object. It is noted that the vector field, $\boldsymbol{\theta} = (\theta_x, \theta_y)$, defined on the xy-plane is irrotational (i.e., integrable) everywhere. This is obvious from the previous discussion where we have derived Lorentz deflections as derivatives of a scalar function [Eq. (7)] and also directly from

$$\frac{\partial \theta_x}{\partial y} - \frac{\partial \theta_y}{\partial x} \propto \int_{-L}^{0} \mathrm{d}z \left(\frac{\partial B_x}{\partial x} + \frac{\partial B_y}{\partial y} \right) = 0 \qquad (\rightarrow \nabla \cdot \mathbf{B} = 0). \tag{13}$$

Accordingly, any line integral of $\boldsymbol{\theta}$ from one point, $P_0(x_0, y_0)$, to the other, $P_1(x_1, y_1)$, both on the same xy-plane, should give the magnetic increment of the phase function,

$$\begin{aligned} \phi(P_1) - \phi(P_0) &= \frac{e}{\hbar} \int_{-L}^{0} \mathrm{d}z A_z(x_1, y_1, z) - \frac{e}{\hbar} \int_{-L}^{0} \mathrm{d}z A_z(x_0, y_0, z) \\ &= \frac{\sqrt{2m|e|V_{\text{ac}}}}{\hbar} \int_{P_0}^{P_1} \mathrm{d}\ell \cdot \boldsymbol{\theta}. \end{aligned} \tag{14}$$

This suggests that the magnetically distorted electron phase function can be reconstructed by using the information on small Lorentz deflections across the area.

Next we consider how the magnetic object affects the electron current density distribution. A portion of the electron beam that traversed the magnetic object at point (x, y, z) with $-L \leqslant z \leqslant 0$ reaches an xy-plane away from the object by F at point (x', y', F) where, as shown in Fig. 2,

$$\begin{cases} x' = x + F\theta_x \\ y' = y + F\theta_y \end{cases}. \tag{15}$$

The current density at (x', y', F) then becomes, within first order to F,

$$\begin{aligned} J^{(D)} = J^{(0)} \frac{\partial(x, y)}{\partial(x', y')} &\cong J^{(0)} \left[1 - F \left(\frac{\partial \theta_x}{\partial x} + \frac{\partial \theta_y}{\partial y} \right) \right] \\ &= J^{(0)} \left[1 - F \sqrt{\frac{|e|}{2mV_{\text{ac}}}} \int_{-L}^{0} \mathrm{d}z (\nabla \times \mathbf{B})_z \right]. \end{aligned} \tag{16}$$

Under the static conditions where most Lorentz microscopy is performed, the current density associated with the specimen (but not that with the probe electron beam!), $\mathbf{i}$, and time derivative of the electric displacement, $\partial \mathbf{D}/\partial t$, vanish in the general relation

$$\nabla \times \mathbf{B} = \nabla \times \mathbf{M} + \mathbf{i} + \frac{\partial \mathbf{D}}{\partial t}, \tag{17}$$

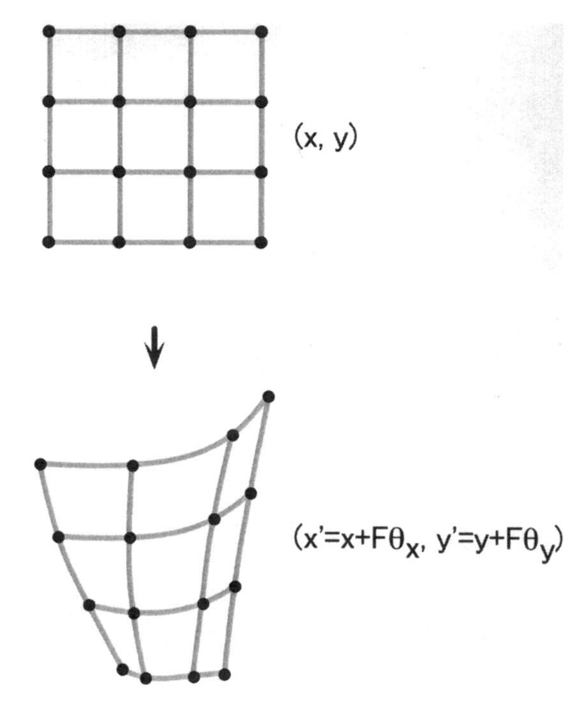

FIG. 2. Lorentz deflection and current density variation (schematically exaggerated).

and only the rotation of the magnetization, $\nabla \times \mathbf{M}$, remains on the right-hand side, giving

$$\nabla \times \mathbf{B} = \nabla \times \mathbf{M}. \tag{18}$$

Accordingly, the current density of the probe electron beam in the static case varies only when there is a finite in-plane magnetization rotation in the object:

$$J^{(D)} \cong J^{(0)} \left[1 - F \sqrt{\frac{|e|}{2mV_{\mathrm{ac}}}} \int_{-L}^{0} \mathrm{d}z (\nabla \times \mathbf{M})_z \right]. \tag{19}$$

This current density variation gives rise to the Fresnel contrast in Lorentz microscopy on conventional TEM, where F corresponds to the defocus used. What is inferred here is that the Fresnel technique is not appropriate for the observation of either magnetization without in-plane rotation or the magnetic field distribution in static-free space where, $\mathbf{M} = 0$. Conversely, the technique extracts information on magnetization distributions from the

Lorentz deflections that bear information on both magnetization and magnetic field.

In the Lorentz STEM, information on the position (x, y) and that on Lorentz deflections $F\theta_{x,y}$ are separated; the former is provided as a scanning signal, whereas only the latter is detected as a shift of the beam spot on the detector. This allows the detection of the magnetic induction itself in the form of its in-plane components even without any rotation:

$$\int_0^L dz\, B_{x,y}(x, y, z) = \mp \sqrt{\frac{2mV_{ac}}{|e|}}\, \theta_{y,x}(x, y). \tag{20}$$

We emphasize here a couple of points about the nature of the information obtained by Lorentz STEM. One is that there is no *a priori* way to distinguish whether the Lorentz deflection is caused by the magnetization in the material or by the magnetic field. In the observation of magnetically hard materials, for instance, they are almost equal in terms of the contribution to Lorentz deflections. If there is some knowledge available beforehand on the overall magnetic structures, one can estimate the relative contribution of magnetization and magnetic field to the image. In the case of recorded magnetic recording films, a numerical procedure has been proposed to determine the ratio of magnetization and magnetic field contribution under the particular observation condition used [292].

The numerical image processing to generate the rotation of magnetic induction could be used to eliminate the effect of the magnetic field, since the magnetic field is free from rotation in the static case. The images generated in this way look similar to corresponding Fresnel images, as they should [291].

The other point worth mentioning here is that the magnetic induction obtained from Lorentz deflections is the quantity integrated along the electron beam path. Therefore, the specimen, or the regions of interest, should be thin enough to render concrete meaning to the in-plane projected information.

7.2.2 Some Features of STEM Essential for Lorentz Microscopy

In this section, we discuss some features of STEM essential for its application to Lorentz microscopy. For simplicity, we use a one-dimensional description of each optical plane, which is sufficient to understand why we can do Lorentz STEM.

In a typical electron optical configuration used in Lorentz STEM (Fig. 3), a point electron source and an appropriate lens system prepares a plane

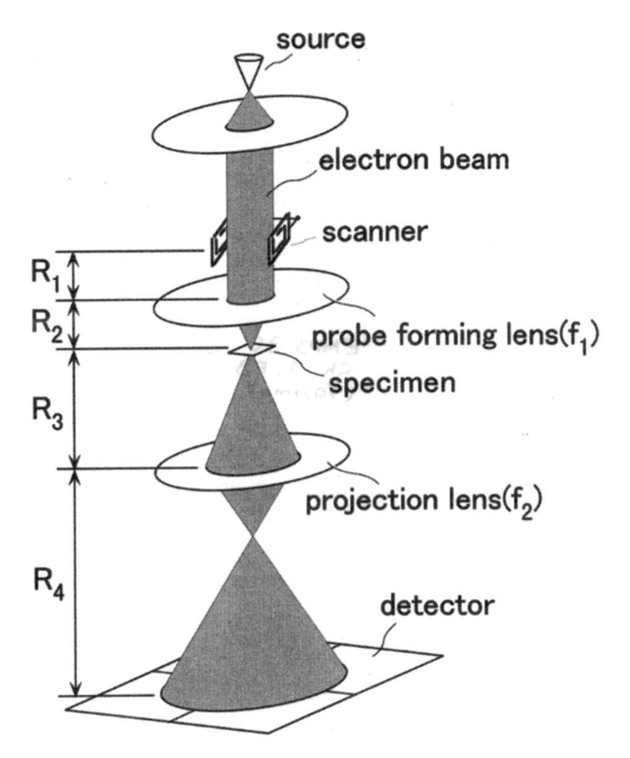

FIG. 3. Electron optical configuration for Lorentz STEM.

wave (with amplitude unity for convenience). This plane wave passes through a scanner/aperture plane where there are the beam deflector for raster scanning and an aperture (not shown in Fig. 3). Then the wave reaches the probe-forming lens. The focused fine probe formed by the probe-forming lens runs through the specimen, then runs through a projection lens, and finally reaches the detector.

7.2.2.1 Probe Formation. By using the point source propagation function in the small angle approximation,

$$p(x;\ R) = \frac{1}{\sqrt{i\lambda}} \frac{e^{(2\pi i/\lambda)R}}{R}\ e^{(\pi i/\lambda)(x^2/R)}, \tag{21}$$

with λ being the electron wavelength and R the propagation distance, the transfer function associated with the scanner/aperture plane, $t(x;\ x_s)$, and the transfer function of a thin lens with a focal length f,

$$l(x;\ f) = e^{-(\pi i/\lambda)(x^2/f)}, \tag{22}$$

the wave amplitude on the specimen plane is expressed as

$$\phi_s(x) = \{[t(x;\ x_s) \otimes p(x;\ R_1)]l(x;\ f_1)\} \otimes p(x;\ R_2), \tag{23}$$

with $\otimes$ representing the convolution integral and R_1, R_2, and f_1 denoting, respectively, the distance between the scanner/aperture plane and the probe-forming lens, that between the probe-forming lens and the specimen plane, and the focal length of the probe-forming lens. The transfer function of the scanner/aperture plane is the product of an aperture transfer function,

$$a(x) = \begin{cases} 1 & (x \leqslant d/2) \\ 0 & \text{elsewhere} \end{cases}, \tag{24}$$

with a constant phase gradient factor describing the probe deflection, and it is written as

$$t(x;\ x_s) = a(x)\ e^{(2\pi i/\lambda)(x_s/R_1 + R_2 - R_1 R_2/f_1)x} \tag{25}$$

Here, the phase gradient factor, expressed as shown for later convenience, leads to the probe deflection of $x_s/(R_1 + R_2 - R_1R_2/f_1)$ radians.

Writing Eq. (23) explicitly with $Q \equiv R_1 + R_2 - R_1R_2/f_1$, one gets

$$\phi_s(x) = \frac{1}{\sqrt{i\lambda}} \frac{e^{2\pi i/\lambda(R_1+R_2)}}{\sqrt{R_1R_2}\sqrt{Q}} \times e^{(\pi i/\lambda)(1-R_1/f_1/Q)x^2} \int du\, a(u)\, e^{(\pi i/\lambda)(1-R_2/f_1/Q)u^2}\, e^{-(2\pi i/\lambda)(x-x_s/Q)y}. \tag{26}$$

If the specimen is situated on the focal plane of the probe-forming lens, $R_2 = f_1 = Q$, this reduces to

$$\begin{aligned} \phi_s(x;\ R_2 = f_1) &= \frac{1}{\sqrt{i\lambda}} \frac{e^{2\pi i/\lambda(R_1+f_1)}}{\sqrt{R_1 f_1}} e^{\pi i/\lambda(1-R_1/f_1)/f_1} \int du\, a(u)\, e^{-(2\pi i/\lambda)(x-x_s/f_1)u} \\ &= \sqrt{\frac{\lambda}{i}} \frac{e^{2\pi i/\lambda(R_1+f_1)}}{\pi\sqrt{R_1}} e^{\pi i/\lambda[(1-R_1/f_1)/f_1]x^2} \frac{\sin\left(\frac{\pi d}{\lambda}\frac{x-x_s}{f_1}\right)}{x-x_s}, \end{aligned} \tag{27}$$

where the explicit form of $a(x)$ Eq. (24) has been used. The probe intensity distribution on the specimen plane is therefore

$$|\phi_s(x;\ R_2 = f_1)|^2 = \frac{\lambda}{\pi^2 R_1} \frac{\sin^2\left(\frac{\pi d}{\lambda}\frac{x-x_s}{f_1}\right)}{(x-x_s)^2}. \tag{28}$$

This shows that a fine probe of about $2\lambda f_1/d$ diameter is formed on the specimen plane, where it moves by x_s when it is deflected by x_s/f_1 radians on the scanner/aperture plane.

7.2.2.2 Lorentz Deflection Detection. On the detector plane, the wave amplitude becomes

$$\phi_d(x) = \{[\phi_s(x)s(x) \otimes p(x;\ R_3)]l(x;\ f_2)\} \otimes p(x;\ R_4), \qquad (29)$$

where $s(x)$ is the specimen transfer function, R_3 and R_4 are the distances between the specimen plane and the projection lens and that between the projection lens and the detector, respectively, and f_2 is the focal length of the projection lens. Then, by defining $E \equiv R_3 + R_4 - R_3R_4/f_2$,

$$\begin{aligned}\phi_d(x) = \frac{1}{i\lambda} & \frac{e^{2\pi i/\lambda(R_1+R_2+R_3+R_4)}\, e^{\pi i/\lambda[(1-R_3/f_2)/E]x^2}}{\sqrt{R_1R_2R_3R_4}\sqrt{Q\times E}} \\ & \times \int du\, t(u;\ x_s)\, e^{\pi i/\lambda[(1-R_2/f_1)Q]u^2} \\ & \times \int d\nu\, s(\nu)\, e^{\pi i/\lambda[(1-R_1/f_1)/Q} + {}^{[(1-R_4/f_2)/E]\nu^2}\, e^{-(2\pi i/\lambda)(u/Q+x/E)\nu}. \end{aligned} \qquad (30)$$

Here we consider the case

$$\begin{aligned} & \frac{1-R_1/f_1}{Q} + \frac{1-R_4/f_2}{E} \\ & = \frac{(1-R_1/f_1)(1-R_4/f_2)}{Q\times E}\left[R_2 + R_3 - \left(\frac{1}{1/f_1 - 1/R_1} + \frac{1}{1/f_2 - 1/R_4}\right)\right] = 0 \end{aligned} \qquad (31)$$

This is the condition necessary for the image of an object R_1 away from a thin lens with focal length f_1 to be formed on a plane R_4 away from another thin lens with focal length f_2 situated coaxially with a distance $R_2 + R_3$. By adjusting the focal length of the projection lens so as to satisfy

$$f_2 = R_4 \frac{(f_1 - R_1)(R_2 + R_3) + f_1R_1}{(f_1 - R_1)(R_2 + R_3 + R_4) + f_1R_1}, \qquad (32)$$

one can achieve this condition without affecting the probe-forming conditions. We assume hereafter that this condition is always satisfied. Also, we define the following shorthand notations, which will be used throughout this chapter:

$$
\begin{cases}
A \equiv \dfrac{1}{R_1 R_2 R_3 R_4} \left| \dfrac{1 - R_1/f_1}{1 - R_4/f_2} \right|; \\
B \equiv \dfrac{1 - R_1/f_1}{1 - R_4/f_2}; \\
C \equiv Q(1 - R_2/f_1); \\
D \equiv \dfrac{(1 - R_1/f_1)(1 - R_2/f_1)}{1 - R_4/f_2};
\end{cases}
\tag{33}
$$

All of these parameters as well as Q and E defined earlier are determined by the optical parameters of the microscope.

Given the specimen transfer function that describes the Lorentz deflection of x_L/E radians,

$$
s(x) = e^{(2\pi i/\lambda)(x_L/E)x}, \tag{34}
$$

the wave amplitude on the detector plane becomes

$$
\begin{aligned}
\phi_d(x) &= \frac{1}{i\lambda} \frac{e^{2\pi i/\lambda(R_1 + R_2 + R_3 + R_4)}\, e^{(\pi i/\lambda)[(1 - B_3/f_2)/E]x^2}}{\sqrt{R_1 R_2 R_3 R_4}\sqrt{Q \times E}} \\
&\quad \times \int du\, t(u; x_s)\, e^{(\pi i/\lambda)[(1 - R_2/f_1)/Q]u^2}\, \delta\left\{ \frac{1}{Q\lambda} \left[u + \frac{Q}{E}(x - x_s) \right] \right\} \\
&= \sqrt{A}\, e^{2\pi i/\lambda(R_1 + R_2 + R_3 + R_4)} \\
&\quad \times e^{(\pi i/\lambda)(1/E)[(1 - R_3/f_2)x^2 - D(x - x_L)^2 - 2x_s(x - x_L)]}\, a[B(x - x_L)].
\end{aligned}
\tag{35}
$$

Hence, the intensity distribution on the detector plane,

$$
|\phi_d(x)|^2 = A a^2[B(x - x_L)], \tag{36}
$$

is B times the magnified image of the aperture. As stated at the end of the previous section, this image on the detector is independent of the probe position at the specimen and moves solely in proportion to the Lorentz deflection θ by

$$
x_L = E\theta, \tag{37}
$$

where E corresponds to F of the previous section. From Eq. (36) as well as the condition in Eq. (31), $BE = -Q$, we see that the Lorentz deflection θ is equivalent to the *virtual* shift of the aperture on the scanner/aperture plane by $-Q\theta$ that would cause the motion of the aperture image on the detector by x_L. The probe defocus, $R_2 = f_1$, does not affect these features on the detector although it does have much influence on the probe size at the specimen.

We examine here a numerical example. If $R_1 = 20$ mm, $R_2 = f_1 = 10$ mm, $R_3 = 15$ mm, and $R_4 = 1000$ mm, the condition of Eq. (33) requires $f_2 \approx 5$ mm. With an aperture width of 50 μm and a 200 keV electron beam ($\lambda = 2.508$ pm), one expects a probe size, which determines the spatial resolution, of about 1 nm on the specimen plane. The size of the aperture image on the detector becomes $200 \times d = 10$ mm, and the Lorentz shift there is $2000\,\text{mm} \times \theta$ (in radians), that is, typically of the order of 10^{-1} mm.

One thing to note is that the Lorentz STEM is often performed on TEM designed only for STEM and accordingly without any projection lens system. In this case, the electron beam is appropriately descanned after passing through the specimen so that the probe scanning does not affect the intensity distribution on the detector.

7.2.2.3 Spatially Varying Lorentz Deflections.

The Lorentz deflection discussed so far has been constant everywhere. Next we extend the discussion to spatially varying Lorentz deflections. The specimen transfer function in this case could be written as

$$s(x) = e^{-i(\ell/\lambda)\theta_0 \cos[2\pi/\ell(x-x_0)]}, \tag{38}$$

which describes the Lorentz deflection changing periodically,

$$\theta = \theta_0 \sin\left[\frac{2\pi}{\ell}(x - x_0)\right]. \tag{39}$$

Assuming that $(\ell/\lambda)\theta_0$ is sufficiently small, this transfer function can be approximated as

$$\begin{aligned} s(x) &= 1 - i\frac{\ell}{\lambda}\theta_0 \cos\left[\frac{2\pi}{\ell}(x - x_0)\right] \\ &= 1 - i\frac{\ell}{2\lambda}\theta_0[e^{2\pi i/\ell(x-x_0)} + e^{-(2\pi i/\ell)(x-x_0)}]. \end{aligned} \tag{40}$$

Then we obtain the following for the wave amplitude at the detector plane:

$$\phi_d(x) = \sqrt{A}\, e^{2\pi i/\lambda(R_1+R_2+R_3+R_4)}\, e^{(\pi i/\lambda)(1/R_3+R_4-R_3R_4/f_2)[(1-R_3/f_2-D)x^2-2x_s x]}$$
$$\times\left(a(Bx) - i\,\frac{\ell}{2\lambda}\theta_0\, e^{\pi i(\lambda/\ell^2)C}\left\{e^{2\pi i/\ell[Dx+(x_s-x_0)]}\, a\left[B\left(x-\frac{\lambda}{\ell}E\right)\right]\right.\right.$$
$$\left.\left. + e^{-(2\pi i/\ell[Dx+(x_s-x_0)]}\, a\left[B\left(x+\frac{\lambda}{\ell}E\right)\right]\right\}. \tag{41}$$

Next, one can write the intensity profile on the detector up to the order $(\ell/\lambda)\theta_0$:

$$|\phi_d(z)|^2 = Aa^2(Bx) + Aa(Bx)\,\frac{\ell}{\lambda}\,\theta_0\cos\left[\frac{2\pi}{\ell}(x-x_0)\right]$$
$$\times\left\{\sin\left(\frac{\pi\lambda}{\ell^2}C+\frac{2\pi}{\ell}Dx\right)a\left[B\left(x-\frac{\lambda}{\ell}E\right)\right]\right.$$
$$\left.+\sin\left(\frac{\pi\lambda}{\ell^2}C-\frac{2\pi}{\ell}Dx\right)a\left[B\left(x-\frac{\lambda}{\ell}E\right)\right]\right\}$$
$$+ Aa(Bx)\frac{\ell}{l}\theta_0\sin\left[\frac{2\pi}{\ell}(x-x_0)\right]$$
$$\times\left\{\cos\left(\frac{\pi\lambda}{\ell^2}C+\frac{2\pi}{\ell}Dx\right)a\left[B\left(x-\frac{\lambda}{\ell}E\right)\right]\right.$$
$$\left.-\cos\left(\frac{\pi\lambda}{\ell^2}C-\frac{2\pi}{\ell}Dx\right)a\left[B\left(x+\frac{\lambda}{\ell}E\right]\right\}. \tag{42}$$

This expression describes the effect of the spatial variation of Lorentz deflections as an overlap of the unshifted aperture function with those shifted by $\pm(\lambda/\ell)E$. Obviously there would be no overlap when

$$\frac{\lambda}{\ell}|BE| = \frac{\lambda}{\ell}|Q| > d. \tag{43}$$

In the in-focus case, $R_2 = f_1$, this gives

$$\frac{\lambda f_1}{d} > 1, \tag{44}$$

indicating that the Lorentz deflections varying with a period shorter than the probe size of $\sim 2\lambda f_1/d$ do not affect the intensity profile on the detector.

In Eq. (42), the deflection-dependent part is broken down into two components: the component symmetric in x and that antisymmetric in x.

Terms with $\cos[2\pi(x_s - x_0)/\ell]$ and with $\sin[2\pi(x_s - x_0)/\ell]$ in Eq. (42) correspond to symmetric and antisymmetric components, respectively [note $a(x) = a(-x)$]. Because of these x_s-dependent factors, both components depend on the probe position on the specimen. In particular, the x_s dependence of the antisymmetric component is proportional to the Lorentz deflection at x_s. Also of note is the difference between these components in terms of the dependence on the defocus, $R_2 - f_1$. Since $C, D \to 0$ as $R_2 - f_1 \to 0$, the symmetric component becomes negligible under near-focus conditions and vanishes at in-focus, whereas the antisymmetric component remains finite and more or less constant, and the in-focus intensity distribution becomes

$$|\phi_d(z)|^2 = Aa(Bx)\left\{a(Bx) + \frac{\ell}{\lambda}\theta_0 \sin\left[\frac{2\pi}{\ell}(x_s - x_0)\right] \times\left[a\left(Bx + \frac{\lambda}{\ell}f_1\right) - a\left(Bx - \frac{\lambda}{\ell}f_1\right)\right]\right\}. \tag{45}$$

Next we compare the two descriptions, namely, the constant deflection description discussed previously and the spatially varying deflection description just presented. For simplicity, we again discuss the in-focus case only. The intensity profile arising from the Lorentz deflection of θ is described as

$$|\phi_d(z)|^2 = \begin{cases} Aa^2(Bx + f_1\theta) & \text{constant deflection description} \\ Aa(Bx)\left\{a(Bx) + \frac{\ell}{\lambda}\theta\left[a\left(Bx + \frac{\lambda}{\ell}f_1\right) - a\left(Bx - \frac{\lambda}{\ell}f_1\right)\right]\right\} & \\ \quad \text{spatially varying deflection description.} & \end{cases} \tag{46}$$

In the constant deflection description, the Lorentz deflection appears as a shift of the aperture image by $f_1\theta$. In the spatially varying deflection description, on the other hand, the image does not move, but its intensity distribution changes antisymmetrically so that the intensity imbalance becomes equivalent to the image shift in the constant deflection description (Fig. 4).

7.2.2.4 Detector Performance. In any description used, the Lorentz deflection gives rise to an asymmetric change in the intensity distribution on the detector. Therefore, the detector should also have an asymmetric response in order to produce a signal. In Lorentz STEM, two types of detectors are used frequently; one is the split detector whose response

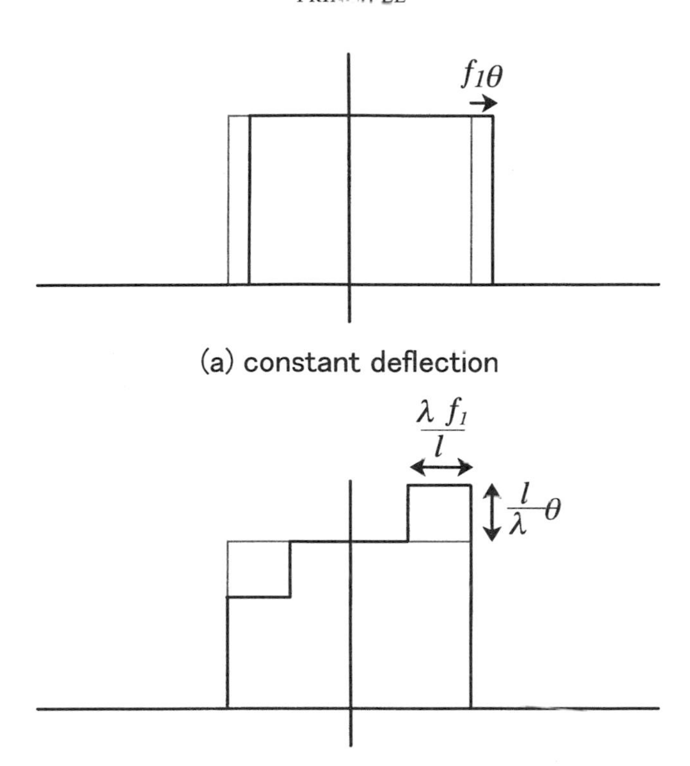

FIG. 4. Intensity imbalance on the detector caused by Lorentz deflections: (a) constant deflection description and (b) spatially varying deflection description.

function has the form

$$K_s(x) = \begin{cases} -1 & (x < 0) \\ 0 & (x = 0)\,, \\ 1 & (0 < x) \end{cases} \tag{47}$$

and the other type is the so-called modified split detector whose response function is

$$K_{ms}(x) = \begin{cases} -1 & (x < -kw) \\ 0 & (-kw \leqslant x \leqslant kw), \\ 1 & (kw < x) \end{cases} \tag{48}$$

where $0 < k < 1$ and w is the width of the aperture image on the detector ($= Bd$) [293].

Some geometrical consideration based on Eqs. (47) and (48) allows straightforward estimation of the signal:

$$I_{s,ms} = \int dx\, K_{s,ms}(x)|\phi_d(x)|^2. \tag{49}$$

To estimate the response to the Lorentz deflection angle θ, the constant deflection description applies to give

$$I_s(\theta) = \begin{cases} \dfrac{2f_1}{d}\theta & \left(0 \leqslant \theta \leqslant \dfrac{d}{2f_1}\right) \\ 1 & \left(\dfrac{d}{2f_1} \leqslant \theta\right) \end{cases} \tag{50}$$

and

$$I_{ms}(\theta) = \begin{cases} \dfrac{2f_1}{d}\,\theta & \left[0 \leqslant \theta \leqslant \dfrac{d}{2f_1}\,(1-k)\right] \\ \dfrac{f_1}{d}\,\theta + \dfrac{1-k}{2} & \left[\dfrac{d}{2f_1}\,(1-k) \leqslant \theta \leqslant \dfrac{d}{2f_1}\,(1+k)\right]. \\ 1 & \left[\dfrac{d}{2f_1}\,(1+k) \leqslant \theta\right] \end{cases} \tag{51}$$

Here the signal intensity has been properly normalized. Comparing Eqs. (50) and (51), one sees that these two types of detectors behave in much the same way for small deflections, whereas at larger deflections the modified split detector tends to reduce the signal level and extends the maximum detectable limit of deflections (Fig. 5a). However, this difference is of no importance in practice, because the Lorentz deflection angle is always very small, as stated earlier.

Next we estimate the dependence of the signal on the spatial frequency $(1/\ell)$ of deflection distributions using the spatially varying deflection description. That is,

$$I_s\left(\frac{1}{\ell}\right) = \begin{cases} 1 & \left(0 \leqslant \dfrac{1}{\ell} \leqslant \dfrac{d}{2\lambda f_1}\right) \\ \dfrac{d\ell}{\lambda f_1} - 1 & \left(\dfrac{d}{2\lambda f_1} \leqslant \dfrac{1}{\ell} \leqslant \dfrac{d}{\lambda f_1}\right) \\ 0 & \left(\dfrac{d}{\lambda f_1} \leqslant \dfrac{1}{\ell}\right) \end{cases} \tag{52}$$

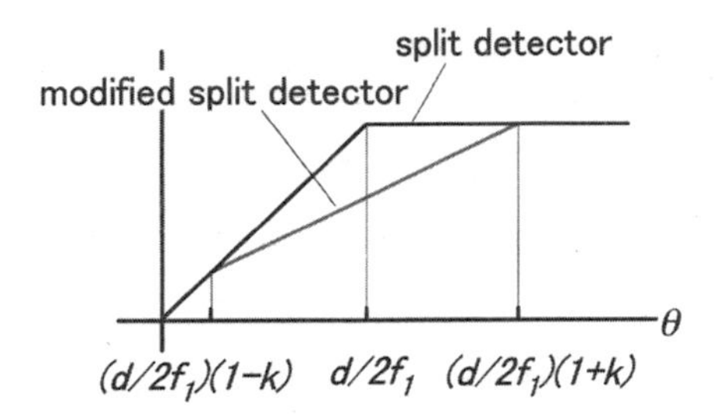

(a) deflection angle dependence

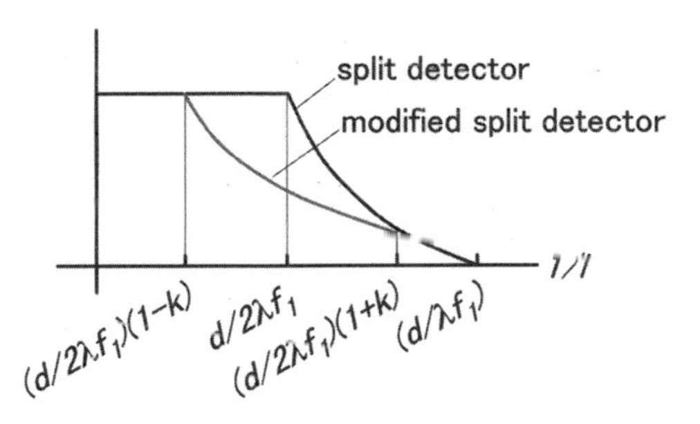

(b) spatial frequency dependence

FIG. 5. Performance of split and modified split detectors: (a) deflection angle dependence and (b) spatial frequency dependence.

and

$$I_{ms}\begin{cases} 1 & \left[0 \leqslant \dfrac{1}{\ell} \leqslant \dfrac{d}{2\lambda f_1}(1-k)\right] \\ \dfrac{d\ell}{2\lambda f_1}(1-k) & \left[\dfrac{d}{2\lambda f_1}(1-k) \leqslant \dfrac{1}{\ell} \leqslant \dfrac{d}{2\lambda f_1}(1+k)\right] \\ \dfrac{d\ell}{\lambda f_1} - 1 & \left[\dfrac{d}{2\lambda f_1}(1+k) \leqslant \dfrac{1}{\ell} \leqslant \dfrac{d}{\lambda f_1}\right] \\ 0 & \left(\dfrac{d}{\lambda f_1} \leqslant \dfrac{1}{\ell}\right) \end{cases}. \tag{53}$$

This shows that, compared with the standard split detector, the modified split detector reduces the signal in the spatial frequency region $(d/2\lambda f_1)$

$(1 - k) < 1/\ell < (d/2\lambda f_1)(1 + k)$ specifically (Fig. 5b). Since this spatial frequency region corresponds to a characteristic length equal to 1–10 times the probe size, the modified split detector works quite effectively for reducing noise in the image.

7.2.2.5 Nonmagnetic Effects. We have seen that the Lorentz deflection gives rise to asymmetry in the intensity profile on the detector plane, and a detector with an asymmetric response is therefore essential for the Lorentz STEM. We point out here that the nonmagnetic amplitude variation across the specimen could also cause the asymmetric intensity profile on the detector and accordingly bring about the unwanted noise signal. Therefore we discuss here the effect of amplitude variation on the intensity profile at the detector in comparison with that due to the Lorentz deflections. The transfer function of the specimen with amplitude variation could be written as

$$s(x) = 1 - g\cos\left[\frac{2\pi}{\ell}(x - x_0)\right] = 1 - \frac{g}{2}\left[e^{2\pi i/\ell(x - x_0)} + e^{-(2\pi i/\ell)(x - x_0)}\right], \tag{54}$$

and the wave amplitude on the detector becomes

$$\begin{aligned} |\phi_d(z)|^2 &= Aa^2(Bx) \\ &+ Aa(Bx)g\cos\left[\frac{2\pi}{\ell}(x_s - x_0)\right] \\ &\times\left\{\cos\left(\frac{\pi\lambda}{\ell^2}C + \frac{2\pi}{\ell}Dx\right)a\left[B\left(x - \frac{\lambda}{\ell}E\right)\right]\right. \\ &\left.\qquad + \cos\left(\frac{\pi\lambda}{\ell^2}C - \frac{2\pi}{\ell}Dx\right)a\left[B\left(x + \frac{\lambda}{\ell}E\right)\right]\right\} \\ &- Aa(Bx)g\sin\left[\frac{2\pi}{\ell}(x_s - x_0)\right] \\ &\times\left\{\sin\left(\frac{\pi\lambda}{\ell^2}C + \frac{2\pi}{\ell}Dx\right)a\left[B\left(x - \frac{\lambda}{\ell}E\right)\right]\right. \\ &\left.\qquad - \sin\left(\frac{\pi\lambda}{\ell^2}C - \frac{2\pi}{\ell}Dx\right)a\left[B\left(x + \frac{\lambda}{\ell}E\right)\right]\right\}. \end{aligned} \tag{55}$$

As in Eq. (42), terms with $\cos[2\pi(x_s - x_0)/\ell]$ and $\sin[2\pi(x_s - x_0)/\ell]$ are even and odd functions of x, respectively. Note, however, that the odd (antisymmetric) term tends to vanish as $R_2 - f_1 \to 0$ (C, $D \to 0$), in marked contrast with the corresponding term in the Lorentz deflection wave amplitude [Eq. (42)]. We infer that the amplitude variation on the specimen does not result in noise signal only when the probe is exactly focused. Hence, it is desirable to carefully adjust the microscope to obtain in-focus images when the specimen would have much structural inhomogeneity, although the Lorentz deflection signal itself is, as seen earlier, rather insensitive to the focusing condition at least in near-focus region. A numerical image processing method to reduce the nonmagnetic contrast from inadvertently defocused Lorentz STEM images has been proposed and proved successful [294].

One thing to merit attention here is that under in-focus conditions, any amplitude change at the specimen, whether it is spatially varying or constant ($\ell \rightarrow \infty$), does not lead to the intensity asymmetry; only the phase change, that is, deflection, does. This is why Lorentz STEM is also called *differential phase contrast* (DPC) Lorentz microscopy.

7.3 Instrument

As a typical example of the microscopes used for Lorentz STEM, we describe here our instrument briefly [295]. We have implemented the Lorentz STEM mode of operation on a 200 kV TEM (Hitachi HF-2000) with a W(310) cold field emission source and a scanning unit. For magnetically nonintrusive observation, a goniometer stage for a conventional side entry specimen holder has been installed halfway between the second condenser and objective lenses. A Permalloy shield around the specimen holder further ensures that the specimen is magnetically intact. The residual field at the specimen stage is well below 10 A/m. An opening placed at a corner of the shield allows the detection of secondary electrons and X-ray emissions, making it possible to analyze surface morphology and composition distributions, respectively. On the goniometer stage, the specimen can be tilted up to 180° to either sense. The second condenser lens is used to form the probe incident on the specimen. With a probe half-angle of 2.5×10^{-3} radian, the measured probe current at the specimen is around 20 pA. Under this condition, bright-field scanning TEM images have confirmed the spatial resolution of 5 nm. The detector is a semiconductor type modified split detector (with $k \approx 0.8$).

In addition to the Lorentz STEM images, conventional Fresnel (defocus) images can also be observed and taken on photographic plates. Besides magnetic measurements, conventional bright-field STEM, energy dispersive X-ray (EDX) analysis, and secondary electron detection are possible on the same specimen stage. The original specimen stage in the objective lens can also be used for standard high resolution TEM observation.

7.4 Applications

7.4.1 Magnetic Induction Mapping

7.4.1.1 Soft Materials. Figure 6 shows Lorentz STEM images of an ion thinned iron single-crystal film [i.e., horizontal (Fig. 6a) and vertical (Fig. 6b) components of in-plane integrated magnetic induction] [295]. We see that the magnetic domains divided by straight domain walls constitute the regularly aligned magnetic structure. In the case of magnetically soft materials such as this example, the image contrast reflects the magnetization distribution because there is virtually no magnetic field around the material.

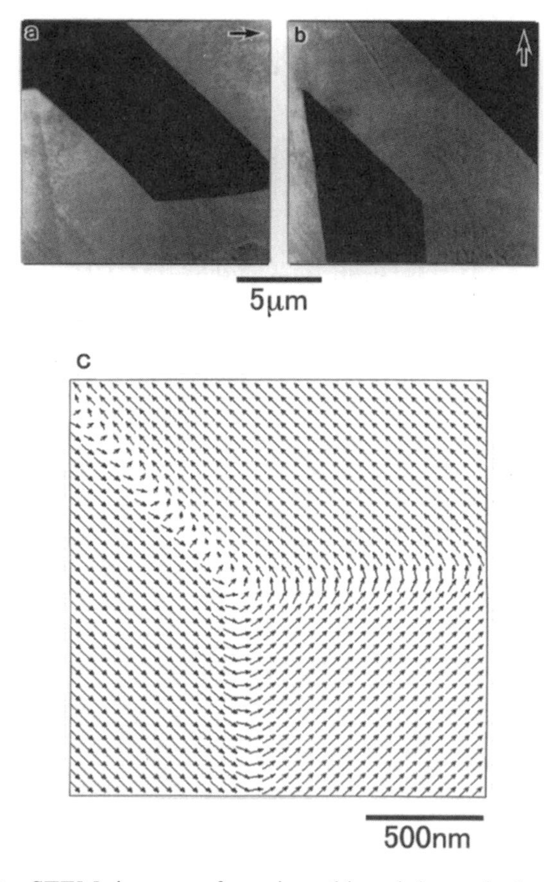

FIG. 6. Lorentz STEM images of an ion thinned iron single-crystal film: (a) horizontal and (b) vertical components of in-plane integrated magnetic induction, and (c) vector plot of a portion of (a) and (b) [295].

A vector map of a portion in Fig. 6a, and b shows the domain wall structure clearly (Fig. 6c). The magnetization in the 90° wall turns in plane (Néel-type wall structure), whereas the magnetization in the 180° wall is likely to have some out-of-plane component, suggesting the out-of-plane rotation (Bloch-type wall structure). The wall energy of a Néel-type wall, σ_N, and that of Bloch-type wall, σ_B, are given by

$$\sigma_N(2\theta) = \sigma_N(180°)(1 - \cos\theta)^2 \tag{56}$$

and

$$\sigma_B(2\theta) = \sigma_B(180°)\sin\theta^2, \tag{57}$$

where $\sigma_N(180°)$ and $\sigma_B(180°)$ are the 180° wall energies of Néel- and Bloch-type walls, respectively, and 2θ is the total change of magnetization direction across the wall [296]. In the case of a 90° wall, they become

$$\sigma_N(90°) = 0.086\sigma_N(180°) \approx 0.1\sigma_N(180°) \tag{58}$$

and

$$\sigma_B(90°) = 0.5\sigma_B(180°). \tag{59}$$

Therefore, the wall structure appearing in Fig. 6c indicates that wall energies in this particular portion of the film might satisfy

$$\sigma_N(90°) < \sigma_B(90°) < 5\sigma_N(90°) \tag{60}$$

and

$$\sigma_B(180°) < \sigma_N(180°) < 5\sigma_B(180°). \tag{61}$$

7.4.1.2 Hard Material. The next example compares the Lorentz STEM image and Fresnel image of recorded bit profiles on a Co-based polycrystalline magnetic recording film (Fig. 7) [295]. The images appear very different. The Lorentz STEM image (Fig. 7a) is substantially affected by the stray magnetic field around the film. In general, the stray magnetic field, although important for information retrieval, tends to cancel out the magnetization in terms of the net magnetic induction experienced by the probe electron beam. Since the spatial distribution of the stray magnetic field around the film is very different from that of magnetization, which is confined of course in the film, the image is strongly dependent on the propagation direction of the probe beam, which can be changed by tilting the specimen [292].

The Fresnel image (Fig. 7b), on the other hand, primarily reflects the distribution of in-plane magnetization rotation. The featherlike pattern, called a ripple structure, in the image changes the direction in accordance with the change of the mean magnetization direction. This example shows

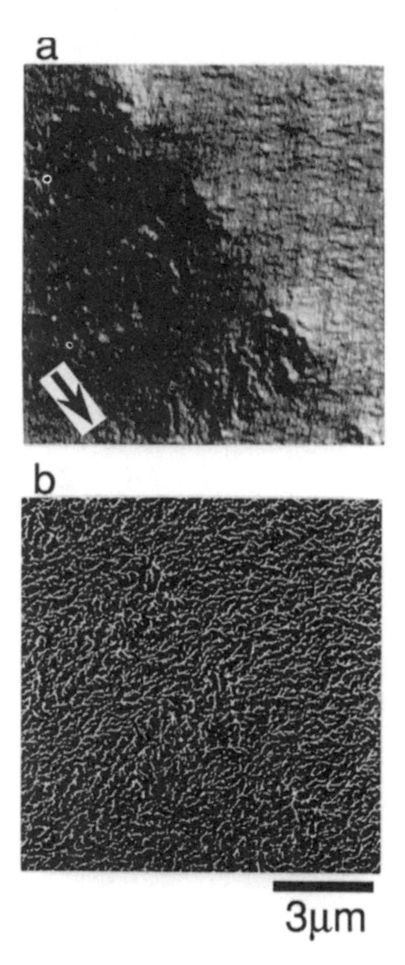

FIG. 7. Recorded bit profiles on a Co-based polycrystalline magnetic recording film: (a) Lorentz STEM image (arrow indicates the direction of the detected component) and (b) Fresnel image [9].

that information obtained by Lorentz STEM is quite different from that given by the Fresnel technique. Hence, they should be used in a complementary manner.

7.4.1.3 Magnetic Field in Free Space. Lorentz STEM images of free space near the edge of a recorded magnetic recording film are shown in Fig. 8. The vector map (Fig. 8c) produced from component images (Fig. 8a, b) clearly depicts the distribution of the magnetic field that emerges at bit boundaries into free space and returns to neighboring boundaries.

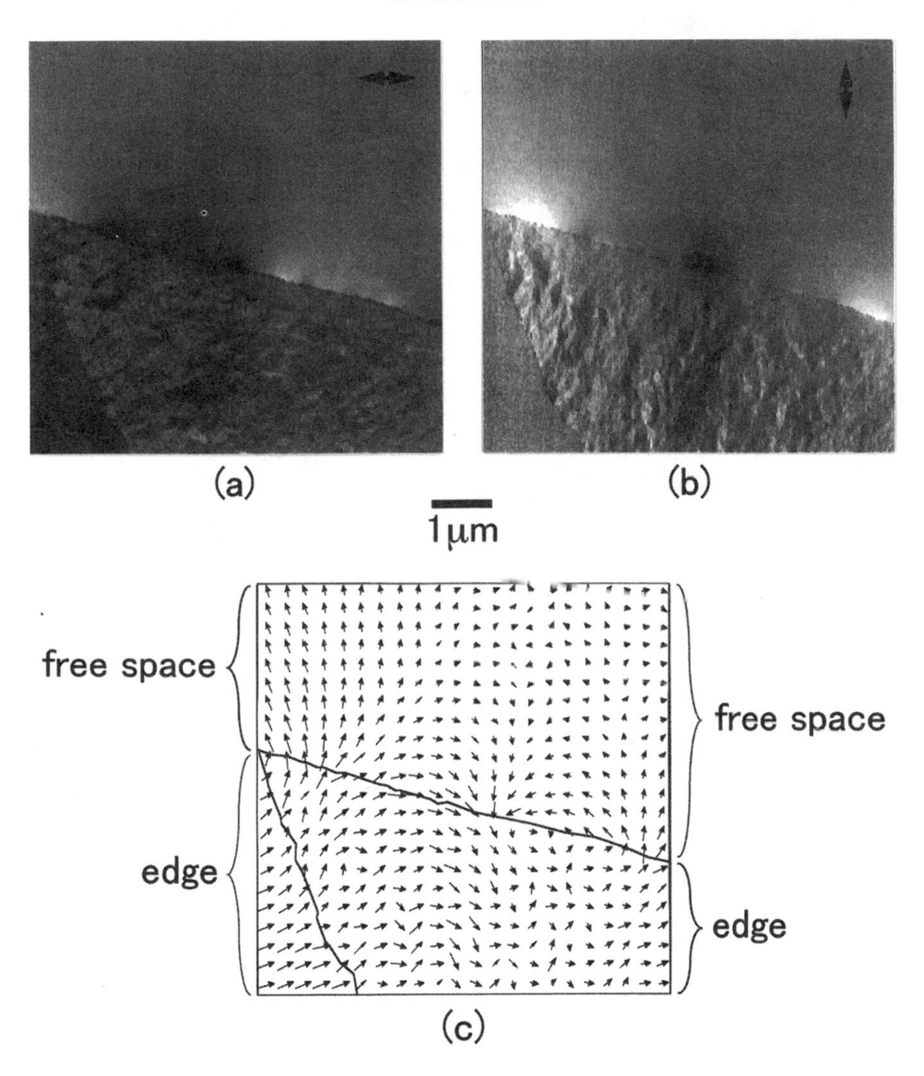

FIG. 8. Lorentz STEM images of free space near the edge of a recorded magnetic recording film: components (a) parallel and (b) perpendicular to the edge, and (c) the vector map.

As presented here, Lorentz STEM is capable of mapping out even the rotation-free, $\nabla \times \mathbf{B} = 0$, magnetic induction distribution, which is hard to observe by the Fresnel method. One must be careful, however, that these maps are in-plane projected images of the magnetic field extending three-dimensionally. Since small Lorentz deflections as obtained by Lorentz

STEM contain highly quantitative information on the magnetic induction well compatible with sophisticated numerical processing, some tomographic method might apply to reconstruct three-dimensional field distributions [297].

7.4.2 Phase Contour (Flux Line) Delineation

Lorentz STEM does not give information on the electron phase as an immediate output. As discussed previously, however, the distribution of small Lorentz deflections is integrable, and by integrating it once, one can generate a scalar function equivalent to the electron phase function [see Eq. (14)] [298].

As in other types of current scanning electron microscopy, the Lorentz deflection data are stored in the form of discrete matrices. Hence, convenient integration paths are the combinations of vertical and horizontal straight paths, such as path 1, $(x_0, y_0) \rightarrow (x_1, y_0) \rightarrow (x_1, y_1)$, and path 2, $(x_0, y_0) \rightarrow (x_0, y_1) \rightarrow (x_1, y_1)$, in Fig. 9.

In principle, any integration path should give the same result. However, to avoid the effect of local noise, it is often useful to average out the results of the integration carried out along various paths. For instance, the averaging of values of integration taken along paths 1 and 2 in Fig. 9 is in practice easy to handle by automatic image processing. Figure 10 shows the vertical (Fig. 10a) and horizontal (Fig. 10b) Lorentz deflection maps of a 30-nm-thick ferromagnetic Ni/Fe alloy film sputtered on a carbon microgrid [301]. Since the effect of the magnetic field on the Lorentz deflection is negligible in this film, the vector plot of the projected magnetic induction distribution (Fig. 10c) shows that this film has a vortexlike magnetization structure.

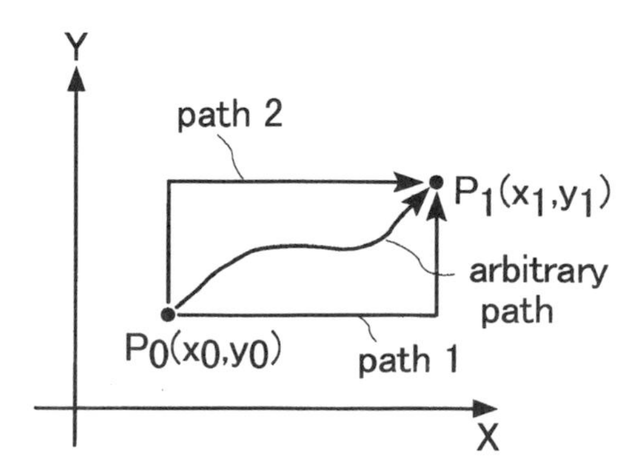

FIG. 9. Possible integration paths.

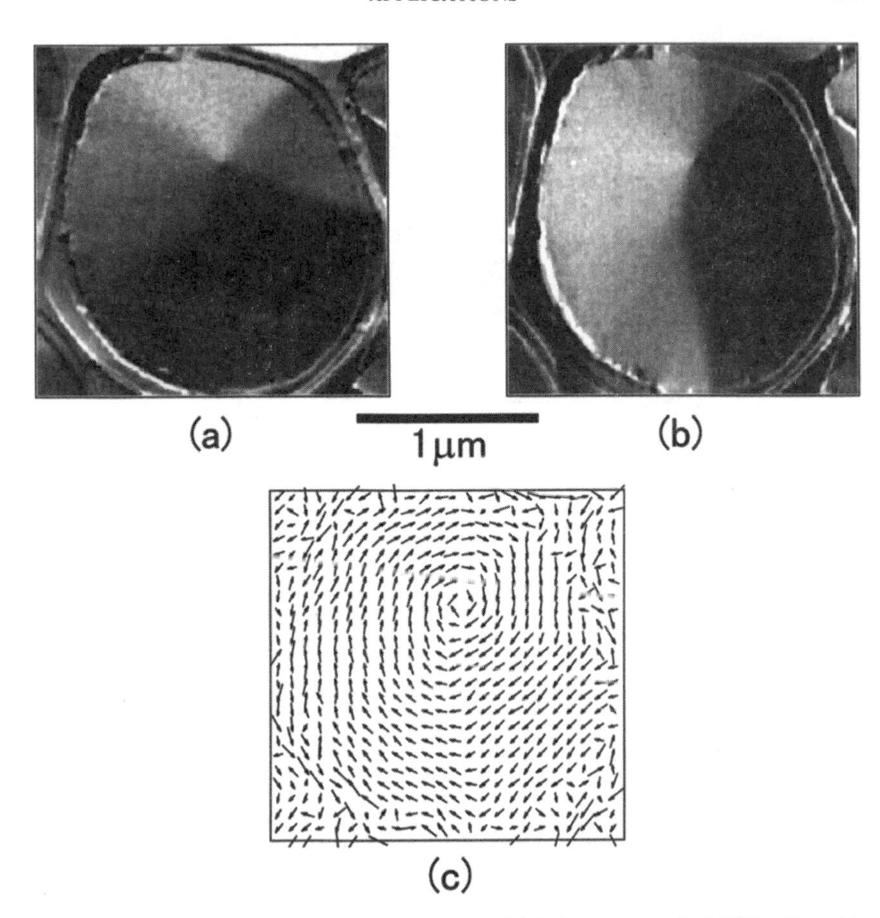

FIG. 10. Lorentz deflection maps of a 30-nm-thick ferromagnetic Ni/Fe alloy film sputtered on a carbon microgrid: (a) vertical and (b) horizontal deflections, and (c) vector plot [301]. (© 1995 IEEE, used with permission).

The result of integration is shown in Fig. 11 as a bird's-eye view. For convenience, the sign of the vertical intensity is taken upside down (upward being negative). It shows the conical deformation of the phase function owing to the effect of the vortex-type magnetization structure. The contour map of the phase function thus obtained is equivalent to the *uncalibrated* flux line image. Figure 12 shows the banded contour (Fig. 12a) as well as solid line contours (Fig. 12b, c) with different spacing of adjacent lines. In Fig. 12c, the line spacing is one-third of that in Fig. 12b, namely, three times higher in density. This demonstrates that once the phase function is constructed, flux lines can be easily mapped out with various conditions. In

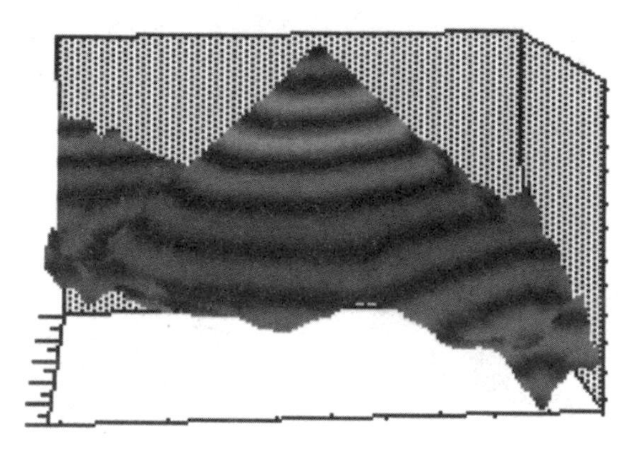

FIG. 11. Magnetically deformed electron phase function. Vertical intensity is taken upside down (upward being negative).

Fig. 12d, for example, flux lines passing over arbitrary points, P and Q in the figure, are delineated.

7.4.3 Combination with Interference Microscopy

Lorentz STEM is highly quantitative in that Lorentz deflection data are precisely proportional to the integrated magnetic induction strengths. For the determination of absolute strengths, however, careful signal calibration is required, and very often we do not dare to do that. On the other hand, interference microscopy in the scanning mode [300] can be used to accurately quantify magnetic flux change in the calibration-free scale of $h/|e|$ [299, 301].

As shown schematically in Fig. 13, scanning-type interference microscopy is performed using a mutually coherent probe pair prepared by splitting the primary probe from a field emission source with an electron biprism. The probe pair runs together, with definite relative distance and direction, through the observation area and then merges again to form interference fringes on a grating-type slit placed just above the detector. The variation of magnetic flux enclosed by the probe paths leads to the relative shift of the probe phases that results in the motion of fringes (Fig. 14). This fringe motion is converted to a periodic change of the net current reaching the detector (Fig. 15). One period of signal change exactly corresponds to the flux change of $h/|e|$. This type of interference microscopy can be performed on a microscope designed for Lorentz STEM just by adding a retractable biprism and slit [302].

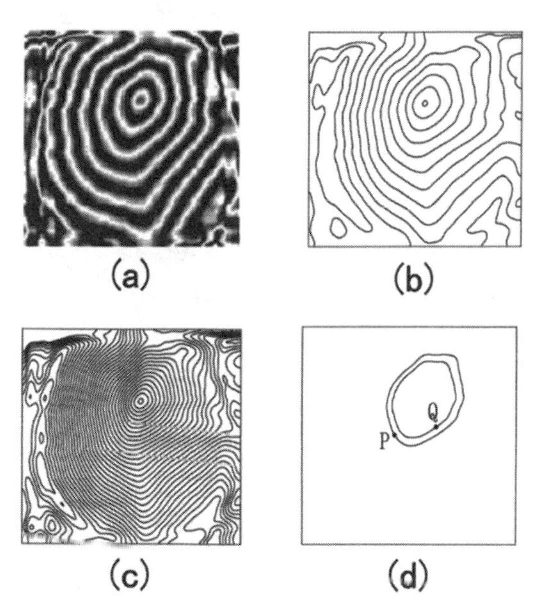

FIG. 12. Contours of the phase function in Fig. 11: (a) banded contour, (b, c) solid line contours with different spacing of adjacent lines [spacing in (b): spacing in (c) = 3:1], and (d) contours passing over arbitrary points, P and Q.

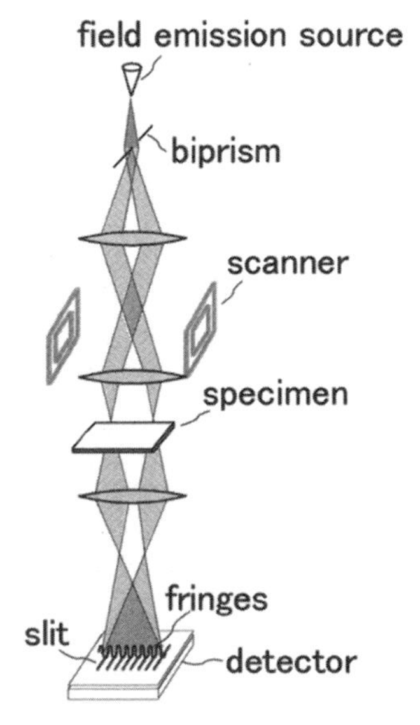

FIG. 13. Electron optical configuration for scanning interference microscopy.

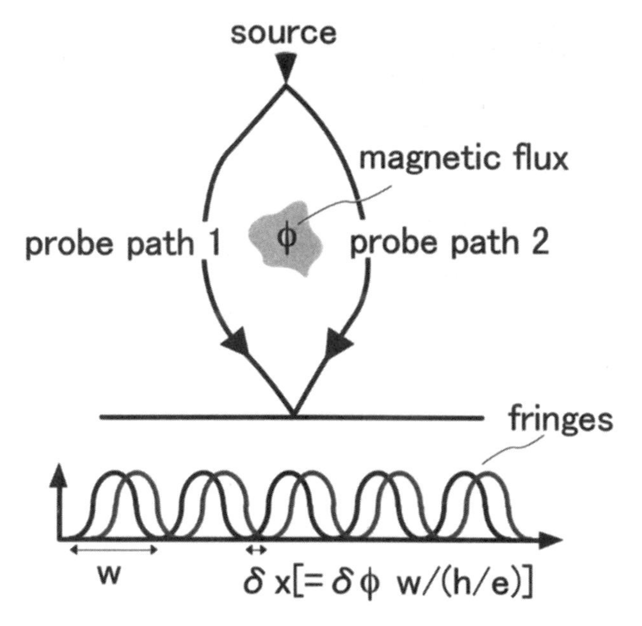

FIG. 14. Magnetic flux and fringe motion.

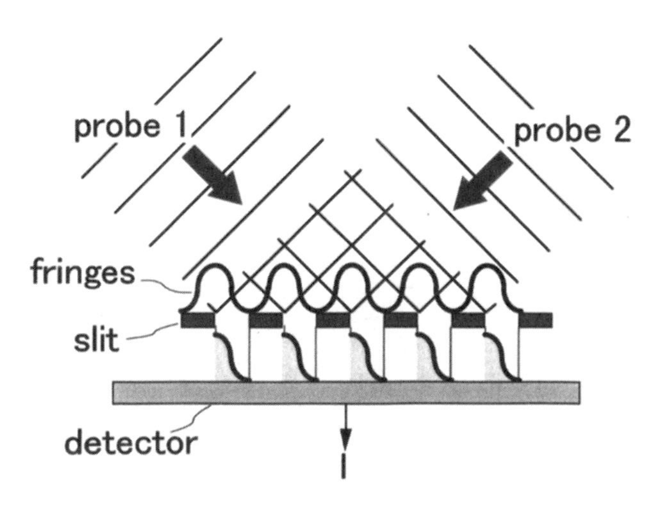

FIG. 15. Conversion of the fringe motion to a periodic signal.

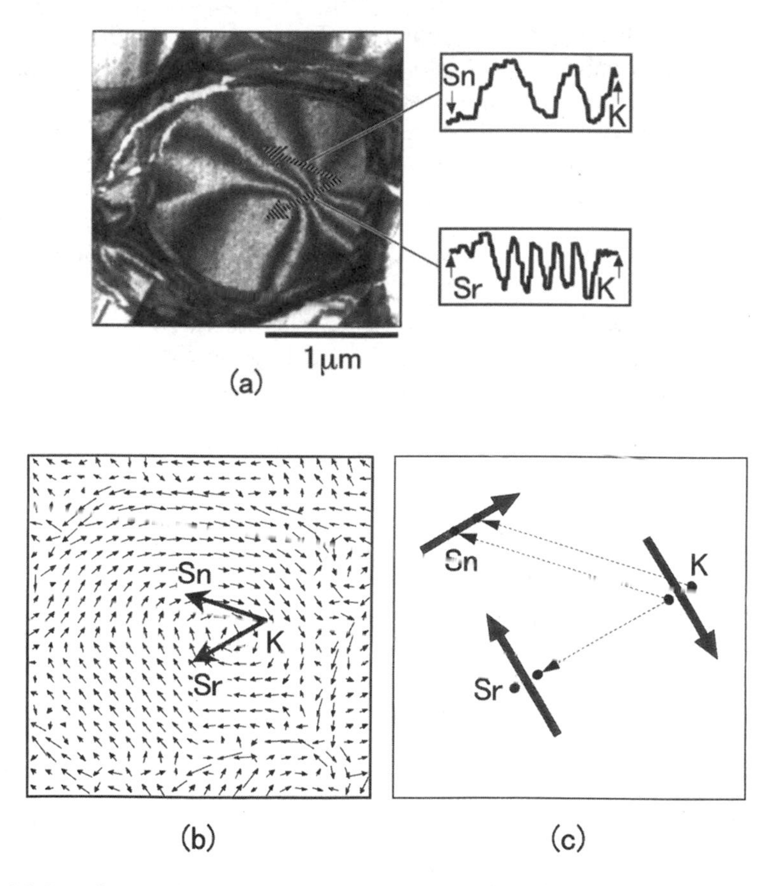

FIG. 16. Interference image (a) of the same Ni/Fe alloy film as in Fig. 10. (© 1995 IEEE [301], used with permission). The vector plot (b) is the same as Fig. 10c. The illustration in (c) shows the directions of magnetization at three particular points, K, S_r, and S_n, and the relative direction of the probe pair.

Figure 16 shows the interference image (Fig. 16a) [301] of the same Ni/Fe alloy film as in Fig. 10 and the same vector plot (Fig. 16b) as Fig. 10c. Also shown is the illustration (Fig. 16c) of the directions of magnetization at three particular points, K, S_r, and S_n, and the relative direction of the probe pair. We see that the direction of magnetization becomes perpendicular to the probe split direction at points K and S_r, and parallel at point S_n. Therefore, in going from point K to point S_r, the net flux enclosed by the probe pair changes by $2Mts$ where M is the magnetization strength, t is the film thickness, and s is the probe separation. Likewise, the enclosed flux changes by Mts in going from point K to point S_n. The number of periodic changes occurring in the image contrast along each path can be counted from the corresponding line profile also shown in the figure (5 times for $K \rightarrow S_r$ and 2.5 times for $K \rightarrow S_n$). Since the flux change along the path is monotonous in either case, the relation $2Mts = 5h/|e|$ must hold for the $K \rightarrow S_r$ path and $Mts = 2.5h/|e|$ for the $K \rightarrow S_n$ path, giving the consistent result $Mts = 2.5h/|e|$. As s is known (400 nm), the result leads to $Mt = 2.5h/(|e|s) = 2.6 \times 10^{-8}$ T·m (=260 G·μm). This completes the procedure for the absolute calibration of the Lorentz STEM signals. In other words, the length of the arrows appearing in the vector plot corresponds to the integrated magnetic induction strength of 2.6×10^{-8} T·m in this particular case. We now understand that interference microscopy in scanning mode gives the contour map of

$$\delta\phi_x = \frac{\hbar}{|e|} \int_{x-s/2}^{x+s/2} dx' \int_{-L}^{0} dz\, B_y(x', y, z) \tag{62}$$

and

$$\delta\phi_y = \frac{\hbar}{|e|} \int_{y-s/2}^{y+s/2} dy' \int_{-L}^{0} dz\, B_y(x', y, z), \tag{63}$$

when the probes are split by s toward the x- and y-directions, respectively (Fig. 17). The signal is therefore dependent on the probe distance, s. If the distance is sufficiently small,

$$\frac{\partial}{\partial y}\delta\phi_x + \frac{\partial}{\partial x}\delta\phi_y = 0 \qquad (\rightarrow \nabla \cdot \mathbf{B} = 0) \tag{64}$$

should be satisfied.

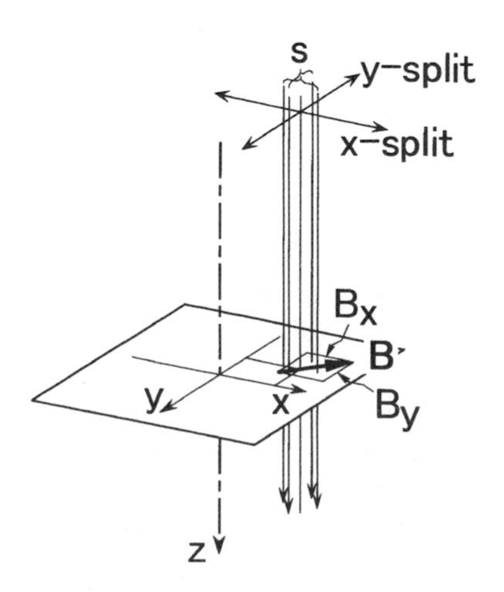

FIG. 17. Probe splitting directions in interference microscopy.

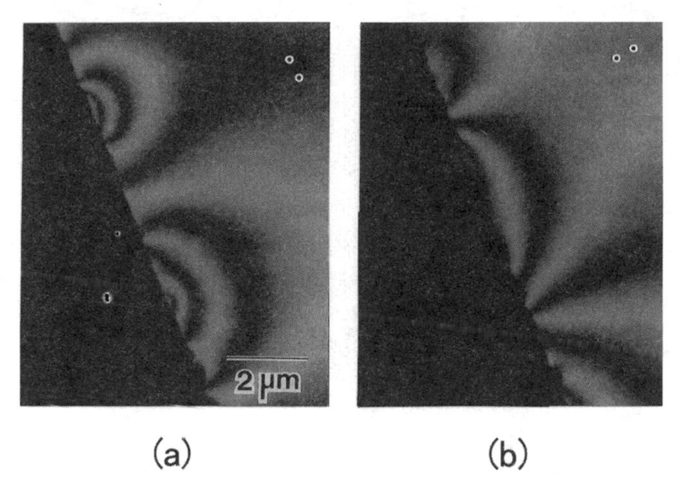

FIG. 18. Interference images near the edge of a recorded magnetic recording film [302]: probes split (a) parallel and (b) perpendicular to the film edge.

And in a static-free space,

$$\frac{\partial}{\partial x}\,\delta\phi_x - \frac{\partial}{\partial y}\,\delta\phi_y = 0 \tag{65}$$

should also be satisfied since $\nabla \times \mathbf{B} = 0$. Equations Eq. (64) and (65) are the Cauchy–Riemann differential equations. Therefore, contours of $\delta\phi_x$ and $\delta\phi_y$ should become mutually orthogonal in a static-free space or in the static material without magnetization rotation, *if* the probe distance is small enough compared with the characteristic length of flux change. Figure 18 shows interference images near the edge of a recorded magnetic recording film [302]. Images obtained with probes split parallel (Fig. 18a) and perpendicular (Fig. 18b) to the film edge show that contours are actually orthogonal. This indicates that the probe distance used (400 nm) is appropriate to reproduce reliable magnetic field distributions.

7.5 Conclusions

Lorentz STEM, the scanning version of Lorentz microscopy, has been proved well applicable to the investigation of magnetic structures in microscopic regions. Setting the temporally random phenomena aside, Lorentz STEM could be used almost anywhere in magnetically oriented material research.

From the exposition presented here, the following points have been made clear. Very quantitative magnetic information is obtained as an immediate output in the form of a temporal sequence of electrical signals, which opens an easy access to various digital image processing techniques. It is also highly compatible with other microscopic and microanalytic techniques.

Acknowledgments

The author would like to thank Dr. Yoshio Takahashi for crucial contributions to the work presented here.

8. MAGNETIC STRUCTURE AND MAGNETIC IMAGING OF $RE_2Fe_{14}B$ (RE = Nd, Pr) PERMANENT MAGNETS

Yimei Zhu and Vyacheslav Volkov

Department of Applied Science
Brookhaven National Laboratory
Upton, New York

8.1 Introduction

$RE_2Fe_{14}B$ compounds (where RE is a rare earth element, Nd or Pr), currently the most powerful permanent magnets, have received much scientific attention because of their great importance in the technology of magnetic materials [303–306]. They have been used for a wide variety of applications, ranging from generators to motors and computer devices, and the market for them is growing rapidly as their magnetic properties and cost-effectiveness are improved. In fact, $Nd_2Fe_{14}B$ and $Pr_2Fe_{14}B$ magnets are quite comparable since their intrinsic properties are nearly identical over their ferromagnetic temperature range. The advantage of the Pr compound is that the easy axis of the magnetization remains aligned along the *c*-axis of its crystal lattice, which may be favorable for low temperature applications. In contrast, the easy axis for the Nd compound undergoes a spin reorientation below 135 K [303, 307–310]. The drawback of the Pr compound is that it is more expensive because praseodymium is less abundant than neodymium.

Considerable efforts were made to optimize the chemistry and processing of these compounds [303, 311]. It was experimentally found that doping with a small amount of impurity elements, such as Co and Ga, and a small excess of Nd, or Pr, over the exact stoichiometric phase composition of $RE_2Fe_{14}B$ (or $RE_{11.43}Fe_{80}B_{5.7}$) can play an important role in achieving high-energy products for the anisotropic magnets [312–314]. In both compounds, the presence of secondary-phase inclusions of approximate composition Nd_3Fe or Pr_3Fe, and of other small nonmagnetic phases or dispersoids, were reported. These impurity phases have contributed to the improvement in the intrinsic coercivity of the hard magnetic materials [315, 316].

Magnetic properties such as coercivity and the energy products of highly anisotropic $RE_2Fe_{14}B$ (RE = Nd, Pr) magnets are known to be very sensitive to microstructure, especially grain size, grain alignment, and

EXPERIMENTAL METHODS IN THE PHYSICAL SCIENCES
Vol. 36
ISBN 0-12-475983-1

ISSN 1079-4042/01 $35.00

secondary phases, is controlled through material synthesis. Two major methods are commercially used for producing RE–Fe–B magnets: the traditional approach of powder metallurgy (sintering) and the rapid-solidification technique of melt spinning (die-upset). In the sintering process, the compounds are prepared by arc melting under an argon atmosphere [317, 318]. The button ingots are crushed and ball milled to form powder grains of 3–6 μm, and these grains then are magnetically aligned and pressed, and finally sintered and rapidly cooled. Such a process generates a strong crystallographic texture in the magnets. A high-energy product with $(BH)_{max} > 45$ MGOe has been reported [319, 320]. In the die-upset process, the magnets are prepared from overquenched ribbons using the melt-spinning technique followed by a hot pressing [321] and die-upsetting procedure [306, 322]. The latter process is used to develop a strongly anisotropic magnet by uniaxial plastic deformation of the compound in a closed die at about 750–800°C, which gives a strong crystallographic texture with the *c*-axes of platelet like grains parallel to the compression direction. The superior quality die-upset magnets produced at General Motors Research had energy products with $(BH)_{max} = 36.38$ MGOe and remanence $B_r = 12.9$ kG (=1.29T in mksa units) [323, 324].

Despite great efforts and undeniable progress in developing commercial magnets based on the tetragonal $RE_2Fe_{14}B$ phase, the energy products thus far achieved are substantially lower than their theoretical upper limits. Our understanding of the role of microstructure in controlling the magnetic structure in hard magnets (e.g., grain alignment and nonmagnetic intergranular phases and their relation to magnetic domain configurations) is still very limited. A complete knowledge of how microstructure is related to magnetic behavior in these compounds may help us to address some major issues of magnetism in materials science, and to optimize the performance of magnetic materials. A primary aspect of research is to understand the factors that limit the coercivity strength, that is, the comparative importance of domain wall depinning and reversed domain nucleation as controlling phenomena [325]. As outlined in the previous chapters, one unique approach to tackle these material problems is the use of Lorentz microscopy, both in the Fresnel and Foucault imaging modes, combined with *in-situ* magnetizing experiments in a field-calibrated transmission electron microscope (TEM). When TEM is compared with other magnetic imaging techniques, the strength of TEM is that the microstructure and magnetic structure of a sample can be examined simultaneously. Real-time observations of local behaviors of magnetic domains and their interactions with structural defects under various applied fields can shed light on the mechanisms of magnetization reversal as well as the structure–properties relationship of the technologically important magnets.

This chapter aims to review the magnetic structures observed in the $RE_2Fe_{14}B$ (RE = Nd, Pr) system using various TEM magnetic imaging techniques. We focus on studies of die-upset Nd-based permanent magnets conducted mainly at Brookhaven National Laboratory in 1993–1999. Investigations on Nd–Fe–B sintered magnets and single crystals as well as on Pr–Fe–B die-upset magnets also will be covered. In Sections 8.2 and 8.3 we review the microstructure, including grain alignment and secondary phases of the materials, and grain boundary structure and composition of the intergranular phase. Section 8.4 is devoted to the domain structure, such as the width of domains and domain walls and the domain wall energy. Monte Carlo simulation of the effects of demagnetization fields is presented in Section 8.5. *In situ* experiments on the dynamic behavior of domain reorientation as a function of temperature, pinning, and grain boundary nucleation related to coercivity under various fields are described in Section 8.6. Finally, in Section 8.7 the correlation between microstructure and properties is discussed.

8.2 Microstructure

The $RE_2Fe_{14}B$ magnets fabricated through a die-upset process are characterized by the presence of a well-developed plateletlike grain texture with a preferential grain orientation of the c-axis parallel to the die-upset direction (hot-press direction), as shown in Fig. 1a in a cross-sectional view. The fine platelets have a dimension smaller than $0.1 \times 0.5\,\mu m^2$ when viewed along a direction perpendicular to the c-axis. Most of them are closely stacked such that their flat facets (a–b planes) are aligned predominantly perpendicular to the die-upset direction. We note that the thickness of these grains does not exceed the critical single-domain grain size (D_c), estimated to be about $0.2\,\mu m$ [326] or $0.3\,\mu m$ [303] for the $Nd_2Fe_{14}B$ phase. Such c-axis-aligned grain platelets and their associated magnetic domain structure directly contribute to the high remanence of the die-upset samples. These plateletlike grains usually have pure twist, or nearly pure twist, interfaces between them, with common c-axes as their rotation axis. These boundaries are often free from intergranular phase.

Figure 1b shows the typical morphology of the same $Pr_{13.75}Fe_{80.25}B_6$ compound shown in Fig. 1a but imaged along the press direction parallel to electron beam. The grains have a polygonal shape with Pr-rich (or Nd-rich for the Nd-based compounds) subnanometer precipitates distributed in the middle of some of the grains. In the regions where the grain boundaries were viewed edge-on, they appear as white lines, nearly even in width ($<5\,nm$). Such line contrast was most visible when the objective lens

FIG. 1. Typical morphology of die-upset $RE_{13.75}Fe_{80.25}B_6$ magnets. (a) The platelet-like grains (RE = Pr) are viewed perpendicular to the press direction, indicated by the arrowhead. (b) The uniaxial grains (RE = Pr) viewed along the press direction. The white-line contrasts at the boundaries are the intergranular phase, and the dot contrast in the center of some grains represents nanometer scale rare earth-rich precipitates. (c) Cross-sectional view of the sample (RE = Nd) showing the mixture of well-aligned regions (marked as layer A) and nonaligned regions (marked as layer B). In the nonaligned B layer, there are areas with a high density of particles of secondary pocket phase (marked as p). The arrow c indicates the press direction.

was underfocused. These spacerlike intergranular phases between the adjacent grains have a significant influence on the magnetic properties of the materials.

In the die-upset RE–Fe–B compounds, the well-aligned grain regions are quasi-periodically separated by nonaligned grain regions, as shown in Fig. 1c, in which they are marked as layer A and B, respectively. The nonaligned grains can have a wide range in size, and they are often mixed with nonmagnetic inclusions (marked as p in Fig. 1c). We shall call these nonaligned areas "defect" layers; they originate from the former flake interfaces between the ribbons and flakes used to form the dense die-upset hard magnets [321, 322]. These inclusions or secondary "pocket phases" concentrate predominantly at the triple junction regions of the $RE_2Fe_{14}B$ grains and have different shapes and sizes. Chemical analysis using energy-dispersive X-ray spectroscopy and electron diffraction reveals that they have a eutectic composition near to Nd_7Fe_3, or Pr_7Fe_3, consistent with the observations of Mishra [313, 327]. In contrast to the well-aligned grains, larger grains over 400 nm (grain size $D > D_c$) with a multidomain structure were also observed in the defect layers. Hence, the microstructure of the die-upset RE–Fe–B compound may be considered as consisting of relatively thick layers of well-aligned grains separated by thin defect layers of nonaligned grains (areas A and B in Fig. 1c). The thickness of these layers was estimated to be 4–6 μm and 0.5–1.0 μm, respectively, with a total quasi-period of about 5–7 μm [328].

8.3 Grain Boundaries in Die-Upset Magnets

8.3.1 The Intergranular Phase

Figure 1b demonstrates that the nontwist grain boundaries in RE–Fe–B often consist of very thin intergranular spacer phase. To reveal the boundary structure at the atomic scale, phase-contrast imaging is often necessary. High-resolution electron microscopy (HREM) on permanent magnets is not a trivial task because the magnetic field generated by the sample can easily distort the beam path in an electron microscope. Thus, realigning the electron beam at an area of interest and correcting axial astigmatism of the microscope lens are crucial for HREM.

We examined several dozens of grain boundaries in RE–Fe–B cut from the center portion of the die-upset buttons to identify the intergranular spacer phase. We found that the interfacial white-line contrasts, shown in Fig. 1b, mostly correspond to an amorphous intergranular phase. However, one should bear in mind that the apparent width of the lines in Fig. 1b (bright-field Fresnel contrast) does not reflect the real thickness of the boundaries because of the out-of-focus imaging conditions. HREM revealed that the average width of the intergranular phase is 8–12 Å for Nd-based magnets (Fig. 2a) and 15–20 Å for Pr-based magnets (Fig. 2b). The grain boundaries are often oriented along the basal plane of one of the grains [the (001) planes of the tetragonal phase; see grain B in Fig. 2(a–c)]. The width of the intergranular phase likely depends on the misorientation of adjacent grains. For basal plane-matched grain boundaries (i.e., pure twist boundaries with rotations around the common *c*-axis), usually no intergranular phase is observed. An example of such a grain boundary in $Pr_{13.75}Fe_{80.25}B_6$ is shown in Fig. 2c (a 36° [001] twist boundary). This suggests that the formation of the intergranular phase at least can be partially attributed to lattice mismatch at the boundaries.

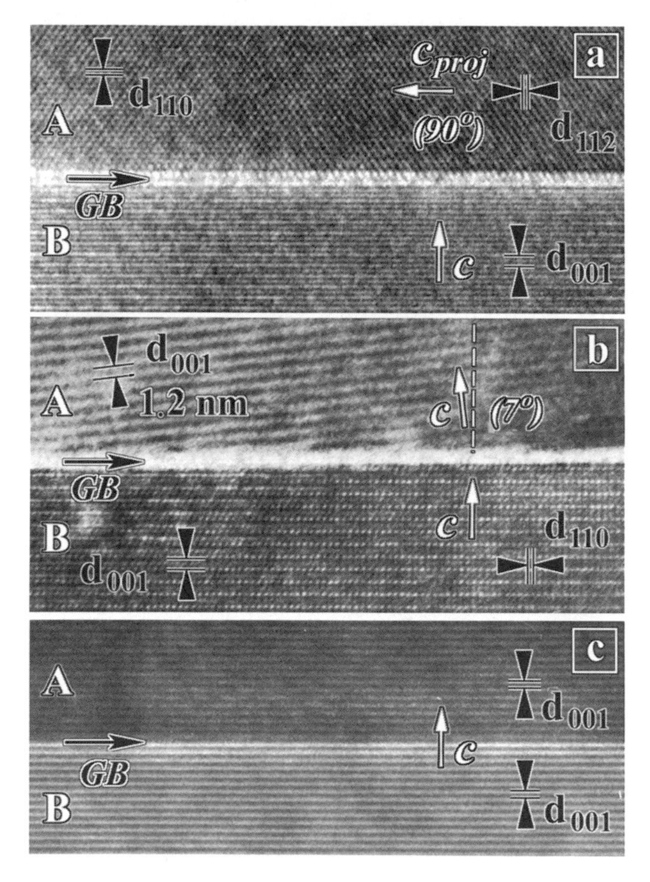

FIG. 2. HREM images of large-angle grain boundaries in $RE_{13.75}Fe_{80.25}B_6$ magnets. (a) RE = Nd, and (b, c) RE = Pr. A crystalline and an amorphous intergranular phase is visible in (a) and (b), respectively, but not in (c). The crystallographic relation of the boundary planes are (a) $(110)_A \| (001)_B$, (b) $(109)_A \| (001)_B$, and (c) $(001)_A \| (001)_B$. GB indicates the position of the grain boundary, while d_{110}, and d_{001}, show the spacing of the corresponding lattice planes. White arrows show the direction of the *c*-axes (easy magnetic directions) in the A and B crystals.

8.3.2 Nanoscale Chemical Analysis

To determine the chemical composition of the thin intergranular phase we examined the spacer phase in die-upset $RE_{13.75}Fe_{80.25}B_6$ magnets using an X-ray energy-dispersive spectrometer attached to a dedicated scanning transmission electron microscope (VG HB501) with an approximately 5 Å probe at liquid nitrogen temperature [329]. The relative change in local composition of RE or Fe was normalized from the average of the integrated peak intensity of the grain interior composition to RE = 13.75%, Fe = 80.25% using the Cliff–Lorimer ratio technique ($C^{Fe}/C^{RE} = kI^{Fe}/I^{RE}$, where k is the Cliff–Lorimer factor determined from standard samples) [330]. Figure 3a shows an example of an excess of iron (after subtraction of the Fe concentration in the matrix) measured across a grain boundary in the Pr-based magnet. The measurement was carried out at ~15 Å intervals with a possible spatial error of 5 Å. At the center of the boundary, the iron concentration was about 89.5% (equivalent to an excess of iron of ~9%), much higher than that 100 Å away from the boundary. Using a simple standard-deviation procedure, we estimated that the measurement error of iron concentration is less than 1.5% in the matrix. On the basis of the "full width at half of maximum" (FWHM) criterion, we found the average iron-rich region was ~20 Å wide, which matches the width of the amorphous intergranular phase, as determined by HREM.

A few dozen grain boundaries in RE–Fe–B samples (RE = Pr, Nd) were examined. Figure 3b gives a histogram for RE = Pr, showing the concentration ratio of iron to praseodymium in the grain boundary region versus that in the grain interior for a sampling of 30 boundaries. Each data point of $(Fe/Pr)_{boundary}/(Fe/Pr)_{interior}$ was the average of twelve measurements, four from the grain boundary regions and eight from the grain interior. The grain boundaries we measured were similar to those shown in Fig. 1(b). We

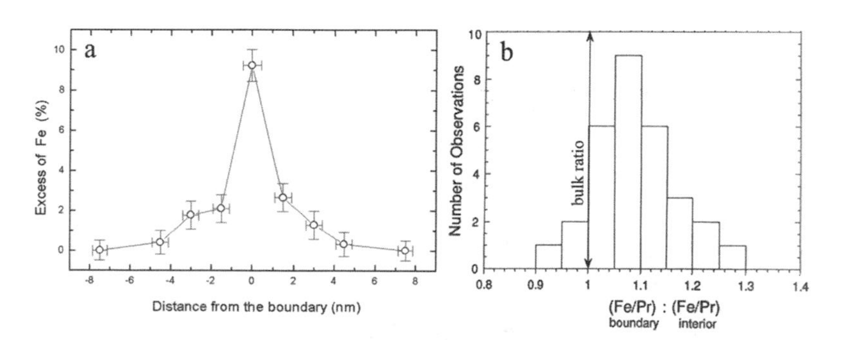

FIG. 3. (a) Excess of Fe as a function of distance from the grain boundary in $Pr_{13.75}Fe_{80.25}B_6$, as measured by nanoprobe (5 Å) X-ray energy-dispersive spectroscopy. The interval distance between the measurements was about 15 Å. (b) Histogram depicting the Fe/Pr concentration ratio for the grain boundary region versus that for the grain interior. The vertical line at $(Fe/Pr)_{boundary}/(Fe/Pr)_{interior} = 1$ indicates the value of the ratio where the concentration of iron at the grain boundary is the same as in the grain interior (bulk ratio).

observed that 90% of the boundaries are rich in iron and that only 10% of them are rich in praseodymium; however, 70% of the former had less than 15% excess iron. The average Fe concentration at the boundaries was 8% more than that in the interior of grains. The findings on the iron-rich grain boundaries in Nd-based samples were very similar to those of the Pr-based samples.

The observations of Fe enrichment at the grain boundaries seem in conflict with those of Mishra [313], who identified the intergranular phase in Nd–Fe–B as Nd_7Fe_3 (i.e., Nd-rich). His work led researchers at General Motors to believe that grain boundary pinning at the nonmagnetic intergranular phase determines the coercivity of Nd-based magnets [331, 332]. We cannot rule out the possibility that this discrepancy is due to differences between the samples or is due to the selection of grain boundaries studied by the two groups. However, it is likely that Mishra was measuring the grain boundary pocket phase Nd_7Fe_3 (several hundreds of angstroms thick, as we discussed in Section 8.2, as is evident from their reported sharp Bragg spots in the diffraction patterns from the Nd_7Fe_3 phase), whereas we were measuring the spacer phase with a width of ~ 20 Å. The Nd_7Fe_3 phase we observed in the die-upset magnets usually is not a thin grain boundary phase. More details on specific role of Fe-rich and Nd_7Fe_3 phases is discussed in Section 8.7.3 and Fig. 25.

8.3.3 Low-Angle Dark-Field Fresnel Imaging

The contrast of the intergranular spacer phase seen in Fig. 1b is a slightly underfocused image in conventional Fresnel imaging. As a complementary technique, low-angle dark-field imaging [333] of the Fresnel contrast can offer chemical information on the boundary at low magnification, disregarding the misorientation of the neighboring grains. This technique, which images the difference between the mean inner potential of two materials at an interface, may be considered as a simpler version of the conventional Fresnel contrast mode [334, 335] because no instrumental parameters are involved, such as the deviation from exact focus. Figure 4a shows schematically the deflection of an electron beam entering a thin specimen parallel to an interface. When no strong Bragg reflections are excited, electrons hitting the specimen in the vicinity of the interface are deflected toward the material with the lower inner coulomb potential (Fig. 4b), generating a streak perpendicular to the interface in the diffraction pattern near the transmitted beam (Fig. 4c). By using the deflected electrons to image the specimen, a bright line is seen at the interface; with a small aperture around the transmitted beam, a black line is present at the interface (Fig. 4d). For an

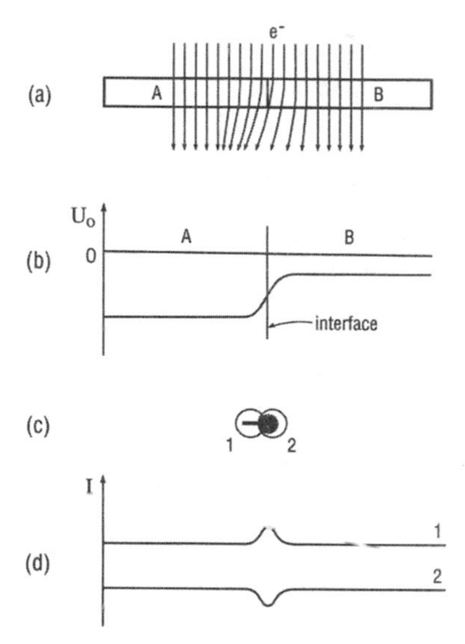

FIG. 4. (a, b) The deflection of a parallel beam of electrons entering a thin specimen parallel to an interface between two materials (A and B) of different mean inner potentials U_0. (c) The streak near the transmitted beam (solid dot) in the diffraction pattern. The open circles represent two aperture positions, 1 and 2, that are placed on the streak (1, dark-field) and on the transmitted beam (2, bright-field), respectively. (d) The corresponding intensity profiles across the interface in low-angle dark-field (trace 1) and bright-field (trace 2) imaging.

intergranular phase present at the grain boundary, we may consider it as two interfaces of opposite sense put together.

Whether the low-angle dark-field imaging method can be used to reveal the mean inner potential of the grain boundary phase will depend on the distance separating the two interfaces. In $Nd_{13.75}Fe_{80.25}B_6$ we observed the contrast variation when the objective aperture was displaced slightly to the left (Fig. 5a). The intensity profile across both vertical grain boundaries from left to the right is seen to be bright–dark, suggesting that the absolute value of the mean inner potential for the intergranular phase is lower (absolute value) than that in the interior of the grains. This is consistent with the observation that the hole in the specimen seen in Fig. 5a is bright on the left side and dark on the right side. Figure 5b schematically illustrates the profile of the mean inner potential and the intensity variation across two grain boundaries and a vacuum region (the line scan in Fig. 5a). The inner potential difference images of the grain boundaries agreed well with our nanoprobe chemical analysis: in general, the boundaries are rich in the

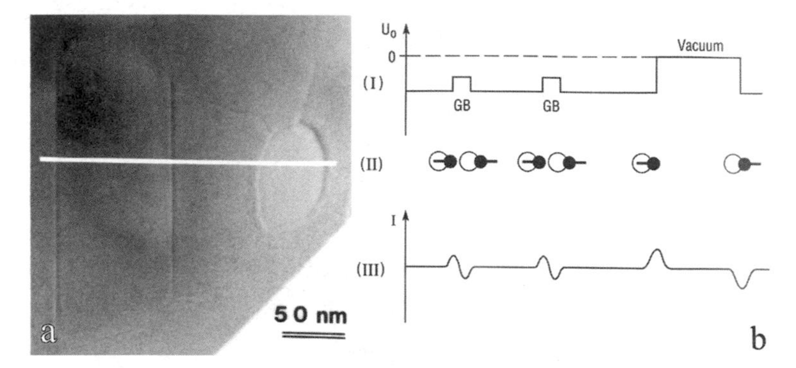

FIG. 5. (a) Low-angle dark-field inner potential difference contrast of two grain boundaries in $Nd_{13.75}Fe_{80.25}B_6$ imaged by placing the aperture on the streak at the left side of the transmitted beam. A hole at the right side of the boundaries also shows the inner potential difference contrast formed by the crystal/vacuum and the vacuum/crystal interfaces. (b) (I) Variation of the mean inner potential along the line in (a). (II) The deflection of the electrons at the different interfaces in the diffraction pattern. The open circles represent the aperture placed to the left of the transmitted beam. (III) The expected dark-field inner potential difference contrast.

low-Z element, Fe, rather than in the high-Z elements, such as Pr and Nd. The dark-field Fresnel imaging method is essentially a phase-imaging technique. In our measurements, we assumed that the local variation in thickness and magnetization across the boundary is negligible.

8.4 Domain Structure

8.4.1 Quantifying Local Magnetization

For a crystalline grain with a size that exceeds the single-domain grain size D_c, multidomain configurations are formed in hard magnets. Such a grain may consist of several ferromagnetic domains separated by 180° Bloch walls of specific thickness, δ, with an antiparallel alignment of single-domain moments, I_s. Figure 6a is a Lorentz Foucault image of a multidomain structure for demagnetized $Nd_2Fe_{14}B$ grains viewed approximately perpendicular to the c-axis. Here, the ratio of domain thickness with magnetic spins up (D_u, black domains) to that of domains with magnetic spins down (D_d, white domains) is close to unity. For partially magnetized or nonuniformly magnetized samples, the local magnetic domain configurations may look like those schematically shown in Fig. 6b, c. The local magnetization of the sample can then be estimated easily by the ratio of the surface areas S_u/S_d

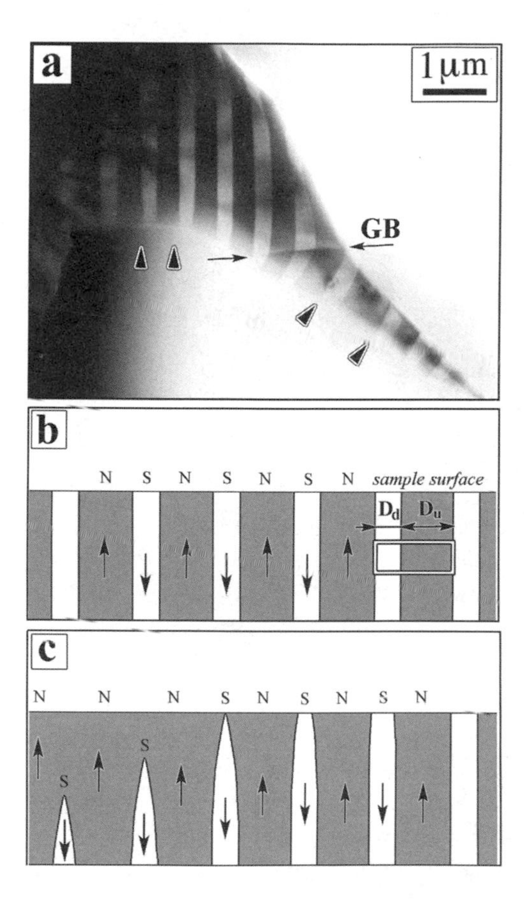

FIG. 6. (a) Foucault image of the multidomain magnetic structure of two $Nd_2Fe_{14}B$ grains. Note the continuity of the domain walls across the grain boundary with the walls parallel along the local easy magnetic direction (*c*-axis) of the grains. (b, c) Schematic drawings of (b) a partially magnetized and (c) a nonuniformly magnetized domain structure. D_d and D_u denote the thickness of the domains with opposite magnetization, which can be used to measure the local magnetization.

(area ratio of white/black domains) defined by the so-called magnetic probe cell with a translation property. Then, the normalized local magnetization I can be determined directly from the Foucault image by a simple relationship:

$$\frac{I}{I_s} \equiv m = \frac{S_u - S_d}{S_u + S_d} \approx \frac{D_u - D_d}{D_u + D_d}, \tag{1}$$

where m and I_s are the relative and saturated magnetization respectively, $S_u = \mathbf{e} \cdot \mathbf{D}_u$ and $S_d = \mathbf{e} \cdot \mathbf{D}_d$ are the surface areas of antiparallel aligned domains which, in turn, are proportional to their widths D_u and D_d ($D_d \neq 0$). Here, $S_d/S_u < 1$ was assumed, and $\mathbf{e}$ is a unit vector along the direction of saturated moment I_s. For instance, the local remanence for the left and right grains in Fig. 6(a) can be estimated to be $m_r =$ 0.20 and 0.17, respectively.

8.4.2 Magnetic Domains and Grain Alignment

Magnetic domains together with the microstructure of the RE–Fe–B magnets can be imaged simultaneously either in the Lorentz Fresnel or in the Lorentz Foucault mode, as demonstrated for RE = Nd in Fig. 7a, b, respectively. The Fresnel contrast, seen as vertical alternating white and black lines, highlights the 180° Bloch domain walls (Fig. 7a), whereas the Foucault contrast, seen as vertical alternating white and black broad stripes, highlights the opposite components in the magnetization of domains themselves. In both images (Fig. 7), the horizontal (perpendicular to the wall direction) black and gray flakelike contrasts of $0.1 \times 0.5\,\mu m^2$ represent the well-aligned platelet grains in the die-upset magnets. Detailed analysis of

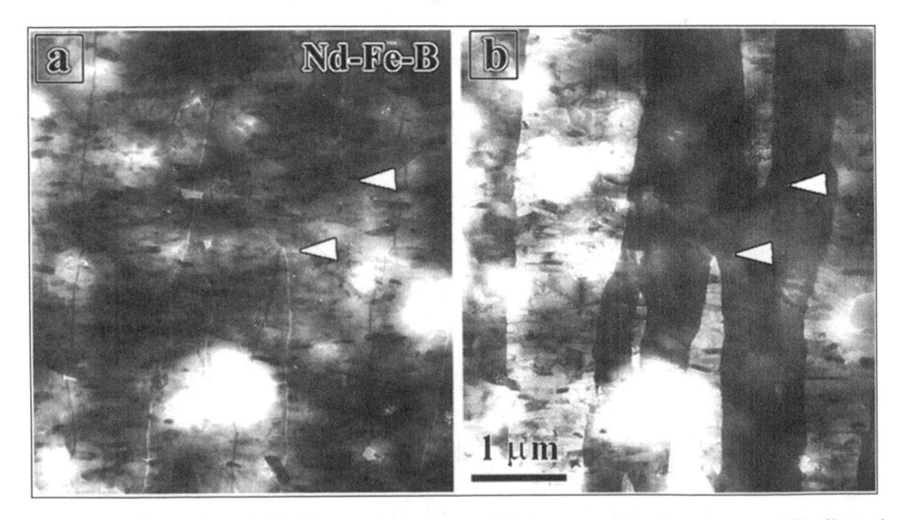

FIG. 7. (a) Fresnel and (b) Foucault images of the same area showing a well-aligned grain area in a die-upset $Nd_{13.75}Fe_{80.25}B_6$ magnet under an almost demagnetized state. The small flakelike contrasts perpendicular to the domain wall are crystalline grains with an average size of $0.1 \times 0.5\,\mu m^2$. Large arrows mark the tip of the domain wall.

Lorentz images at high magnification indicates that these small well-oriented grains are well coupled into bigger grain clusters by ferromagnetic exchange interactions. Therefore, they form large "interacting" domains, running across a large number of small grains along their common *c*-axes as a common easy magnetization direction, without significant interruption at any grain interfaces inside the grain clusters. In a demagnetized anisotropic magnet, the domain clusters with one preferential magnetic direction, say positive, are self compensated by domain clusters with negative magnetization in such a way that the total magnetization m_r remains zero.

As we mentioned earlier (Fig. 1c), die-upset samples have quasi-periodically well-aligned regions separated by nonaligned, or defect, regions. The magnetic domain structure is also quasi-periodically interrupted by these defect layers. Figure 8 is a Foucault image showing the splitting or discontinuity of the magnetic domains in the vicinity of the defect layer B, where the easy magnetic axes of the majority of grains are strongly misoriented. These grains are polygonal and are larger in size by a factor of 3 to 10 compared to the elongated small grains within the well-aligned region (marked as A) above and below in Fig. 8.

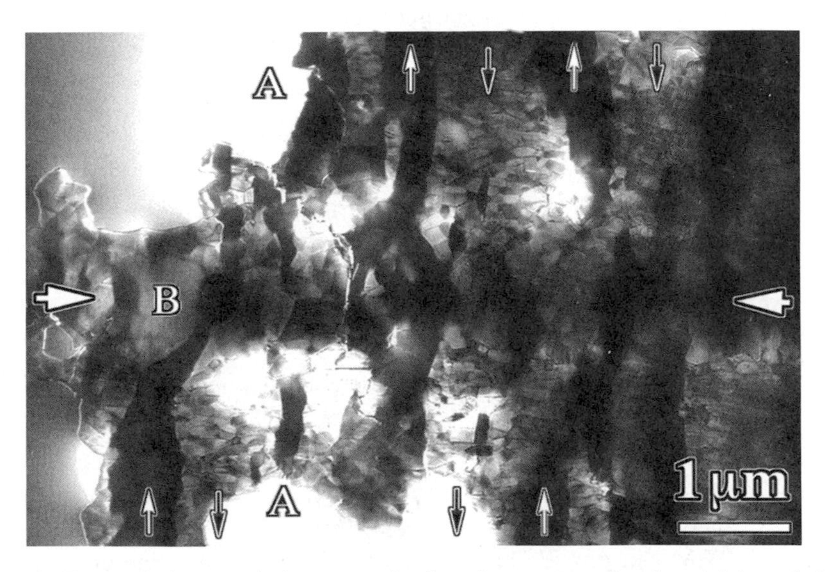

FIG. 8. Foucault image of the magnetic domain structure in the vicinity of the "defect" layer (area B indicated by the pair of arrowheads) in a die-upset $Nd_{13.75}Fe_{80.25}B_6$ magnet. Note the interruption, reversal, and splitting of the magnetic domains when they approach the "defect" layer from the "perfect" well-aligned grain layers. Arrows show the spin orientations within domains of opposite contrast.

For large crystalline grains of 0.5–1 μm, which far exceeds the single-domain grain size $D_c = 0.2–0.3$ μm, a multidomain configuration exists in the remanent state of $Nd_2Fe_{14}B$. Each grain consists of several ferromagnetic domains separated by domain walls, presumably 180° Bloch walls, with antiparallel aligned magnetized moments for each domain. We note that the equilibrium magnetic domain width of different grains varies over a surprisingly wide range, depending on the grain size and crystal thickness. In the demagnetized state, large interacting domains approximately 0.4–0.6 μm wide were often observed. In most cases, their length was limited by the extended "defect" layers (B layers in Fig. 8) or by local defects, which serve as pinning (or nucleating) centers of reversed domains (see domain wall tips marked by the arrowheads in Fig. 7).

Because the grain texture of well-aligned platelet-shaped grains and their magnetic domain structure strongly correlate with the die-upset direction, they directly contribute to the high remanence of the anisotropic die-upset samples. The nonaligned grains with random orientations contribute to it much less (such as area B in Fig. 8). The remanence $I_r/I_s = m_r$ for the hot-pressed sample was found experimentally to be 0.60(4). After the die-upset procedure, its remanence increased to 0.83(4) [324]. Both values can be explained in terms of a much better grain alignment after the die-upset procedure on the basis of Eq. (11) discussed in Section 8.7. This formula gives the relation of remanence, m_r, to the grain orientation distribution, characterized by the texture angle θ_0 of the polycrystalline magnet. For instance, for a hot-pressed sample with $I_r/I_s = 0.60(4)$, we estimated that the maximum deviating angle from the c-axis alignment, θ_0, was 75–85°. This is close to the ideal 90° angle distribution expected for grains that are randomly oriented. Angles smaller than 90° are expected owing to the presence of weak texture or weak remanence enhancement caused by exchange coupling near the grain interfaces.

8.4.3 Domain Width as a Function of Crystal Thickness

Domain size in magnetic materials often varies with sample thickness. However, investigations were limited by using the magneto-optic Faraday and Kerr effect in the past [336, 337] and type I contrast using scanning electron microscopy (SEM) more recently [338–340]. Measuring domain size with TEM can provide high spatial resolution for a range of thickness through which electrons can penetrate, although determining the thickness of a TEM sample is not trivial [341]. Figure 9a is a Lorentz image revealing the configuration of magnetic domains from a $Nd_2Fe_{14}B$ single crystal with its surface normal near the [001] easy axis. The maze pattern, which is formed by the intersection of 180° walls with the sample surface, corresponds to domains with their magnetization directions antiparallel, pointing

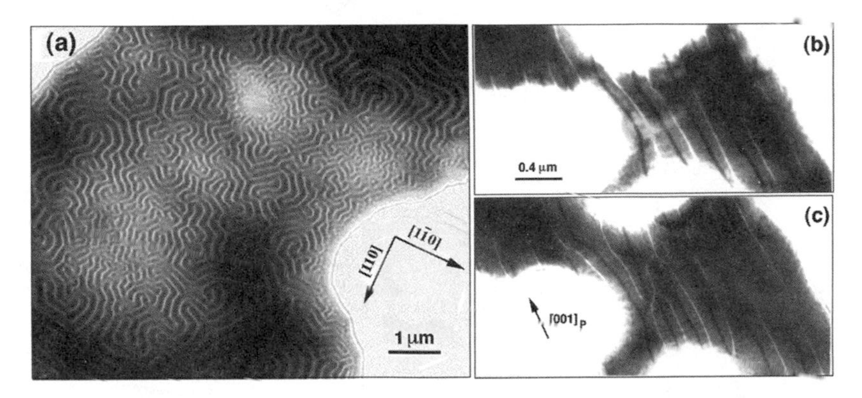

FIG. 9(a) Magnetic domains from a (001) single crystal of $Nd_2Fe_{14}B$ magnetized and viewed along the *c*-axis. The domain width varies with the crystal thickness. (b, c) Magnetic domains from a $Nd_2Fe_{14}B$ polycrystal magnetized with a large [001] component projected into the viewing plane (marked as $[001]_p$). The domains in the thin region (center area of b) split after the sample was heated above T_c, and then cooled back to 300 K (c).

either into or out of the paper. The domain width changes considerably with the area thickness, as evident in the image where the bright and dark background intensities correspond to thin and thick areas, respectively. Although magnetic domains can be visualized in Lorentz mode only when there is at least a small in-plane component of magnetization (i.e., with the Lorentz force pointing to a direction perpendicular to the incident beam), domain size varies little with the viewing direction.

A similar dependence of the domain width on areal thickness was also observed for (100) oriented crystals that consist of strips of parallel domains separated by Bloch walls that lie in the projected [001] direction, as shown in Fig. 9b, c for sintered polycrystalline $Nd_2Fe_{14}B$ [317]. When the samples were heated above the Curie temperature, $T_c = 312°C$ [303], the domains disappeared. After the samples were cooled to room temperature, the domains in the thin regions often split (the initial four domains in the thin region at the center of Fig. 9b split into six, shown in Fig. 9c), while the domains in thicker regions remained unchanged. Since the spacing, or period, of domain walls does not change with the defocus of the objective lens, domain size can be measured directly with reasonable accuracy from Lorentz images. The local crystal thickness *t*, on the other hand, can be determined independently using electron energy-loss spectroscopy based on the relationship

$$t = \lambda \ln\left(\frac{I_t}{I_0}\right), \tag{2}$$

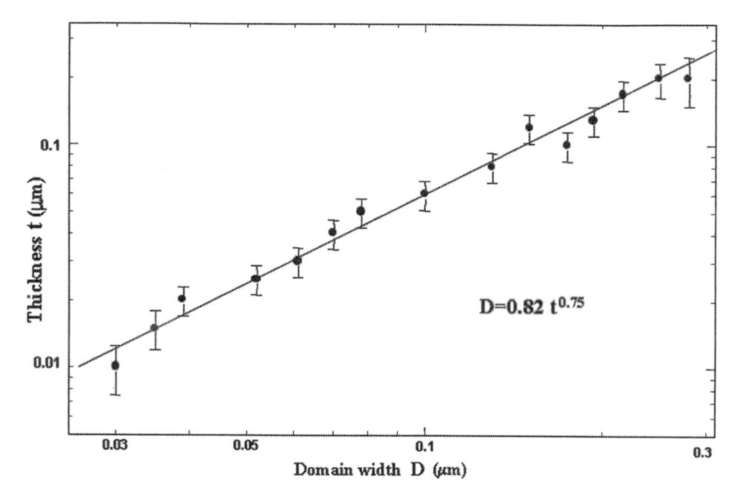

FIG. 10. Width of the magnetic domain as a function of the crystal thickness in $Nd_2Fe_{14}B$. Each data point represents an averaged value from at least three measurements.

where I_0 is the integrated intensity of the zero-loss peak, I_t is the total intensity of the spectrum, and λ is the inelastic mean free path of the crystal. As the inelastic mean free path is unknown for $Nd_2Fe_{14}B$, we used a value of 60 nm from the experimentally determined cross sections for Fe [188]. The domain size D, as a function of sample thickness t, is plotted in Fig. 10, from measurements of more than 50 areas ranging from 10 to 200 nm thick for single-crystal and polycrystalline $Nd_2Fe_{14}B$ samples. The error bars, representing the uncertainty of the measurements, were primarily attributed to the Lorentzian fitting of the shape of the zero-loss peak that has to be separated from the total spectrum. The measurement error is large when the area is extremely thin or thick. Figure 10 suggests that the thickness t has a power dependence of domain width D with an exponent smaller than 1, and a linear relationship in a logarithmic scale. With a least squares fitting procedure, we obtained $D = 0.82t^{0.75}$. This value of the exponent does not significantly differ ($<50\%$) from those observed for much thicker cobalt crystals ranging from 10 μm to 1.9 mm, based on SEM measurements [340].

8.4.4 Measurements of Domain Wall Width

Characteristics of magnetic domain configurations and domain wall motion strongly depend on the thickness, δ, and interfacial energy, γ, of the domain wall separating domains in hard anisotropic magnets. Unlike the magnetic domains, the width of the domain wall is not so sensitive to the crystal thickness. In the past, estimates of δ for the Nd–Fe–B phase were made by various indirect methods based on the equilibrium of minimum energy with respect to domain size, domain wall energy, and saturation magnetization. With TEM magnetic imaging techniques, as outlined in Chapter 2, it is possible to directly measure the domain wall width.

8.4.4.1 Fresnel Method. Although Lorentz Fresnel microscopy is unique in imaging domain walls, its disadvantage is that the walls are imaged under "out-of-focus" conditions. Thus, the width of the black- and white-line contrast from the domain wall cannot be directly used to measure the domain wall width.

Nevertheless, a systematic approach to measuring the domain wall width against different defocus values may allow us to derive the wall width by Fresnel imaging [328]. Figure 11 is a Lorentz Fresnel image taken at a defocus value of 250 μm, showing meandering domain walls. The black- and white-line contrast (on a film positive) corresponds to the so-called convergent and divergent walls with a width W_c and W_d, respectively. Line scans of the domain wall intensity from areas A and B at three different focuses are plotted in Fig. 11b, c. Note that the convergent wall is always sharper than the divergent wall. The wall width can be measured using the intensity profile across the divergent or convergent wall alone, or the difference profiles of the two. We found that the measurement of the divergent wall width is most reliable because it is described by a simple linear relation $W_d(z)$ valid for any positive defocus values z ($z > 0$). Simple analytical expressions for convergent and divergent wall widths versus defocus values were given, for instance, by Wade [342]. The formula for divergent wall width is

$$W_d(z) = \delta_0 + 2z\Psi, \tag{3}$$

where Ψ is the deflection angle owing to the Lorentz force on the electron beam determined by local magnetization. The value δ_0 can be derived as a linear asymptote $W_d(0) = \delta_0$ at zero defocus ($z \rightarrow 0$), in principle, without knowledge of the Ψ value, because the defocus z can be measured in arbitrary units. Here W_d was measured using full width at half of maximum (FWHM) for the intensity profile across the image of the $W_d(z)$ wall (see Fig. 11b, c). Figure 11d summarizes the measurements from the areas A and

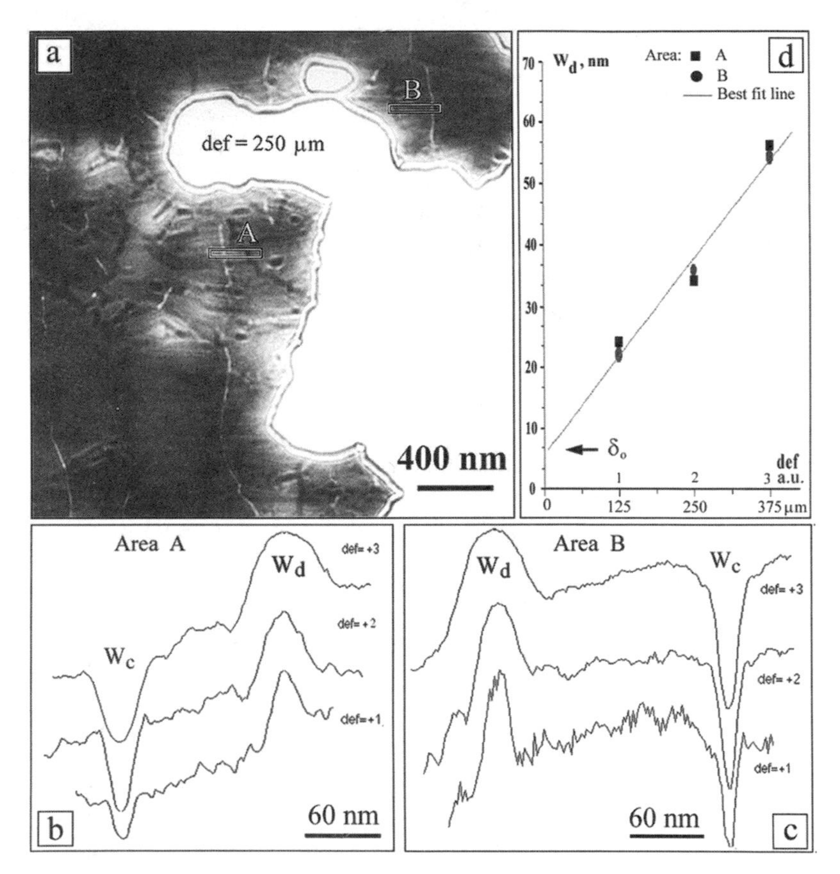

FIG. 11. (a) Fresnel contrast of domain walls in the die-upset magnet. (b, c) Line profiles from area A and area B in (a). The domain wall width δ is determined by extrapolating W_d to the defocus value $z = 0$. (d) Image width of the divergent wall W_d as a function of defocus.

B in Fig. 11a. By extrapolating the FWHM for a $W_d(z)$ from defocus to in focus ($z = 0$), we derived $\delta_0 = 5.8 \pm 2.8$ nm with best-fit parameters $P = 2.6 \times 10^{-4}$, $N = 6$, $R = 0.98$, where P, N, and R are, respectively, the mean square deviation, number of points, and the Student criterion [328]. It is worth noting that the Lorentz deflection angle Ψ can be estimated from the slope of the $W_d(z)$ curve, provided the defocus values are calibrated. In our case (Fig. 11d), we found $\Psi = 0.06$ mrad, which fits well with the expected value obtained by calculating $\Psi(B_s, V)$ at $V = 300$ kV ($B_s = 1.6$ T for $Nd_2Fe_{14}B$). The measured value δ_0 also agrees well, within the experimental error, with the theoretical estimate $\delta = \pi(A/K_1)^{1/2} = 4.2$ nm made by using the exchange constant ($A = 7.7 \times 10^{-12}$ J/m) and the magnetocrystalline anisotropy constant ($K_1 = 4.3 \times 10^6$ J/m^3) for the $Nd_2Fe_{14}B$ phase at $T = 300$ K reported in the literature [343, 344].

8.4.4.2 Holographic Method. Lorentz microscopy is an easy way, but not a straightforward one, for measuring domain wall width. Extrapolating the widths at zero defocus from defocused wall images can be erroneous [313], especially for $Nd_2Fe_{14}B$ where subnanometer spatial resolution is required. A robust way of measuring the domain wall width may be by measuring phase gradients across the domain interfaces using off-axis holography. The underlying principle is based on the fact that electron holography gives phase shifts due to both magnetic and electrostatic potentials of a local area. For a magnetic material, the phase gradient is proportional to the in-plane component of the induction provided that the thickness and potential are constant in the area [184, 345].

Figure 12a is a Lorentz Fresnel underfocus image of $Nd_2Fe_{14}B$ showing curved black/white domain wall contrast. Figure 12b is the x-phase gradient of the phase image reconstructed from a hologram acquired from the boxed area shown in Fig. 12a. The gray scale represents the phase slope in the region. When the electrostatic potential and crystal thickness of the region are constant, the phase gradient image can be used to directly map local magnetic induction. Figure 12c, superimposed on a background contrast of the gradient image, shows an induction map converted from the area outlined in Fig. 12b. The map, consisting of $10 \times 10\,\text{nm}^2$ finite elements, displays the local magnetization configuration at nanometer scale. The arrows represent the projected in-plane (in the plane of projection) components of induction vectors, B ($B = I + \mu_0 H_d$, where μ_0 is the permeability of vacuum and I and H_d are the magnetization and demagnetizing field, respectively), averaged through the sample thickness along the beam direction. The minimum length of the vectors in Fig. 12c is zero, indicating an out-of-plane induction, and the maximum length in the figure corresponds to the in-plane component of ~ 1 T. We note that although the 90° domains in Fig. 12c show a large in-plane component, the domain walls display a variety of characteristics. The walls labeled A and B appear to separate 90° domains in Fig. 12c and show in-plane (Néel) and out-of-plane (Bloch) rotation, respectively. The wall labeled C separates 180° antiparallel domains and has an out-of-plane character.

The width of domain walls can be measured from the regions where the slope of the phase perpendicular to the wall changes abruptly. Figure 12d is an example of a line scan through the gradient of the phase from the region marked on Fig. 12c across the in-plane wall (A, left) and the out-of-plane wall (B, right). Although the gradient profile is noisy, the abrupt change in the phase profile is clear. The wall width measured for the line scan was

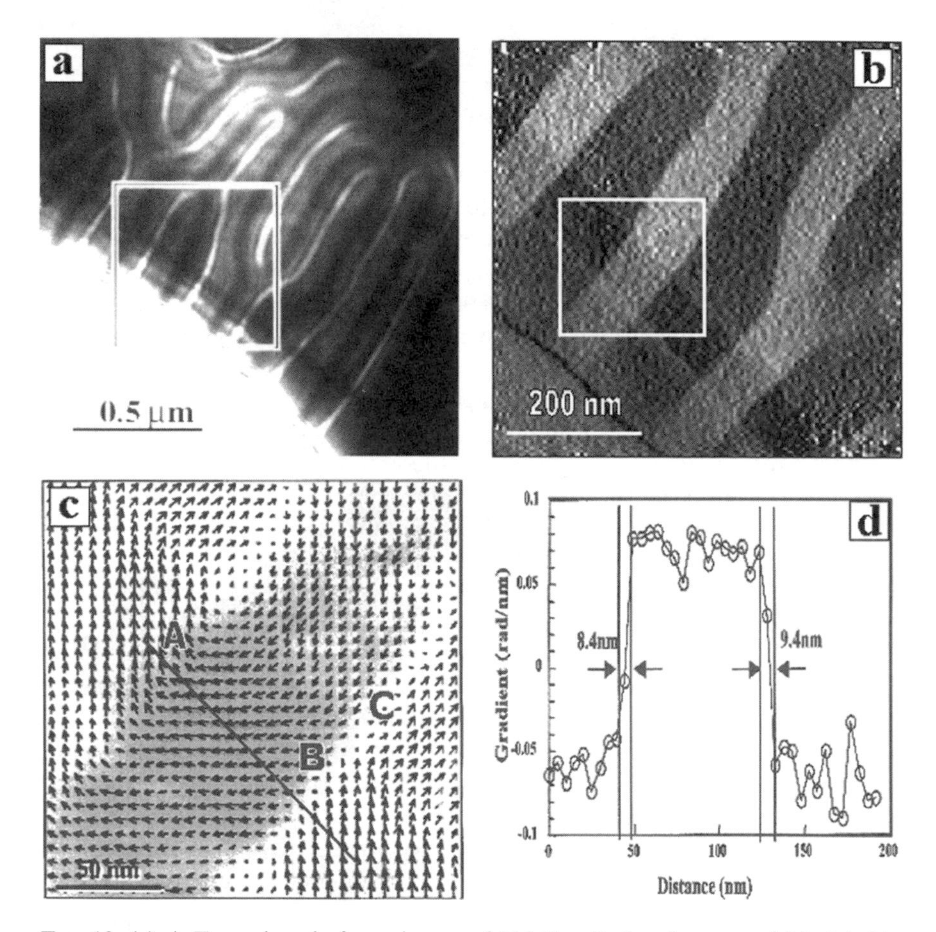

FIG. 12. (a) A Fresnel underfocus image of $Nd_2Fe_{14}B$ showing curved black/white domain wall contrast. (b) x-gradient of a phase image acquired using off-axis holography from the same area of the boxed region in (a). (c) Induction map derived from both x- and y-phase gradients of the region outlined in (b). (d) Line scan across domain walls at indicated regions in (c) showing an abrupt change in the slope of the phase, which can be used to accurately measure the domain wall width.

determined to be about 8.4 nm for the former and 9.4 nm for the latter. Using a similar method, we measured about a dozen domain walls ranging from 7 to 19 nm wide. The wide domain walls we observed may be attributed to the inclination of the interface, since we could only measure the projection of the wall along the direction of the incident beam. The average thickness of the walls was about 9 and 10 nm, respectively, for a 90°

and a 180° wall configuration. There was no significant difference in the width of the wall for these two types. In the following section, we show that the δ_0 determined through the line profile of the phase gradient (Fig. 12d) should be considered as the upper limit of the domain wall width.

8.4.4.3 Comparison of Fresnel and Holographic Methods. For a useful comparison of the domain wall widths measured by Fresnel and holographic methods, we should use the same width definition. For instance, in the Fresnel mode we used the FWHM criterion. The principal difference between these two methods is schematically shown in Fig. 13, which illustrates that the holographic method measures the phase $\phi(x)$ (Fig. 13a), whereas the Fresnel method deals with the analytical signal $d^2\phi(x)/dx^2$ (Fig. 13c). The phase gradient maps (Fig. 12c and Fig. 13b), proportional to the **B**-vector map, can be obtained from Fig. 13a only by numerical differentiation. Here, the regions of the most rapid change in phase gradient, $[d\phi(x)/dx]_{max}$, correspond to domain wall positions of thickness $x(I_{max}) - x(I_{min}) \approx 2\delta$. Thus, the δ value is defined by the usual FWHM for the peak-profile intensity shown in Fig. 13c. Because the Fresnel method does not require any double numerical differentiation, defining the width using the analytical signal $d^2\phi(x)/dx^2$ is more straightforward than defining

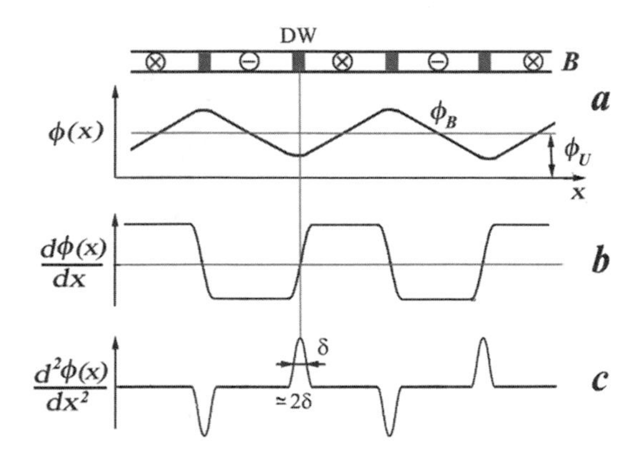

FIG. 13. (a) The ideal phase profile $\phi(x) = \phi_U(x) + \phi_B(x)$ for a magnetic multidomain sample of constant thickness (t) recorded by electron holography (one-dimensional case). Here, $\phi_U(x) \approx U_0 t$ and $\phi_B(x) \approx \pm Bxt$ are the contributions of electrostatic and magnetic vector potentials, respectively. (b) Plot of $d\phi(x)/dx$, the first derivative of the profile in (a) [equivalent to $\nabla\phi(\mathbf{r})$ in the three-dimensional case]. (c) Plot of $d^2\phi(x)/dx^2$, the second derivative of the same profile. The positions of the domain walls in (b) correspond to transition areas $\sim 2\delta$ in thickness with the rapid change of the phase gradient, $[d\phi(x)/dx]_{max}$. The δ value was determined using the FWHM criterion in the Fresnel mode. The same measuring should be employed in the holographic mode to compare the domain wall width.

the holographic phase distribution $\phi(x)$. If we convert the phase gradient profile into a Gaussian distribution measured by FWHM, the domain width δ will be reduced by a factor of 2. In other words, the domain wall width for the 90° and 180° domain walls would be 4–5 nm wide. The result would then fit well with the value of the divergent wall measured by the Fresnel method and the theoretical estimate $\delta = \pi\sqrt{A/K} = 4.2$ nm.

It is also noteworthy that the induction maps we reconstructed using the holographic method for Nd–Fe–B contain not only 180° but also 90° domain walls (Fig. 12c). Care should be taken when we interpret the 90° domain walls because they are not expected in the highly anisotropic uniaxial bulk crystals. In this sense we cannot exclude the possibility that the 90° domain walls are not intrinsic characteristics of the domain configuration, even if the artifacts owing to the variation in the potential and thickness have been corrected. Two factors can contribute to the discrepancy: (1) in TEM, we only probe the in-plane component, which may represent a small fraction of the total local magnetization; and (2) in electron holography, to obtain a reference beam from vacuum, a thin area near the sample is often required. For a thin area, the demagnetization field, or stray field, can be substantial, which may significantly alter the projected induction distribution in the area. We discuss the effects of the stray field on magnetization configuration in Section 8.5.

8.4.4.4 Domain Wall Energy. Knowledge of the domain wall width δ can be used to estimate the specific domain wall energy γ. In a good first-order approximation, by neglecting the long-range interactions and taking into account only the exchange energy and magnetocrystalline anisotropy energy of the domain wall, the total energy per unit area of the wall can be considered as the sum of the exchange energy γ_{ex} and the anisotropy energy γ_{an}:

$$\gamma = \gamma_{ex} + \gamma_{an} = \frac{A\pi^2}{\delta} + K\delta, \tag{4}$$

where A and K are the exchange and anisotropy constants. In the equilibrium state, the minimum value of the total domain wall energy will satisfy the condition $d\gamma/d\delta = 0$, which results in $\gamma_{ex} = \gamma_{an}$. Combining these two equations, we can calculate the domain wall energy in the $Nd_2Fe_{14}B$ phase. The anisotropy field is defined as $H_a = 2K/(4\pi M_s)$ in cgs units, where M_s is the saturation magnetization; $M_s = 1.274$ kG [346] and $H_a = 73$ kOe [347] for the $Nd_2Fe_{14}B$ phase at room temperature. Thus, we derived $K = 4.65$ MJ/m$^3 = 4.65 \times 10^7$ erg/cm^3. For a measured domain wall width of $\delta \approx 5$ nm, $\gamma_{ex} = \gamma_{an} \approx 0.023$ J/m$^2 =$ 23.2 erg/cm^2, $\gamma = 46.4$ erg/cm^2, and $A \approx 1.2 \times 10^{11}$ J/m.

On the basis of the measurements of domain wall width, we can estimate the collective spin rotation across the domain walls. We assume that the direction of spin within the wall rotates evenly and gradually by an angle $\Delta\theta$ at an interatomic distance a, until the total rotation angle θ across the wall region reaches $\pi/2$, or π, for a 90°, or 180°, wall. We estimate that the spin rotation of the nearest neighbor $\Delta\theta$ is 2.8° for a 90° wall, and 5.1° for a 180° wall, using the averaged interplanar distance for the Fe sublattice $a = 0.20$ nm, calculated from the Wyckoff positions for a $Nd_2Fe_{14}B$ compound.

8.5 Effects of Stray Fields on Magnetic Imaging

One of the drawbacks in TEM imaging may be associated with its image projection along the beam direction. This raises a special concern for magnetic imaging because the stray fields, or the demagnetization fields, present on the top and bottom surface of a sample are superimposed into the magnetization within the sample. To understand the induction distribution in a thin TEM sample, we have to understand the effects of stray fields on magnetic imaging.

8.5.1 Contrast from Stray Fields

Figure 14a–c are TEM images in deflecting mode for a thick electron non-transparent sample (thickness $\approx$ 300 nm), showing dark and bright lobe contrast at a specimen edge that reveals stray fields generated by magnetic domains in TEM. The imaging condition of the deflecting mode is very similar to the Foucault mode of Lorentz microscopy, and it is based on positioning the selected area aperture to collect or block electrons displaced by the Lorentz force originating from the sample [348]. The bright and dark lobe contrast in Fig. 14 directly corresponds to the areas with high and low density of the incident electrons displaced by the opposite sign of stray fields at the specimen edge. The images in Fig. 14a, b are almost complementary since the aperture was placed in opposite directions. The size of the lobe contrast is determined by the extension and the strength of the stray field, which may be directly related to the size of the magnetic domain, as shown in Fig. 14d.

For a comparison, Fig. 14e–g shows the contrast of stray fields from a much thicker ($\sim$ 3000 μm) $Nd_2Fe_{14}B$ single crystal at much lower magnification, using secondary electrons in a scanning electron microscope [349]. To view the stray field present at the sample surface three-dimensionally, the (001) surface of the sample was tilted in such a way that there is a shallow angle with respect to the optical axis of the microscope. As shown in Fig.

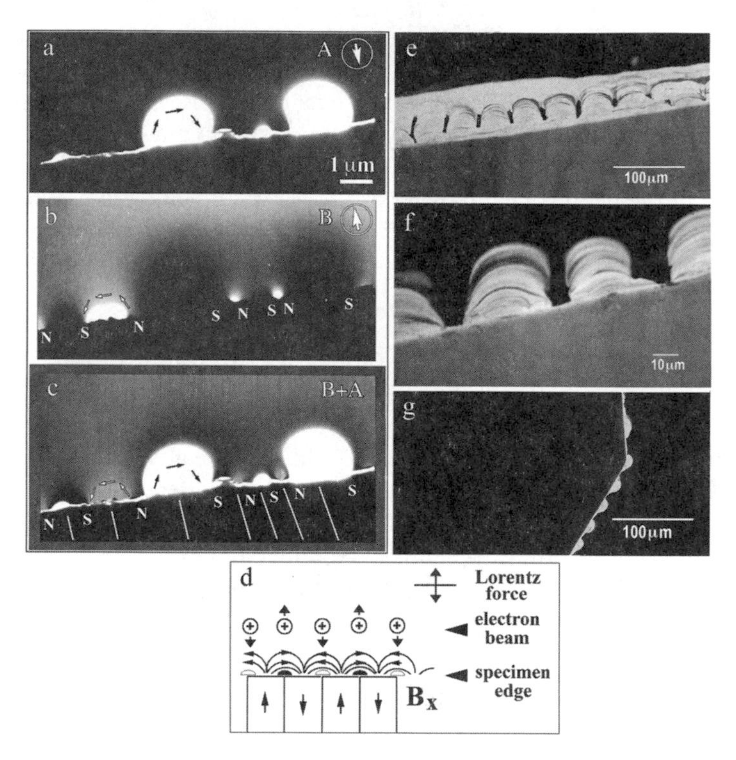

FIG. 14. (a, b) Stray field contrast of a $Nd_2Fe_{14}B$ magnet seen in an opaque sample using the TEM deflection mode, where the black and white band contrast corresponds to the magnetic domain with opposite magnetization. The sample edge was tilted 45° and viewed in opposite deflected beams. (c) A superimposed Foucault image from (a) and (b) of the corresponding area. (d) Schematic drawing of the stray field and the magnetic deflection of the beam by the sample edge. (e–g) Stray field contrast seen in SEM on a bulk $Nd_2Fe_{14}B$ single crystal (by courtesy of American Institute of Physics, from Ref. [349]). (e) A low-magnification secondary electron image with the crystal ~45° away from the viewing direction; (f) enlargement of (e), showing the troughs and hillocks formed by the stray fields; (g) edge-on view of surface flux from two crystallographic facets, similar to that seen in TEM (a).

14e, f, the stray fields appear as parallel row of hillocks separated by narrow troughs, suggesting that the origin of the stray fields are the parallel plate domains that extend through the bulk of the crystal (also see Fig. 14d). We note that the stray fields can extend, as far as the incident electrons can see, to more than 10 μm from the sample edge for a thick SEM sample, but they extend ~1–2 μm in a thin TEM sample. However, the characteristics of the stray field for thin and thick samples are quite similar, as demonstrated in

Fig. 14a, g. Such extended stray fields, present at the top and bottom surface of a magnetic sample, can modify the trajectory of the incident electrons when they pass through it and may give misleading information on the projected final images. In the following, we use micromagnetic simulation techniques to shed light on the effect of the stray fields on magnetic imaging in $Nd_2Fe_{14}B$.

8.5.2 Micromagnetic Simulations of Domain Structures

A thin magnetic sample has unique characteristics in its magnetic structure. It generally does not have net magnetization because of the existence of magnetic domains with different magnetization directions. There are three energy terms to describe the magnetic structure of a system in the absence of an external magnetic field, namely, E_{an}, E_{ex}, and E_{dip}. E_{an} *is the anisotropy energy* representing the magnetocrystalline anisotropy, E_{ex} *is the exchange energy* representing the ferromagnetic coupling between neighboring spins aligning them parallel within the domain, and E_{dip} is the *dipole–dipole interaction energy* representing long-range interactions between the spins. These three energies can be expressed as

$$E_{an} = K \sum_i |\mathbf{m}_i \times \mathbf{c}|^2; \tag{5}$$

$$E_{ex} = -J \sum_{i,j} \mathbf{m}_i \cdot \mathbf{m}_j; \tag{6}$$

$$E_{dip} = -\frac{\mu_0}{4\pi} \sum_{ij} \left[\frac{\mathbf{m}_i \cdot \mathbf{m}_j}{r_{ij}^3} - \frac{3(\mathbf{m}_i \cdot \mathbf{m}_j)(\mathbf{m}_j \cdot \mathbf{r}_{ij})}{r_{ij}^5} \right], \tag{7}$$

where i and j represent the ith and jth dipole in the system, $\mathbf{m}$ is the magnetic moment of the dipole, $\mathbf{r}$ is the distance vector between the dipoles, and $\mathbf{c}$ is the unit vector along the easy axis. K and J are coefficients for anisotropy and exchange energy, respectively [350]. The dipole–dipole interaction energy is an essential part of total magnetostatic energy of a magnetic sample. The competition between the exchange and anisotropy energies is the source for spin rotation during the reorientation phase transition of magnetization. The balance between the long-range dipolelike magnetostatic interactions and strong short-range interactions (E_{ex} and E_{an}) defines the final domain structure. The actual magnetic structure of a system depends on the minimization of the total energy E,

$$E = E_{an} + E_{ex} + E_{dip}, \tag{8}$$

and its minimization strongly depends on the shape of the sample.

In our simulation we simplify the energy expression as

$$E_{\text{eff}} = \frac{E}{KM} = \sum_i (\hat{\mathbf{m}}_i \cdot \mathbf{c})^2 + J_{\text{eff}} \sum_{\langle ij \rangle} \hat{\mathbf{m}}_i \cdot \hat{\mathbf{m}}_j + A_{\text{eff}} \sum_{i \neq j} \left[\frac{\hat{\mathbf{m}}_i \cdot \hat{\mathbf{m}}_j}{r_{ij}^3} - \frac{3(\hat{\mathbf{m}}_i \cdot \mathbf{r}_{ij})(\hat{\mathbf{m}}_j \cdot \mathbf{r}_{ij})}{r_{ij}^5} \right], \tag{9}$$

where M is the magnitude of the magnetic moment of each dipole, $\mathbf{m} = M\hat{\mathbf{m}}$, $J_{\text{eff}} = J/K$, and $A_{\text{eff}} = \mu_0/(4\pi K)$. As we see, the parameter A_{eff} determines the relative strength of the anisotropic energy and the magnetostatic energy. The larger A_{eff} is, the larger is the magnetostatic energy compared to the anisotropy energy.

The magnetic domain structure of the model system can be obtained by minimizing the total energy E_{eff} using Monte Carlo methods to solve the problem in statistical mechanics. The thermodynamic average of any mechanical variable A is defined as

$$\langle A \rangle = \frac{\sum_\mu A_\mu e^{-E_\mu/k_B T}}{\sum_\mu e^{-E_\mu/k_B T}} \tag{10}$$

where E_μ represent the energy eigenvalues. Because only a few states μ that are close to the most probable values contribute significantly to the average, it is sufficient to generate only the important states by assigning enhanced probabilities to them.

A possible implementation of the Monte Carlo method is to start with some initial configuration and then repeat again and again the following four steps, which simulate the real thermal fluctuations [351]: (1) choose an arbitrary trial configuration, (2) compute the energy difference ΔE between the initial and trial configuration, (3) calculate the "transition probability" $P = e^{-\Delta E/k_B T}$ for that change, and (4) generate a random number ε $(0 < \varepsilon < 1)$. If $P > \varepsilon$, the new configuration is accepted; otherwise, the system remains in the old state. Having obtained the low-energy magnetic configuration from the simulation, we then calculate the projection of the magnetization parallel to the film $B_\parallel$, which can be used to compare to the magnetization distribution obtained using TEM Lorentz microscopy or electron holography.

8.5.3 Effects of Crystal Orientation

We model magnetic films with an array of dipoles of size $16 \times 16 \times L_z$, where L_z corresponds to the thickness of the sample. On a workstation, one

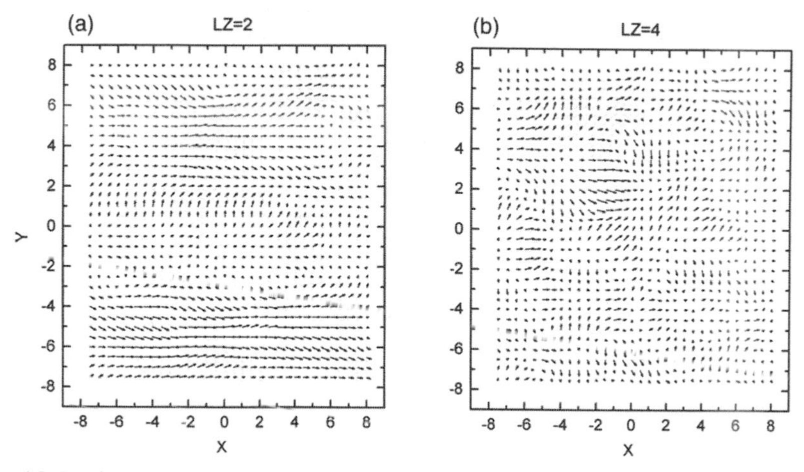

FIG. 15. Projected in-plane magnetization $B_{\|}$ of the sample for different thicknesses L_z with the easy axis oriented 45° to the film surface, obtained via Monte Carlo simulation. (a) $L_z = 2$ and (b) $L_z = 4$. The size of $B_{\|}$ in (a) is scaled by one-third.

calculation cycle took about 10 hours. Figure 15 shows the simulated magnetization map; the direction and the length of the vector represent the direction and magnitude of the local magnetization. The magnetization distribution $B_{\|}$ was projected along the incident beam direction for thicknesses $L_z = 2$ and $L_z = 4$ with the easy axis oriented 45° relative to the film surface. For thin samples (Fig. 15a), there are fewer domain walls, and the magnetic moments mostly lie in the plane owing to the strong demagnetization field caused by the shape anisotropy. In contrast, in thick samples (Fig. 15b), we see a vortexlike magnetic configuration with most of the magnetization parallel to the surface normal. This is because in a thin sample, the excess magnetostatic energy caused by stray fields dominate the domain wall energy. The vortex structure minimizes the magnetostatic energy at the expense of excess domain wall energy owing to the increase of the domain wall area. On the other hand, the calculated magnetization distributions for samples with their easy axis parallel to the surface normal show a vortexlike magnetic configuration in a thin sample but not in a thick one. It is evident that because the easy axis has a substantial component parallel to the sample surface, when the *c*-axis is tilted away from the surface normal (Fig. 15), it facilitates the conversion of the out-of-plane $B_{\perp}$ to in-plane $B_{\|}$. As sample thickness increases, the effects of the demagnetization field on the sample decreases; thus, the magnetization perpendicular to the film surfaces, $B_{\perp}$, increases.

8.5.4 Effects of Anisotropy

Figure 16 shows the competition effect between magnetostatic and anisotropy energy in magnetization distribution using different values of the constant A_{eff} inversely proportional to anisotropy coefficient K. If the relative contribution of anisotropy energy is much stronger than the

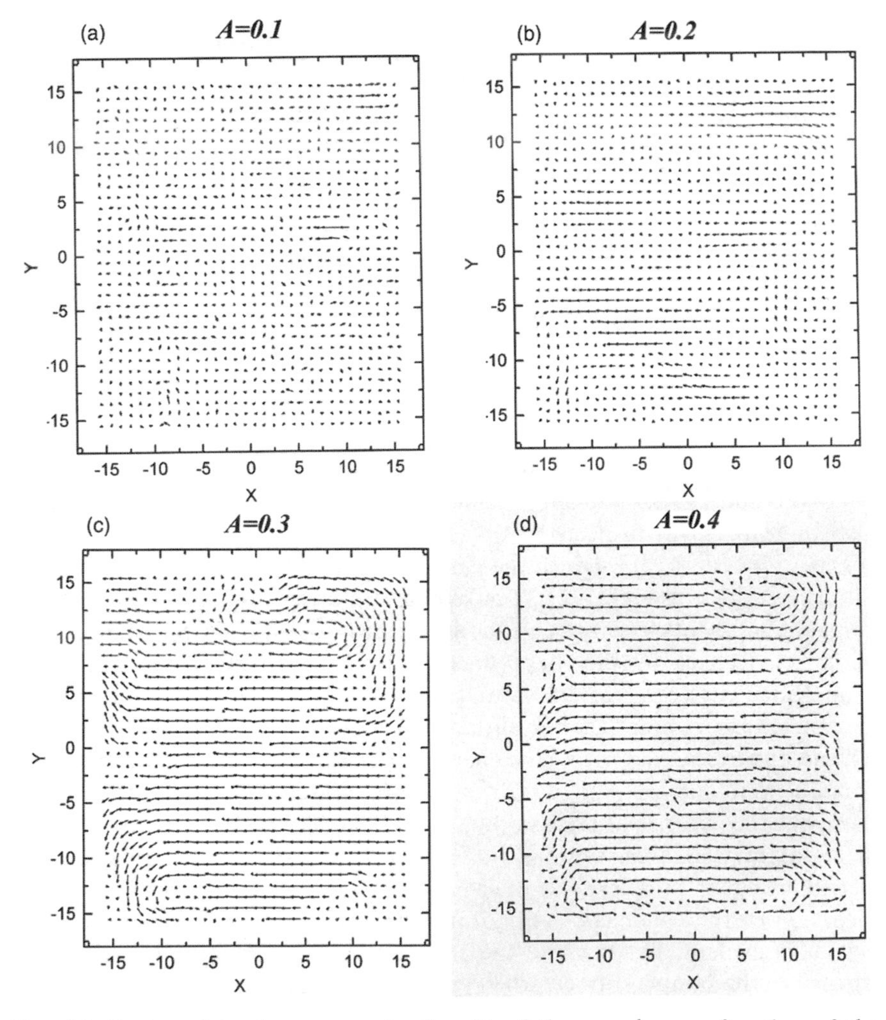

FIG. 16. Projected in-plane magnetization $B_{\parallel}$ of the sample as a function of the anisotropy constant A_{eff} associated with the dipole–dipole interaction energy E_{dip}, obtained via Monte Carlo simulation. The easy axis of the sample is parallel to the film surface normal. (a) $A_{eff} = 0.1$, (b) 0.2, (c) 0.3, and (d) 0.4.

magnetostatic one (hard magnet approximation), the effective constant A_{eff} is small enough, let say, $A_{eff} = 0.1$. In this case, almost all the $\mathbf{B}_{\|}$ run perpendicular to the sample surface with negligible in-plane components (Fig. 16a). For $A_{eff} = 0.2$ (Fig. 16b), about 10% of the magnetization is in-plane, in either the positive or negative x-direction. A remarkable difference was observed when A_{eff} reaches 0.3 (Fig. 16c). About 80% of **B** reoriented from lying perpendicular to the sample surface to lying parallel to it. Domain walls are visible near the center, while the out-of-plane magnetization or vortex configuration is mainly distributed near the edge or at the corner of the sample. This type of magnetization distribution is similar to closure domain structures observed in several soft magnets, for which the anisotropy energy is small enough and, hence, A_{eff} is expected to be high. No significant reorientation was observed when the constant A_{eff} changed from 0.3 to 0.4 (Fig. 16d). Only an additional ~10% magnetization converted from out-of-plane to in-plane, mainly eliminating domain boundaries from the center and out-of-plane component at the edge of the sample. Our simulation suggests that the effects of stray fields strongly depend on the sample thickness, crystal orientation, as well as material parameters, such as the anisotropy coefficient. A stray field can significantly alter the local magnetization configuration of a sample.

8.6 *In Situ* Experiments

The dynamic behavior of domain structure, such as orientation dependence and interaction with defects, can be studied *in situ*. For magnetic imaging, samples are normally examined in a microscope when the main objective lens is either switched off or slightly exited. The latter case is used for *in situ* magnetizing experiments. Because the direction of the field, parallel to the incident beam, is fixed, the magnitude of the in-plane magnetizing field, which contributes to magnetic imaging of a sample with respect to the crystal orientation, can be changed continuously by tilting the sample back and forth to the maximum permissible angles. Demagnetization of the sample can be realized by either reversing the objective-lens current or by heating the sample above the Currie temperature in a heating stage. In our *in situ* experiments, JEM 2000FX and JEM 3000F microscopes were used. The magnetic field in a sample area (B_0) versus objective-lens potential (U) for the latter instrument was carefully calibrated using a $Nd_2Fe_{14}B$ test sample measured with a superconducting quantum interference device (SQUID) magnetometer following a procedure described by Volkov *et al.* [352], as well as by direct Hall probe measurements with a Bell 610 gaussmeter [353]. As plotted in Fig. 17, the magnetic fields in

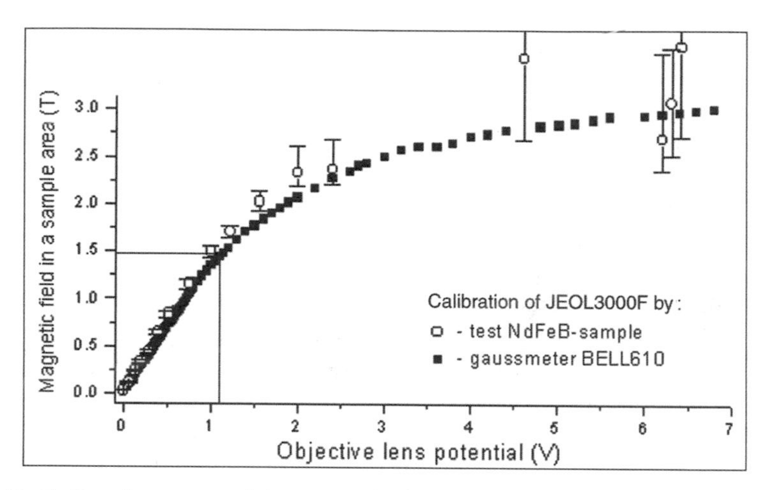

FIG. 17. Calibration curve of the magnetic field B_0 in the specimen area of the Jeol 3000 microscope as a function of the potential (U) of the objective lens. The open circles represent the measurements using a SQUID magnetometer with a Nd–Fe–B test sample, whereas the black squares correspond to the direct Hall sensor measurements.

sample area of the JEM 3000FEG microscope can be varied from 0.02 to 3 T.

8.6.1 Domain Reorientation under Thermal Cycles

The experimentally observed domain configuration in an anisotropic $Nd_2Fe_{14}B$ hard magnet strongly depends on the viewing direction with respect to the easy axis (c-axis) of the sample. Figure 9 serves as a good example, showing a maze pattern when viewed near the c-axis (Fig. 9a) but a strip pattern when viewed along a direction nearly perpendicular to the c-axis (Fig. 9b, c). In both cases, the major components of magnetization direction are antiparallel to the c-axis to form 180° domain walls. The domain reorientation and orientational dependence can be continually monitored during magnetization and demagnetization using (001) and (100) oriented tetragonal single crystals.

Figure 18 is a set of domain images of the same area recorded on video at different stages of a thermal cycle. The crystal was originally magnetized with an applied field of $\sim$2 T parallel to its surface normal and the [001] easy axis, and then the field was removed. The domain configuration was mainly isotropic along the [001] direction (Fig. 18a), similar to that in Fig. 9a. The crystal was then tilted $\sim$27° about the [110] axis with its surface

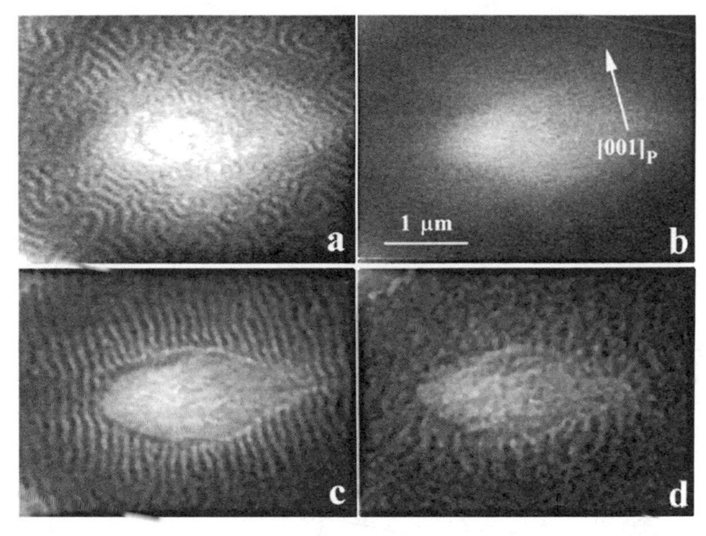

FIG. 18. A series of images of the same area captured from a videotape recorded during an *in situ* experiment: (a) original configuration at room temperature with the easy axis nearly parallel to the viewing direction; (b) at 350°C; (c) cooled back to room temperature; and (d) after a field of ~2 T was applied and then removed. The sample orientations in (b–d) are the same with the viewing direction near [1$\bar{1}$2], about 27° away from (a). The projected [001] direction is indicated in (b).

normal near the [1$\bar{1}$2] axis to project a large [001] component in the viewing plane. No change in domain configuration was observed. When the sample was demagnetized at above T_c (Fig. 18b), the domain structure vanished completely. When the sample was subsequently cooled back to room temperature, a new configuration of domains was formed with their walls well aligned along the projected [001] direction (Fig. 18c), similar to those in Fig. 9b, c), owing to the residual field along the [1$\bar{1}$2] axis that gives rise to an in-plane field $[001]_p$. *In situ* observations revealed that these new domains might be nucleated at the surface or the edge of the single crystal. When the ~2 T external field was applied again while the crystal remained tilted (i.e., with the applied field along the [1$\bar{1}$2] rather than the original [001] axis), the alignment was destroyed immediately. Only a small portion of the alignment was restored when the large field was removed. Lorentz images showed that the majority of the parallel domain walls were broken apart and rearranged themselves with the magnetization direction close to the field direction. With an increase of the magnetic field, the domains became much coarser (Fig. 18d), owing to alternate domain walls moving in opposite directions and being annihilated under the applied field. The origin of the motion of the domain wall is the competing force inside the domains between the demagnetizing field and the applied field. The magnetization directions in the sample are dominated by the anisotropy, which resists the rotation of magnetization caused by an external field.

In situ observations also showed that the domain configuration in thin areas was most susceptible to alteration, and it often was reversible in the residual field ($\sim$200 G) but irreversible in a large field ($>$1 T). Domain nucleation and growth often exhibited unpredictable, irregular behavior under the residual field when the samples were magnetized and demagnetized through thermal cycles. The movement of the domain walls is often delayed by imperfection in the crystal so that they are not free to return to their original positions. Such behaviors are attributed to inhomogeneities in samples, including local structural defects and even impurity atoms and internal stress that cause fluctuations in the domain wall energy, since after a thermal cycle, defects and internal stress may be redistributed. For example, the center of the image in Fig. 18a is a perforated hole that is amorphous at its edge. During heating, this nonmagnetic amorphous region grew aggressively from the edge of the hole, as demonstrated by the retreat of the newly formed domain walls from the edge after two thermal cycles (Fig. 18c) although the hole size remained the same.

8.6.2 Pinning Centers

As mentioned in the introduction (Section 8.1), an important issue in studying RE–Fe–B hard magnets is understanding whether the nucleation or pinning of the structural defects determines the coercivity. Dynamic magnetizing experiments combined with Lorentz imaging can help to identify the nature of the defects.

Figure 19 shows two Foucault images taken in the remanent state (applied field $\sim$2 T) of $Nd_2Fe_{14}B$ from the same area, magnetized along the easy (Fig. 19a) and hard (Fig. 19b) magnetic directions. The in-plane magnetizing component of the field was proportional to the projection of the tilt angle, which is given in both images. The drastic change in local remanence with the direction of the applied field is clear, and it can be quantified by measuring the domain ratio of opposite magnetization [black (D_u) to white (D_d) regions, as described earlier by Eq. (1)]. For the major in-plane component applied along the hard axis (Fig. 19b), the domain ratio D_u/D_d is close to unity, suggesting that the $B_{r\perp}$ component of the remanence $\mathbf{B}_r$ and magnetic hysteresis must be small along the hard magnetic plane in the material. In contrast, the well-aligned grain region (area A in Fig. 19a) remains well saturated when the field was applied along the easy axes and then removed, while the defect layer (area B in Fig. 19a) is far from such

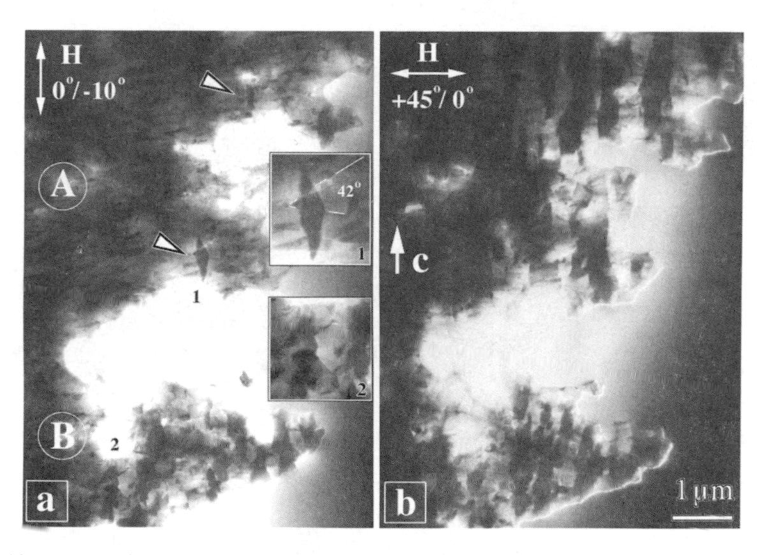

FIG. 19. Foucault images of domain structure for the same area in die-upset Nd–Fe–B under remanent state after *in situ* magnetization saturation with the in-plane component of an external field applied along the easy (a) and hard (b) magnetization axis by properly tilting the specimen. The tilt angles in the x- and y-direction are also included. The area A, composed of well-aligned grains, remains magnetically saturated in remanence (a), whereas area B (defect layer) remains unsaturated. The insets are enlarged areas showing reversal domains. The white lines in inset 1 indicate the large-angle grain boundaries.

saturation. Hence, only aligned grains directly contribute to the high remanence of the die-upset magnet ($B_{r\|} \approx B_s$ in area A where B_s refers to the saturation remanence); in the nonaligned grains (area B in Fig. 19) the associated domains essentially cancel each other out in the absence of a magnetic field. This suggests that a higher local remanence of the die-upset magnet can be reached only along the die-upset direction, even with a small magnetic field applied.

A few very small reversal domains, marked by arrowheads, were observed in area A (see inset 1, Fig. 19a). Their presence is controlled by local imperfections of the crystal, such as inclusions or misalignments of grains. In the latter case, the reversal domains, whose shape is sensitive to the angle of grain misalignment, grow slowly under the increasing negative field until they nucleate a c-axis-aligned reversal domain at the nearest grain boundary. Then, the magnetization reversal quickly becomes an irreversible process. Such domains can often easily expand throughout the area of well-aligned grains in a negative field exceeding a critical value, as clearly

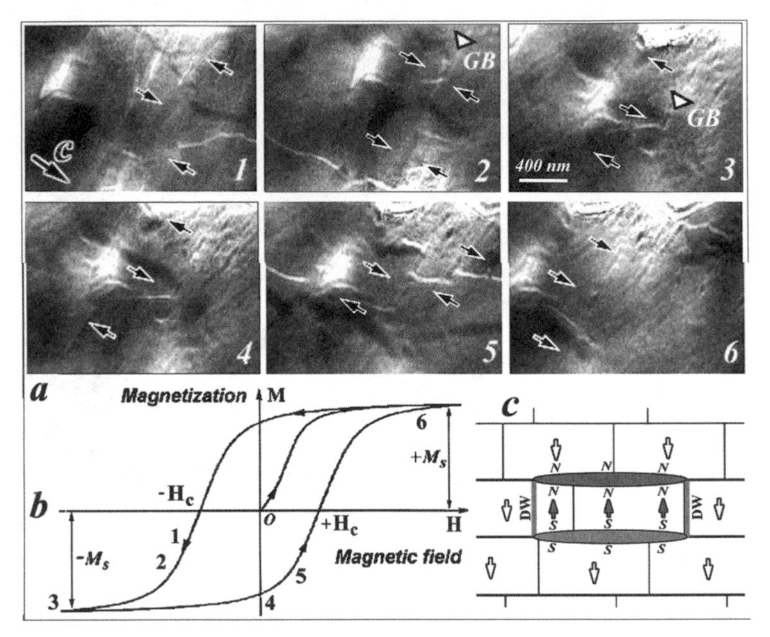

FIG. 20. Successive Fresnel images recorded *in situ* (1–6). (a) Evolution of local domain structure in a die-upset magnet at different points of the magnetization loop [(b), where the number corresponds to the number of the image frames]. Evidence of strong pinning centers that did not vanish under a strong negative field ($\mu_0 H \approx -1\,T$) is seen in image 3. (c) A simple structural model of the pinning center, showing how the domain of opposite magnetization is trapped by the low-energy defect formed by the grains magnetically decoupled from the outer area by a couple of thick nonmagnetic boundary layers and domain walls (DW).

demonstrated in Fig. 20 by a series of successive images captured *in situ.* Once the reversal domain is nucleated, only the pinning of the domain walls at grain boundaries (Fig. 20) controls its further expansion. Thus, the leading mechanism in controlling the magnetization reversal seems to be the nucleation of reversal domains, preferentially at misaligned grain interfaces or sample surfaces where the demagnetization field is the largest.

Frame 3 in Fig. 20a shows an example of a strong "double-boundary" pinning center, where the domain trapped by the boundaries remains positively magnetized under the extreme negative applied field $H < -H_c$ (Fig. 20b). This center was found to be responsible for the nucleation of positive domains approximately at $H \geqslant +H_c$ for the fourth quadrant of hysteresis loop (frames 4–6 of Fig. 20a), starting from negative magnetization. Figure 20c is a schematic structural model of such a double-boundary pinning center. In this model, a single-domain grain with magnetization opposite to that of the matrix is uncoupled along the easy magnetic direction owing to a thick nonmagnetic layer or pocket phase at the boundaries. The presence of a nonmagnetic phase at the boundaries is the necessary condition for the pinning. This is consistent with the observations that the higher the saturation field, the higher is the coercive field can be reached. A stronger magnetizing field can gradually remove the nuclei of reversed domains existing at such grain boundaries.

8.6.3 Dipole Domains

So far, we have discussed mainly the die-upset hard magnets which feature small grains. Our *in situ* experiments suggest that remagnetization processes in sintered Nd–Fe–B magnets, characterized by large grains (3–6 μm), essentially differs from those described earlier (Section 8.6.2) for die-upset samples, especially when the applied field decreases slowly from magnetic saturation. In this case, a new mechanism of magnetization reversal starts to develop [354], as shown in Fig. 21a. Magnetic imaging reveals how narrow so-called dipole domains of negative magnetization start to propagate and grow in the interiors of the grains when the applied field decreases slowly from a positive saturation. This mechanism contributes to the reversible part of the total remagnetization process. We note that the grains with *c*-axes tilted away from the applied field direction are more likely to form reversal dipole domains. When the applied field decreases, the dipole domains tend to align more closely to the *c*-axis of the grains because this

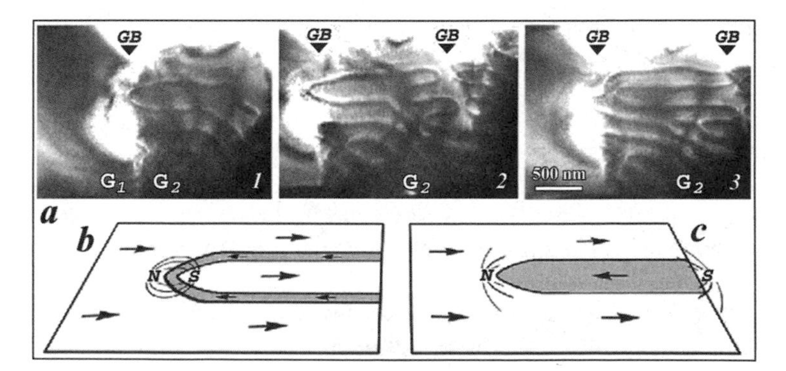

FIG. 21. (a) A set of Fresnel images, captured from video, showing the expansion of dipolelike domains in sintered Nd–Fe–B under a decreasing magnetic field. A simple model of such a dipole domain (b) is compared with the ordinary magnetic domain (c). Note that the dipole domain requires lower magnetostatic energy change.

reduces the energy of the domain walls. Figure 21b is a schematic drawing of the dipole domain motion. These domains can be considered as a pair of ordinary reversal domains, enclosed one into another. In this case, the total area of a dipole domain, sandwiched between two parallel domain walls and magnetized with an opposite sign to that of the matrix, remains very narrow. Such a dipole domain may only slightly change the local magnetization. On the other hand, it facilitates an easy and almost reversible motion for the coupled domain walls, since their movement in a dipole domain configuration does not require a big change of magnetostatic (or dipole) energy by comparison with traditional single-domain motion (Fig. 21c). At the tips of the reverse dipole domain, the magnetic charges, or the N/S poles, are magnetostatically coupled and remain coupled during their movement (Fig. 21b) until they reach a large-angle grain boundary.

8.6.4 Domain Nucleation near Grain Boundaries

When the applied field decreases further and becomes negative, the slow dipole domain motion begins to be replaced with another faster and irreversible demagnetization process in a sintered magnet. This process takes place usually at the grain boundaries, where the nucleation of reversal (negative) domains occurs via several sudden splittings of positive domains [354]. The nucleation transforms a "positive" domain into a positive–negative–positive domain configuration, enclosing a pair of newly generated 180° domain walls. This nucleation process is essentially irreversible, and it may be considered as a cascadelike discharge of oversaturation of magnetic charge (or poles) at the boundary. The formation of reversal (or negative) domains through this process reduces the magnetostatic energy of each boundary by self-compensating the newly created N/S poles at the grain boundaries. The frames in Fig. 22a show the consecutive images of the nucleation process. Figure 22b shows the enlarged images of the same domain area for frames 4–6 recorded with a rate of 24 frames per second. (Here the black and white arrows denote domains with opposite magnetization, and GB indicates a grain boundary position.) Finally, the most active phase of the remagnetization process in sintered magnets is completed very fast by simple lateral expansion of many newly created domain walls within each big grain, as shown in Fig. 22.

It is interesting to note that in the case of curved grain boundaries with a large interfacial misorientation, high-field nucleated domains, which we call "GB domains," prefer to propagate along the grain boundary (GB). Figure 23a shows an example where the two adjacent grains across the boundary have a misorientation of the *c*-axes about 84° and are viewed in

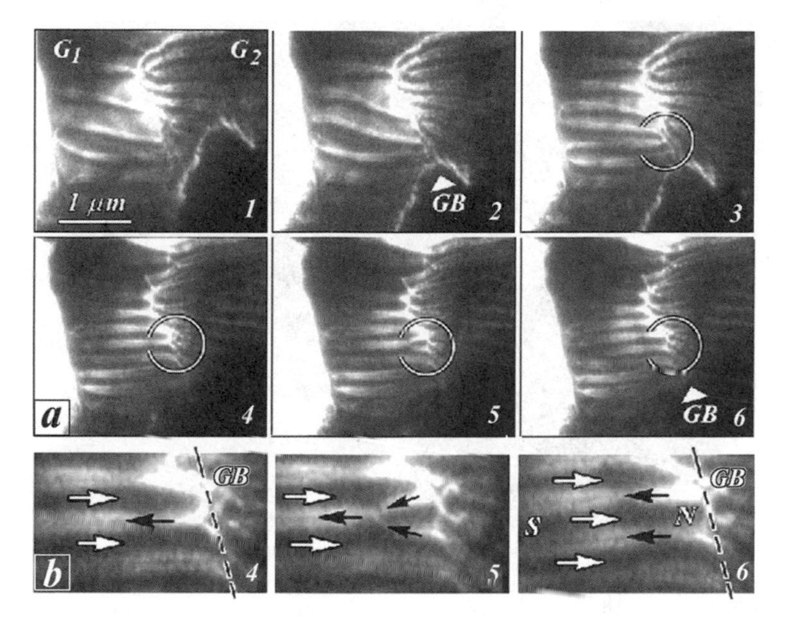

FIG. 22. (a) Successive Fresnel images captured from video showing *in situ* irreversible remagnetization in a sintered Nd–Fe–B magnet under a decreasing magnetic field. The process takes place by the nucleation of new domains at the grain boundary (GB), marked with dashed line in (b) via splitting of single domains of the same magnetization toward the grain interior, as shown in the enlarged images (b).

a such way that their projected c-axes, marked as c_{proj} in Fig. 23a, appear to be parallel in the image plane. Therefore, the in-plane components of the magnetization of the two grains (G_1, G_2) cause opposite magnetic contrast in the Foucault image in Fig. 23a. The local magnetization direction of this nucleated "GB domain" is determined by the boundary plane, and it does not follow any easy magnetic axis of the grains separated by the boundary. Thus, the formation of the GB domain reduces the local boundary anisotropy. The width of the GB domain is field sensitive and usually does not exceed 100 nm. The presence of a GB domain eases magnetization reversal through the grain boundary at low fields (Fig. 23b). Such domains may reduce the density of interfacial magnetic charges through the nucleation and expansion of the magnetic domain at the boundary at lower magnetic field. Thus, this process is somewhat similar to the nucleation of Néel spike domains that occurs near the holes in some other ferromagnetic materials.

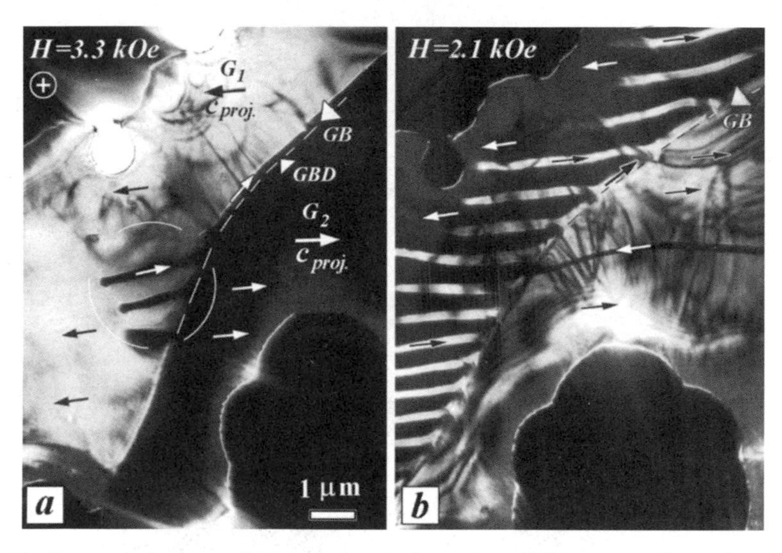

FIG. 23. Foucault images of the same grain boundary (GB) area under a decreasing magnetic field. (a) Nucleation and (b) expansion of reverse magnetic domains in grain G_1 occur via the formation of a specific grain boundary domain (GBD), which lowers the local magnetic anisotropy along the GB line (a). Here, the GB is traced by the dashed line, and the local in-plane direction of magnetization for magnetic domains in (a) and (b) is shown by the fine arrows. The magnetization directions are opposite for grains G_1 and G_2, as evident from the black/white contrast of these domains. Diffraction analysis confirmed that grains G_1 and G_2 have a parallel in-plane c-axis component (c_{proj}) and enclose a misorientation angle of about 84°.

8.7 Structure and Properties Correlation

So far, we have mostly discussed the TEM observations on the microstructure and magnetic structure of die-upset hard magnets containing the $RE_2Fe_{14}B$ hard phase. Now, we briefly summarize these observations and describe how they may be related to microstructure-sensitive parameters of the permanent magnets, such as their remanence ($I_r = I_{H=0}$) and coercivity field ($H_c = H_{I=0}$).

8.7.1 Remanence

As we discussed in the previous sections, grain alignment is of primary importance for the high remanence, I_r, of a hard anisotropic magnet. All high-energy product commercial RE–Fe–B magnets feature the grain texture, no matter which fabrication process was involved (i.e., sintering, die-upset, or other methods of synthesis). For the die-upset RE–Fe–B, the materials consist of platelet-shaped grains stacked along the press (die-upset) direction [312, 313], resulting in a strong magnetic anisotropy of magnetic properties. The typical aspect ratio of the grains is about 1/4 to 1/6. The thickness of the grains ranges from ~0.07 to 0.15 μm, with an average value of ~0.1 μm, which is less than the single-domain grain size $D_c = 1.4\gamma_w/M_s^2$ (where γ_w is the domain wall energy and $4\pi M_s = 16.1$ kOe is the saturated magnetization) estimated as 0.2 μm [326] or 0.3 μm [303] for $Nd_2Fe_{14}B$ magnets. Under such conditions ($D < D_c$), magnetic domains are expected to run across several grains (Fig. 7) if they are coupled with ferromagnetic exchange interactions.

Let us explore the relation of remanence for such an anisotropic magnet of easy c-axis type to the grain texture of a polycrystalline magnet. In a first approximation, we presume that the remanence may be defined by the average of the random distribution $M(\theta)$ of grain magnetic moments within a certain angle range $0 \leqslant \theta \leqslant \theta_0$ around the common easy magnetic texture direction. Here, θ_0 defines the maximal cone angle of the grain texture around the common easy direction of the "uniaxial" magnet. Each grain will contribute to the remanence, with $M(\theta) = M_s \cos\theta$. By neglecting the contribution arising from the exchange coupling of grains, the relative remanence m_r can be derived by averaging as follows:

$$\frac{I_r}{I_s} = m_r = \frac{2\pi \displaystyle\int_0^{\theta_0} \cos\theta \sin\theta \, d\theta}{2\pi \displaystyle\int_0^{\theta_0} \sin\theta \, d\theta} = \frac{1}{2}(1 + \cos\theta_0). \tag{11}$$

Here a random azimuthal distribution of single-grain moments was assumed. Therefore, the number of atomic moments in the unit cone layer about the press direction within the angle between θ and $\theta + d\theta$ must be proportional to $2\pi \sin\theta \, d\theta$. This formula provides a general description of the remanence as function of grain texture misorientation angle θ_0 for any anisotropic polycrystalline magnet. Indeed, assuming completely random grain orientations ($0° \leqslant \theta \leqslant 90°$), the remanence m_r in Eq. (11) equals 0.5, in exact accordance with the Stoner–Wohlfarth model [142] for isotropic magnets. In contrast, for a perfectly aligned grain texture, defined by $\theta_0 = 0°$, we get a trivial solution $m_r = 1$ for a fully anisotropic magnet, for which $I_r = I_s$. Real anisotropic magnets always occupy some intermediate position.

The less trivial solution of Eq. (11) can be derived when a complex microstructure or the real microstructure of a permanent magnet is taken into account [328]. For example, let us make some estimates for the 50%

die-upset anisotropic magnet, discussed earlier (see Fig. 8). The experimental remanence of this magnet was found to be $m_r = 0.84$ (relative) and $I_r = 4\pi M_r = 12.9$ kG (absolute). The grain microstructure of this magnet (Fig. 8) can be approximated with a quasi-periodic grain stucture along the easy magnetic axis, consisting of alternating well-aligned layers (A) and not well-aligned "defect" layers (B). Hence, for this type of microstructure (A layers with $0 \leqslant \theta \leqslant \theta_a \leqslant 90°$ of relative thickness $x_a = 1 - x_b$, and B layers with $0 \leqslant \theta \leqslant \theta_b \approx 90°$ of relative thickness x_b, where the subscripts a and b denote the A and B layers, respectively), Eq. (11) can be rewritten as

$$\frac{I_r}{I_s} = m_r = \frac{1}{2}(1 + \cos\theta_a)(1 - x_b) + \frac{1}{2}(1 + \cos\theta_b)x_b$$

$$\approx \frac{1}{2}(1 + \cos\theta_a)(1 - x_b) + \frac{1}{2}x_b. \qquad (12)$$

Our TEM observations revealed that the ratio of the thickness of the B to A layers in the 50% die-upset magnet manufactured by General Motors was within (0.5–1.0) μm/(4.0–6.0) μm = 0.083–0.25. In other words, the maximal relative thickness of the B layer has to be about $x_b = 0.25/(1+0.25) = 0.2$. By substituting $x_b = 0.2$ and $m_r = 0.84$ into Eq. (12), we get

$$\frac{I_r}{I_s} = m_r = 0.84 = \frac{1}{2}[(1 + \cos\theta_a)(1 - 0.2) + 0.2], \qquad (13)$$

from which an estimate of the expected texture angle for the A layers gives $\theta_a = 31.8°$. This value of the texture angle is consistent with our TEM observations and fits very well to the "surprisingly large" texture angle $\theta = 32.3°$ reported [355] by direct X-ray texture measurement [FWHM of the rocking curve of the (006) reflection] for a 50% die-upset $Nd_2Fe_{14}B$ magnet. Figure 24 is a schematic explanation of the magnetic microstructure in terms of a "quasi-periodic brick wall" model (Fig. 24a) to show the specific role of defect B layers both for the remanence and for the magnetization reversal process (successive steps I–II–III in Fig. 24a, b) of textured die-upset magnets.

8.7.2 Coercivity Field

Micromagnetic theory predicts that the magnetization reversal of a single-domain grain by the rotation of magnetic moment will require a minimum energy with a nucleation field $H_N^{min} \approx \frac{1}{2}H_a = K_1/I_s$ [43, 142, 356] for a grain tilted at 45° with respect to an applied magnetic field. For the $Nd_2Fe_{14}B$ phase, this amount is equivalent to $H_N^{min} = \frac{66.3}{2}$ kOe = 33.2 kOe [356], which is still much higher than the experimental results. Such

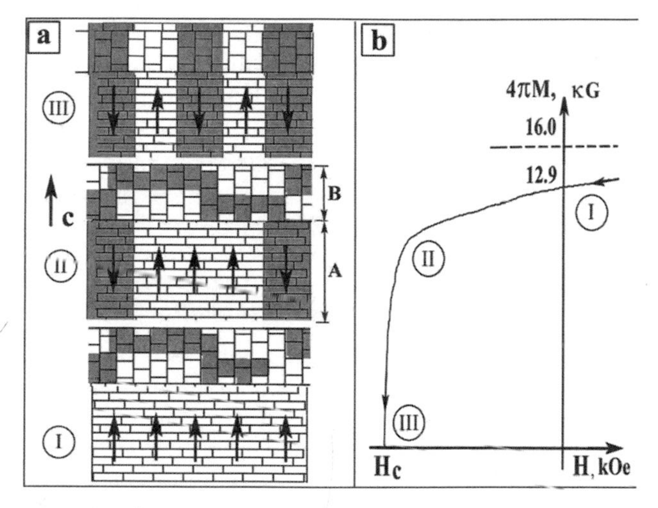

FIG. 24. (a) Schematic drawing of the proposed structural model for the well-aligned (A layer) and nonaligned (B layer) regions used to explain the role of defect layers in magnetization reversal of die-upset magnets. I, II, and III denote successive steps in magnetization reversal along a demagnetizing curve shown in (b). The thick bricks represent the defect layer, while the thin bricks represent the well-aligned grain platelets.

nonaligned 45° grains were indeed observed by TEM within the defect B layers of a die-upset magnet, which also contained a high fraction of nonmagnetic phases. Hence, such defect layers should generate strong local demagnetizing fields owing to the magnetostatic (dipole) energy of the free magnetic poles [43]. To estimate the coercivity field we assume that the nonmagnetic defects in these layers are similar to platelet cavities (such as those shown in Fig. 20c). Then, a good approximation for the demagnetizing factor is $N_{\text{eff}} = 1$ [43]. The external field H_{ext} (equal to the coercivity field H_c) needed to reverse the magnetization of grains 45° tilted will be defined as

$$H_{\text{ext}} \approx H_c = H_N^{\min} - N_{\text{eff}} \frac{I_s}{\mu_0}. \tag{14}$$

The numerical estimate from Eq. (14) in cgs units gives $H_c = (33.2-16.1)$ kOe = 17.1 kOe, which is close to the experimental value of the coercive field $H_c =$ 18.9 kOe found in the best hard die-upset Nd–Fe–B samples. Therefore, we may assume that the high-coercivity mechanism occurs via delayed nucleation of reverse domains near the 45°-tilted grains of defect B layers that exhibit a magnetization reversal owing to the rotation of the local magnetic moment under a reversal magnetic field [328]. A similar nucleation process within the A layers seems less likely, because of the smaller texture angle $\theta_a = 31.8° < 45°$ and, hence, the higher activation critical field H_N within the A layer. Any further expansion of the reversal domains along the easy texture axis is limited by the nearest defect layer until the next delayed nucleation of reversal domains occurs, as schematically shown by steps I, II, and III in Fig. 24a, b.

8.7.3 Correlation of Die-Upset and Sintered Magnets

Two major classes of permanent magnets with very high-energy products are available in the market: the sintered and the die-upset Nd–Fe–B compounds. Their different synthesis processes, namely, the powder metallurgy and melt-spun quenching techniques (see Section 8.1), result in their having quite distinct microstructures. In particular, there is a big difference in the shape and sizes of the grains of the Nd–Fe–B hard magnetic phase, and there is a big variation in the characteristics of the distribution of the major nonmagnetic Nd-rich secondary phase and other nonmagnetic precipitates in the composition-optimized hard magnets. Furthermore, grain boundary chemical analysis suggests that a small excess of Fe ($\sim 8\%$) is present in the major part of the grain boundaries in die-upset samples (Section 8.3), in contrast to the rare earth-rich grain boundary phase (presumably composed of Nd_7Fe_3) in sintered magnets. Despite the differences in microstructure, the magnetic properties of the sintered and die-upset magnets are surprisingly similar, including those of high remanence, moderate or high coercivity, and the similar shape of hysteresis loops. Because remanence and coercivity are known to be very sensitive to microstructure in anisotropic magnets, different microstructures are expected to generate quite different magnetic properties. To fully understand the correlation between structure and properties of the RE–Fe–B magnets, we must able to explain the paradox of the similar magnetic parameters yet very different microstructures of these two types of magnets.

We assume there are some "effective" microstructural parameters, common to both types of magnets and responsible for their similar properties. By comparing the characteristics of the microstructures of the two types, two structural models can be proposed, as shown in Fig. 25a and Fig. 25b for die-upset and sintered magnets, respectively, depicted under the same scaling factor. There are several interesting findings. First, for the die-upset magnets, we found that the "large interacting domains," consisting of many small grains coupled by ferromagnetic exchange interactions, are only present within well-aligned A layers. Hence, the small excess of Fe found at the majority of grain boundaries (see Section 8.3) can help to couple such

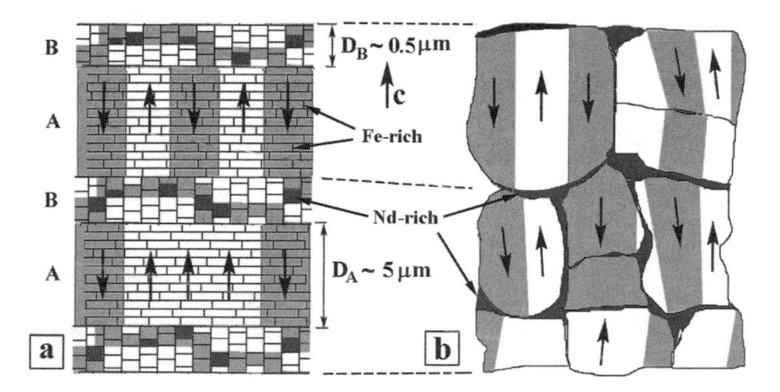

FIG. 25. Schematic of the microstructural models for die-upset (a) and sintered (b) permanent magnets. The well-aligned region (A layers) and "defect" region (B layers; the thickness of the B layer is exaggerated to show the microstructural detail) are highlighted in (a). Here, the nonmagnetic Nd-rich phase is concentrated mostly within B layers as a pocket phase in (a), and as the intergranular phase in (b). The Fe-rich grain boundaries are believed to concentrate within the A layers. The distinct microstructures presented here for the die-upset and sintered RE–Fe–B magnets are likely to be responsible for their similar magnetic properties.

small grains into so-called magnetic clusters [354] of 5–7 μm. In sintered magnets, on the other hand, similar clusters can comprise only a few big grains of the same lateral size (Fig. 25b). Second, these clusters appear to be magnetically decoupled along the easy magnetic direction by nonmagnetic spacer layers, such as the Nd-rich grain boundary phase observed in the sintered magnets (Fig. 25b), which is mirrored in the Nd_7Fe_3 nonmagnetic precipitates in the defect layers in die-upset magnets. Hence, despite the difference in the microstructures of the two types of magnets, there are similar common elements in terms of "magnetic clusters" and "nonmagnetic spacer" layers. The former provide a high remanence, whereas the latter act as a pinning barriers and, therefore, are responsible for higher coercivity.

Acknowledgments

The authors would like to thank their collaborators M. R. McCartney, Z.-X. Cai, L. H. Lewis, J. Tafto, J.-Y. Wang, D. O. Welch, D. Crew, M. De Graef, R. K. Mishra, and C. D. Fuerst for contributions and useful discussions. The work at Brookhaven National Laboratory was supported by the Division of Materials Sciences, U.S. Department of Energy, under Contract No. DE-AC02-98CH10886.

REFERENCES

1. G. Prinz, and K. Hathaway, *Phys. Today*, **48**, 24, (1995)
2. W. F. Brown, Jr., *Micromagnetics*, Wiley (Interscience), New York, (1963)
3. W. F. Brown, Jr., *Magnetostatic Principles in Ferromagnetism*, North-Holland, Amsterdam, (1962)
4. E. H. Frei, S. Shtrickman, and D. Treves, *Phys. Rev.* **106**, 446 (1957)
5. S. Shtrikman and D. Treves, *Magnetism*, Vol. 3, Chap. 8, Academic Press, New York, (1963)
6. J.-G. Zhu and H. N. Bertram, *J. Appl. Phys.* **63**, 3248 (1988)
7. M. Mansuripur and R. Giles, *IEEE Trans. Magn.* **MAG-24**, 2326 (1988)
8. J. L. Blue and M. R. Scheinfein, *IEEE Trans. Magn.* **MAG-27**, 4778 (1991)
9. R. D. McMichael and M. J. Donahu, *IEEE Trans. Magn.* **33**, 4167 (1997)
10. T. Schrefl, *J. Magn. Magn. Mater.*, **207**, 45 (1999)
11. J.-G. Zhu, Y. Zheng, and G. A. Prinz, *J. Appl. Phys.*, in press (2000)
12. J.-G. Zhu, *Interactive in Magnetic Thin Films*, Ph.D. Thesis, University of California at San Diego, 1989
13. N. H. Bertram and J.-G. Zhu, in *Solid State Phys.* (H. Ehrenreich and D. Turnbull, eds.). Vol. 46, p. 271, 1992
14. Y. Zheng and J.-G. Zhu, *J. Appl. Phys.* **85**, 4776 (1999)
15. T. Schrefl, J. Fidler, K. J. Kirk, . and J. N. Chapman, *J. Appl. Phys.*, **85**, 6169 (1999)
16. T. Fang and J.-G. Zhu, *IEEE Trans. Magn.* **36**, 2000
17. Y. Guo and J.-G. Zhu, *IEEE Trans. Magn.* **28**, 2919 (1992)
18. T. Chang, M. Lagerquist, J.-G. Zhu, J. Judy, P. Fisher, and S. Chou, *IEEE Trans. Magn.* **28**, 3139 (1992)
19. R. Madabhushi, R. D. Gomez, E. R. Burke, and I. D. Mayergoyz, *IEEE Trans. Magn.* **32**, 4147 (1996)
20. D. K. Cheng, *Fundamentals of Engineering Electromagnetics*, Addison-Wesley, Reading, Massachusetts, 1994
21. J. N. Chapman and M. R. Scheinfein, *J. Magn. Magn. Mater.* **200**, 729–740 (1999)
22. K. J. Kirk, J. N. Chapman, and C. D. W. Wilkinson, *J. Appl. Phys.*, **85**, 5237 (1999)
23. J. Gadbois and J.-G. Zhu, *IEEE Trans. Magn.* **31**, 3802 (1995)

24. Y. Zheng, *Micromagnetic Study of Magnetic Random Access Memory*, Ph.D. Thesis, Carnegie Mellon University, 1999
25. R. J. Celotta, J. Unguris, M. H. Kelley, and D. T. Pierce, *Methods in Materials Research: A Current Protocols Publication*, (E. N. Kaufman, ed.). Wiley, New York, 2000.
26. T. Chen, *IEEE Trans. Magn.* **MAG-17**, 1181 (1981)
27. J. G. Zhu and N. H. Bertram, *IEEE Trans. Magn.* **MAG-24**, 2706 (1988)
28. T. Chen, *IEEE Trans. Magn.* **MAG-24**, (1988)
29. T. Yogi, C. Tsang, T. A. Nguye, K. Ju, G. L. Gorman, and G. Gastillo, *IEEE Trans. Magn.* **MAG-26**, 2271 (1990)
30. K. E. Johnson, P. R. Ivett, D. R. Timmons, M. Mirzamaani, S. E. Lambert, and T. Yogi, *J. Appl. Phys.*, **67**, 4686 (1990)
31. J.-G. Zhu, *IEEE Trans. Magn.*, **MAG-29**, 195 (1993)
32. T. Min and J.-G. Zhu, *J. Appl. Phys.*, **75**, 6129 (1994)
33. M. Mirzamaani, C. V. Jahanes, and M. A. Russak, *J. Appl. Phys.*, **69**, 5169 (1991)
34. M. Futamoto, M. Suzuki, N. Inaba, A. Nakamura, and Y. Honda, *IEEE Trans. Magn.*, **MAG-30**, 3975 (1994)
35. J. Ding and J.-G. Zhu, *IEEE Trans. Magn.* **MAG-30**, 3978 (1994)
36. X.-G. Ye and J.-G. Zhu, *IEEE Trans. Magn.* **MAG-28**, 3087 (1992)
37. E. D. Dahlberg and J.-G. Zhu, *Physics Today* **48**, 34 (1995)
38. T. J. Silva and T. M. Crawford, *IEEE Trans. Magn.* **35**, 671 (1999)
39. C. E. Patton, Z. Frait, and C. H. Wilts, *J. Appl. Phys.* **46**, 5002 (1975)
40. D. C. Jiles, *J. Phys. D*, **27**, 1–11 (1994)
41. K. Ullakko, J. K. Huang, C. Kantner, R. C. O'Handley, and V. V. Kokorin, *Appl. Phys. Lett.* **69**, 1966–1968 (1996)
42. M. Q. Huang, Y. Zheng, and W. E. Wallace, *J. Appl. Phys.* **75**, 6280–6282 (1994)
43. S. Chikazumi, *Physics of Magnetism.* Wiley, New York, 1964.
44. J. Dooley and M. De Graef, *Ultramicroscopy* **67**, 113–132 (1997)
45. T. Suzuki, C. H. Wilts, and C. E. Patton, *J. Appl. Phys.* **39**, 1983–986 (1968).
46. T. Suzuki and C. H. Wilts, *J. Appl. Phys.* **40**, 1216–1217 (1969).
47. D. Wohlleben, in *Electron Microscopy in Material Science* (U. Valdrè, ed.), pp. 713–757. Academic Press, New York, 1971.
48. H. Gong and J. N. Chapman, *J. Magn. Magn. Mater.*, **67**, 4–8 (1987)
49. J. N. Chapman, E. M. Waddell, P. E. Batson, and R. P. Ferrier, *Ultramicroscopy* **4**, 283–292 (1979)
50. Y. Aharonov, and D. Bohm, *Phys. Rev.* **115**, 485–491 (1959)
51. S. Olariu and S. I. Popescu, *Rev. Mod. Phys.* **57**, 339–436, 1985

52. A. Tonomura, N. Okasabe, T. Matsuda, T. Kawasaki, J. Endo, S. Yano, and H. Yamada, *Phys. Rev. Lett.* **56**, 792–795 (1986)
53. L. Reimer, *Transmission Electron Microscopy*, Springer-Verlag, Berlin, 1993
54. J. D. Jackson, *Classical Electrodynamics*, 2nd Ed. Wiley, New York, 1975
55. J. C. H. Spence, *Experimental High Resolution Electron Microscopy*, 2nd Ed. Oxford Univ. Press, New York, 1988
56. E. J. Kirkland, *Advanced Computing in Electron Microscopy*, Plenum, New York, 1998
57. C. Salling, S. Schultz, I. McFadyen, and M. Ozaki, *IEEE Trans. Magn.* **27**, 5184–5186 (1991)
58. C. Salling, R. O'Barr, S. Schultz, I. McFadyen, and M. Ozaki, *J. Appl. Phys.* **75**, 7989–7992 (1994).
59. M. De Graef, N. T. Nuhfer, and M. R. McCartney, *J. Microsc.* **194**, 84–94 (1999)
60. G. Y. Fan and J. M. Cowley, *Ultramicroscopy* **21**, 125–130 (1987)
61. M. Mansuripur, *J. Appl. Phys.* **69**, 2455–2464 (1991)
62. R. James and D. Kinderlehrer, *Philos. Mag. B* **68**, 237–274 (1993).
63. J. N. Chapman, P. E. Batson, E. M. Waddell, and R. P. Ferrier, *Ultramicroscopy*, **3**, 203–214 (1978).
64. J. N. Chapman, *Mater. Sci. Eng. B* **3**, 355–358 (1989).
65. A. Daykin, and A. Petford-Long, *Ultramicroscopy* **58**, 365–380 (1995)
66. J. Dooley and M. De Graef, *Micron* **28**, 371–380 (1997)
67. T. E. Gureyev, A. Roberts, and K. A. Nugent, Phase retrieval with the transport of intensity equation: Matrix solution with use of zernike polynomials. *J. Opt. Soc. Am. A* **12**, 1932–1941 (1995).
68. T. E. Gureyev, A. Roberts, and K. A. Nugent, Partially coherent fields, the transport-of-intensity equation, and phase uniqueness. *J. Opt. Soc. Am. A* **12**, 1942–1946 (1995).
69. T. E. Gureyev and K. A. Nugent, Phase retrieval with the transport of intensity equation II: Orthogonal series solution for non-uniform illumination. *J. Opt. Soc. Am. A* **13**, 1670–1683 (1996).
70. D. Paganin and K. A. Nugent, Non-interferometric phase imaging using partially coherent light. *Phys. Rev. Lett.* **80**, 2586–2589 (1998).
71. M. R. Teague, Deterministic phase retrieval: A Green's function solution. *J. Opt. Soc. Am.* **73**, 1434–1441 (1983).
72. M. De Graef, *Advanced Hard and Soft Magnetic Materials*, (M. Coey, L. H. Lewis, B.-M. Ma, T. Schrefl, L. Schultz, J. Fidler, V. G. Harris, R. Hasegawa, A. Inoue, and M. E. McHenry, eds.), *Mater. Res. Soc. Symp. Proc.* **577**, 519–530 (1999)
73. P. Grutter, E. Meyer, H. Heinzelmann, *et al.*, *J. Vacuum Sci. Technol.* **A6**, 279–282 (1988)
74. D. Rugar, H. J. Mamin, P. Guethner, *et al.*, *J. Appl. Phys.* **68**, 1169 (1990)

75. U. Hartmann, T. Goddenmhenrich, and C. Heiden, , *J. Magn. Magn. Mater.* **101**, 263–270 (1991)

76. Y. Martin, and H. K. Wickramasinghe, *Appl. Phys. Lett.* **50**, 1455–1457 (1987)

77. P. C. D. Hobbs, D. W. Abraham, and H. K. Wickramasinghe, *Appl. Phys. Lett.* **55**, 2357–2359 (1989)

78. P. Grutter, T. Jung, and H. Heinzelmann, *et al.*, *J. Appl. Phys.* **67**, 1437–1441 (1990)

79. Digital Instruments, *Interleave Mode Implementation.*

80. E. Meyer, and H. Heinzelmann, Scanned force microscopy. In *Scanning Tunneling Microscopy II*, (R. Wiesendanger, and H.-J. Guntherodt, eds.) Vol. 28, pp. 99–149. Springer-Verlag, Hamburg, 1992.

81. J. N. Israelachvili, *Intermolecular and Surface Forces*, Academic Press, New York, 1985.

82. G. Meyer, and N. M. Amer, *Appl. Phys. Lett.* **53**, 1045–1047 (1988)

83. A. J. den Boef, *Appl. Phys. Lett.* **56**, 2045–2047 (1990)

84. R. Allenspach, H. Salemink, and A. Bischof, *et al.*, *Z. Phys. B—Condensed Matter* **67**, 125–128 (1987)

85. J. Moreland, and P. Rice, *J. Appl. Phys.* **70**, 520 (1991)

86. R. D. Gomez, A. A. Adly, I. D. Mayergoyz, *et al.*, *IEEE Trans. Magn.* **29**, 2494–2498 (1993)

87. J. J. Saenz, N. Garcia, and J. C. Slonczewski, *Appl. Phys. Lett.* **53**, 1449–1451 (1988)

88. G. Binnig, C. F. Quate, and C. Gerber, *Phys. Rev. Lett.* **56**, 930–933 (1986)

89. Y. Martin, C. C. Williams, and H. K. Wickramasinghe, *J. Appl. Phys.* **61**, 4723–4725 (1987)

90. R. Erlandsson, G. M. McClelland, C. M. Mate, *et al.*, *J. Vacuum Sci. Technol. A* **6**, 266–270 (1988)

91. D. Rugar, H. J. Mamin, R. Erlandsson, *et al.*, *Rev. Sci. Instrum.* **59**, 2337–2340 (1988).

92. C. Schonenberger and S. F. Alvarado, *Rev. Sci. Instrum.* **60**, 3131–3134 (1989).

93. T. R. Albrecht, P. Grutter, D. Horne, *et al.*, *J. Appl. Phys.* **69**, 668–673 (1991).

94. Y. Martin, C. C. Williams, and H. K. Wickramasinghe, *J. Appl. Phys.* **61**, 4723–4729 (1987).

95. A. DiCarlo, M. R. Scheinfein, and R. V. Chamberlin, *Appl. Phys. Lett.* **61**, 2108–2110 (1992).

96. P. Grutter, H. J. Mamin, and D. Rugar, Magnetic force microscopy. In *Scanning Tunneling Microscopy II* (R. Wiesendanger and H.-J. Gunterodt, eds.) Vol. 28, pp. 151–207. Springer-Verlag, Berlin, 1992.

97. C. Schonenberger and S. F. Alvarado, *Z. Phys. B.* **80**, 373–378 (1990).

98. C. Schonenberger and S. F. Alvarado, *Z. Phys. B Condensed Matter* **80**, 373–383 (1990).

99. U. Hartmann, *Phys. Lett. A* **137**, 475–478 (1989).
100. H. J Mamin, D. Rugar, J. E. Stern, *et al.*, *Appl. Phys. Lett.* **53**, 1563–1565 (1988).
101. E. R. Burke, R. D. Gomez, and I. D. Mayergoyz, *J. Appl. Phys.* **75**, 5759–5761 (1993).
102. U. Hartmann, *J. Vacuum Sci. Technol. A* **8**, 411–415 (1990).
103. A. Wadas and P. Grutter, *Phys. Rev. B* **39**, 12013–12017 (1989).
104. P. Grutter, D. Rugar, H. J. Mamin, *et al.*, *J. Appl. Phys.* **69**, 5883–5885 (1991).
105. P. Bryant, S. Schultz, and D. R. Fredkin, *J. Appl. Phys.* **69**, 5877–5879 (1991).
106. D. Streblechenko, M. R. Scheinfein, M. Mankos, *et al.*, *IEEE Trans. Magn.* **32**, 7954–7956 (1996).
107. B. G. Frost, N. F. van Hlulst, E. Lunedei, *et al.*, *Appl. Phys. Lett.* **68**, 1865–1867 (1996).
108. K. Sueoka, F. Sai, K. Parker, *et al. J. Vacuum Sci. Technol. B* **12**, 1618–1622 (1994).
109. M Ruhrig, S. Porthun, J. C. Lodder, *et al.*, *J. Appl. Phys.* **79**, (1996).
110. P. B. Fischer, M. S. Wei, and S. Y. Chou, *J. Vacuum Sci. Technol. B* **11**, 2570 (1993).
111. G. D. Skidmore and E. D. Dahlberg, *Appl. Phys. Lett.* **71**, 3293–3295 (1997).
112. H. J. Mamin, D. Rugar, J. E. Stern, *et al.*, *Appl. Phys. Lett.* **55**, 318–320 (1989).
113. T. Goddenhenrich, M. Anders, U. Hartmann, *et al.*, *J. Microsc.* **152**, 527 (1988).
114. R. D. Gomez, M. C. Shih, R. M. H. New, *et al.*, *J. Appl. Phys.* **80**, 342–346 (1996).
115. P. Grutter, Y. Liu, P. LeBlanc, *et al.*, *Appl. Phys. Lett.* **71**, 279–281 (1997).
116. S. L. Tomlinson and E. W. Hill, *J. Magn. Magn. Mater.* **161**, 385–396 (1996).
116a. Mayergoyz *et a.*.
116b. Hug *et al.*
117. T. Goddenheinrich, H. Lemke, M. Muck, *et al.*, *Appl. Phys. Lett.* **57**, 2612–2614 (1990).
118. K. Babcock, V. Elings, M. Dugas, *et al.*, *IEEE Trans. Magn.* **30**, 4503–4505 (1994).
119. L. Kong and S. Chou, *J. Appl. Phys.* **81**, 5026–5028 (1997).
120. K. Babcock, V. Elings, J. Shi, *et al.*, *Phys. Rev. Lett.* **69**, 705 (1996).
121. S. Huo, J. E. Bishop, J. W. Tucker, *et al.*, *IEEE Trans. Magn.* **33**, 4056 (1997).
122. G. P. Heydon, A. N. Farley, S. R. Hoon, *et al.*, *IEEE Trans. Magn.* **33**, 4059–4061 (1997).
123. R. Proksch, G. Skidmore, E. D. Dahlberg, *et al.*, *Appl. Phys. Lett.* **69**, 2599–2601 (1996).
124. R. D. Gomez, A. O. Pak, A. J. Anderson, *et al.*, *J. Appl. Phys.* **83**, 6226–6228 (1998).
125. A. N. Campbell, E. I. Cole, Jr., B. Dodd, *et al.*, *IEEE/IRPS*, 168–177 (1993).

126. R. S. Eliot, *Electromagnetics*. McGraw-Hill, New York, 1966.

127. G. A. Gibson, J. F. Smyth, and S. Schultz, *IEEE Trans. Magn.* **27**, 5187–5189 (1991).

128. R. D. Gomez, E. R. Burke, and I. D. Mayergoyz, *J. Appl. Phys.* **78**, 6441–6446 (1996).

129. R. D. Gomez, I. D. Mayergoyz, and E. R. Burke, *IEEE Trans. Magn.* **31**, 3346–3348 (1995).

130. H. N. Bertram, *Theory of Magnetic Recording*. Cambridge Univ. Press, Cambridge, 1994.

131. J. F. Smyth, S. Schultz, D. R. Fredkin, *et al. J. Appl. Phys.* **69**, 5262–5266 (1991).

132. J.-G. Zhu, Y.-F. Zheng, and X.-D. Lin, *J. Appl. Phys.* **81**, 4336–4341 (1997).

133. T. Schrefl, J. Fidler, K. J. Kirk, *et al., J. Magn. Magn. Mater.* **175**, 193–204 (1997).

134. D. R. Fredkin and T. R. Koehler, *J. Appl. Phys.* **67**, 5544 (1990).

135. D. R. Fredkin, T. R. Koehler, J. F. Smyth, *et al., J. dAppl. Phys.* **69**, 5272–5274 (1991).

136. S. J. Hefferman, J. N. Chapman, and S. McVitie, *J. Magn. Magn. Mater.* **95**, 76–84 (1991).

137. R. D. Gomez, T. V. Luu, A. O. Pak, *et al. J. Appl. Phys.* **85**, 6163–6165 (1999).

138. B. Khamsehpour, C. D. W. Wilkinson, J. N. Chapman, *et al., J. Vacuum Sci. Technol. B* **14**, 3361–3366 (1996).

139. K. Runge, Y Nozaki, Y. Otani, *et al., J. Appl. Phys.* **79**, 5075–5077 (1996).

140. M. Donahue, The object oriented micromagnetic framework (oommf) project at itl/nist. NIST, 1999.

141. T. V. Luu, *Domain Characteristics of Small Permalloy Elements Observed Using Magnetic Force Microscopy*. Master's Thesis, University of Maryland, 1999.

142. E. C. Stoner and E. P. Wohlfarth, *Philos. Trans. R. Soc. London, A* **240**, 599–642 (1948).

143. H. Koo, R. D. Gomez, and V. Metlushko, *J. Appl. Phys.* **87**, in press (2000).

144. R. L. White, *Data Storage* **4**, 58 (1997).

145. S. Y. Chou, M. S. Wei, P. R. Krauss, *et al., J. Appl. Phys.* **76**, 6673–6675 (1994).

146. R. M. H. New, R. F. W. Pease, and R. L. White, *J. Vacuum Sci. Technol. B* **13**, 1089–1094 (1995).

147. R. M. H. New, R. F. W. Pease, and R. L. White, *IEEE Trans. Magn.* **31**, 3805–3807 (1995).

148. S. Ganesan, R. L. White, H. C. Koo, *et al.*, (1900).

149. J. P. Jakubovics, In *Electron Microscopy, Part IV* (U. Valdre and E. Ruedl, eds.), p. 1303. Commission of the European Communities, Brussels, 1976.

150. J. N. Chapman, The investigation of magnetic domain structures in thin foils by electron microscopy. *J. Phys. D: Appl. Phys.* **D17**, 623–647 (1984).

151. H. Lichte, in *Advances in Optical and Electron Microscopy* (T. Mulvey and C. J. R. Sheppard, eds.), Vol. 12, pp. 25–91. Academic Press, New York, 1991.

152. A. Tonomura, in "Electron Holography", Vol. 70 of *Springer Series in Optical Sciences*. Springer-Verlag, Heidelberg, 1993.

153. J. M. Cowley, *Ultramicroscopy* **41**, 335–348 (1992).

154. G. Möllenstedt and H. Düker, *Naturwissenschaften*, **42**, 41 (1955).

155. M. E. Haine and T. Mulvey, *J. Opt. Soc. Am.* **42**, 763 (1952).

156. A. Tonomura, T. Matsuda, and J. Endo, *Jpn. J. Appl. Phys.* **18**, 1373 (1979).

157. A. Tonomura, "Applications of Electron Holography." *Rev. Mod. Phys.* **59**, 639–669 (1987).

158. W. J. de Ruijter, *Micron* **26**, 247 (1995).

159. E. Völkl and M. Lehmann, Chap. 6, pp. 125–151. 1999.

160. W. J. de Ruijter and J. K. Weiss, *Ultramicroscopy* **50**, 269–283 (1993).

161. D. J. Smith and M. R. McCartney, in *Introduction to Electron Holography* (E. Völkl, L. F. Allard, and D. C. Joy, eds.), Vol. 00, Chap. 4, pp. 87–106. Kluwer Academic, New York, 1999.

162. D. Gabor, *Proc. R. Soc. London* **A**197, 454 (1949).

163. A. Tonomura, T. Matsuda, J. Endo, H. Todokoro, and T. Komoda, *J. Electron Microsc.* **28**, 1–11 (1979).

164. H Lichte, *Ultramicroscopy* **20**, 293 (1986).

165. A. Orchowski, W. D. Rau, and H. Lichte, *Phys. Rev. Lett.* **74**, 399–402 (1995).

166. W. D. Rau and H. Lichte, In *Introduction to Electron Holography* (E. Völkl, L. F. Allard, and D. C. Joy, eds.), Vol. 00, Chap. 9, pp. 201–229, Kluwer Academic, New York, 1999.

167. S. Frabboni, G. Matteucci, G. Pozzi, and M. Vanzi, *Phys. Rev. Lett.* **55**, 2196–2199 (1985).

168. B. G. Frost, L. F. Allard, E. Völkl, and D. C Joy, In *Electron Holography*, (A. Tonomura, L. F. Allard, G. Pozzi, D. C. Joy, and Y. A. Ono, eds.), pp. 169–179. Elsevier, Amsterdam, 1995.

169. M. R. McCartney and M. M. Gajdardziska-Josifovska, *Ultramicroscopy* **53**, 283–298 (1994).

170. W. D. Rau, P. Schwander, F. H. Baumann, W. Hoppner, and A. Ourmazd, Two-dimensional mapping of the electrostatic potential in transistors by electron holography. *Phys. Rev. Lett.* **82**, 2614–2618 (1999).

171. V. Ravikumar, R. P. Rodrigues, and V. P. Dravid, *Phys. Rev. Lett.* **75**, 4063–4066 (1995).

172. N. Osakabe, K. Yoshida, Y. Horiuchi, T. Matsuda, H. Tanabe, T. Okuwaki, J. Endo, H. Fujiwara, and A. Tonomura, *Appl. Phys. Lett.* **42**, 792–794 (1983).

173. J. Bonevich, K. Harada, T. Matsuda, H. Kasai, T. Yoshida, G. Pozzi, and A. Tonomura, *Phys. Rev. Lett.* **70**, 2952–2955 (1993).

174. J. W. Cowley, M. Mankos, and M. R. Scheinfein, Greatly defocused, point-projection, off-axis electron holography. *Ultramicroscopy* **63**, 133–147 (1996).

175. J. N. Chapman, R. P. Ferrier, L. J. Heyderman, S. McVitie, W. A. P. Nicholson, and B. Bormans, *Inst. Phys. Conf. Ser.* **138**, 1–8 (1993).

176. M. Mankos, A. A. Higgs, M. R. Scheinfein, and J. M. Cowley, *Ultramicroscopy* **58**, 87 (1995).

177. M. R. McCartney, D. Smith, R. Farrow, and R. Marks, Off-axis electron holography of epitaxial FePt films. *J. Appl. Phys.* **82**, 2461–2465 (1997).

178. T. Hirayama, J. Chen, Q. Ru, K. Ishizuka, T. Tanji, and A. Tonomura, *J. Electron Microsc.* **43**, 190–197 (1994).

179. M. R. McCartney, P. Kruit, A. H. Buist, and M. R. Scheinfein, *Ultramicroscopy* **65**, 179 (1996).

180. J. M. Zuo, M. R. McCartney, and J. C. H. Spence, *Ultramicroscopy* **66**, 35–47 (1997).

181. M. Gajdardziska-Josifovska, M. R. McCartney, W. J. de Ruijter, D. J. Smith, J. K. Weiss, and J. M. Zuo, *Ultramicroscopy* **50**, 285–299 (1993).

182. L. Reimer, *Transmission Electron Microscopy*. Springer-Verlag, Berlin, 1989.

183. H. Lichte, H. Banzhof, and R. Huhle, in *Electron Microscopy 98* (H. A. Calderon Benavides and M. C. Yacaman, eds.), Vol. 1, pp. 559–560. IOP, Bristol, 1998.

184. M. R. McCartney and Y. Zhu, *Appl. Phys. Lett.* **72**, 1380–1382 (1998).

185. H. Kronmuller, in *Science and Technology of Nanostructured Materials* (G. C. Hadjipanayis and G. Prinz, eds.), p. 657. Plenum, New York, 1990.

186. J. F. Herbst and J. J. Croat, *J. Magn. Magn. Mater.* **100**, 57 (1991).

187. R. K. Mishra and R. W. Lee, *Appl. Phys. Lett.* **48**, 733–735 (1986).

188. P. A. Crozier, *Philos. Mag. B* **61**, 311–336 (1990).

189. B. O. Cullity, *Introduction to Magnetic Materials*. Addison-Wesley, New York, 1972.

190. D. G. Streblechenko, Ph.D. Thesis, Arizona State University, Tempe, 1999.

191. G. Matteucci, G. Missiroli, E. Nichelatti, A. Migliori, M. Vanzi, and G. Pozzi, *J. Appl. Phys.* **69**, 1853–1842 (1991).

192. G. Lai, T. Hirayama, A. Fukuhara, K. Ishizuka, T. Tanji, and A. Tonomura, *J Appl. Phys.* **75**, 4593–4598 (1994).

193. T. Hirayama, J. Chen, T. Tanji, and A. Tonomura, *Ultramicroscopy* **54**, 9–14 (1994).

194. R. E. Dunin-Borkowski, M. R. McCartney, R. B. Frankel, D. A. Bazylinski, M. Posfai, and P. R. Buseck, *Science* **282**, 1868–1870 (1998).

195. M. Mankos, J. M. Cowley, and M. R. Scheinfein, *Phys. Stat. Sol. a* **154**, 469–504 (1996).

196. R. Dunin-Borkowski, M. R. McCartney, D. J. Smith, and S. Parkin, Towards electron holography of magnetic thin films using in-situ magnetisation reversal. *Ultramicroscopy* **74**, 61–73 (1998).

197. R. E. Dunin-Borkowski, M. R. McCartney, B. Kardynal, and D. J. Smith, *J. Appl. Phys.* **84**, 374–378 (1998).

198. R. E. Dunin-Borkowski, M. R. McCartney, B. Kardynal, D. J. Smith, and M. R. Scheinfein, *Appl. Phys. Lett.* **75**, 2641–2643 (1999).

199. D. J. Smith, R. E. Dunin-Borkowski, M. R. McCartney, B. Kardynal, and M. R. Scheinfein, *J. Appl. Phys.* in press, 2000.

200. M. R. McCartney, R. E. Dunin-Borkowski, R. R. Scheinfein, D. J Smith, S. Gider, and S. S. P. Parkin, *Science* **286**, 1337–1340 (1999).

201. M. Mankos, Z. J. Yang, M. R. Scheinfein, and J. M. Cowley, *IEEE Trans. Magn.* **30**, 4497–4499 (1994).

202. J. Bonevich, G. Pozzi, and A. Tonomura, in *Introduction to Electron Holography* (E. Völkl, E. F. Allard, and D. C. Joy, eds.), pp. 153–181. Kluwer Academic, New York, 1999.

203. J. Chen, T. Hirayama, G. Lai, T. Tanji, K. Ishizuka, and A Tonomura, *Opt. Rev.* **2**, 304–307 (1994).

204. E. Madelung, Quantentheorie in hydrodynamischer form, *Z. Phys.* **40**, 322–326 (1926).

205. E. C. Kemble, *The Fundamental Principles of Quantum Mechanics with Elementary Applications.* Dover Publ., New York, 1937.

206. M. R. Teague, Irradiance moments: Their propagation and use for the unique retrieval of phases. *J. Opt. Soc. Am.* **72**, 1199–1209 (1982).

207. F. Roddier, Curvature sensing and compensation: A new concept in adaptive optics. *Appl. Opt.* **27**, 1223–1225 (1988).

208. F. Roddier, Wavefront sensing and the irradiance-transport equation. *Appl. Opt.* **29**, 1402–1403 (1990).

209. C. Roddier and F. Roddier, Wave-front reconstruction from defocused images and the testing of ground-based optical telescopes. *J. Opt. Soc. Am. A* **10**, 2277–2287 (1993).

210. T. E. Gureyev and K. A. Nugent, Rapid quantitative phase imaging using the transport of intensity equation. *Opt. Commun.* **133**, 339–346 (1997).

211. K. A. Nugent, T. E. Gureyev, D. Cookson, D. Paganin, and Z. Barnea, Quantitative phase imaging using hard x-rays. *Phys. Rev. Lett.* **77**, 2961–2964 (1996).

212. S. Bajt, A. Barty, K. A. Nugent, M. R. McCartney, M. Wall, and D. Paganin, Quantitative phase-sensitive imaging in a transmission electron microscope. *Ultramicroscopy* **83**, 67–73 (2000).

213. E. Schroedinger, Quantisierung als eigenwertproblem I, *Ann. Phys.* **79**, 361–376 (1926).

214. A. Messiah, *Quantum Mechanics*, Vol. 1. North-Holland, Amsterdam, 1961.

215. E. Freenberg, *The Scattering of Slow Electrons in Neutral Atoms*. Ph.D. Thesis, Harvard University, 1933.

216. I. Bialynicki-Birula, M. Cieplak, and J. Kaminski, *Theory of Quanta*. Oxford Univ. Press, New York, 1992.

217. Morse and H. P. M. Feshbach, *Methods of Theoretical Physics*, Part 1. McGraw-Hill, New York, 1953.

218. A. V. Efimov, Y. G. Zolotarev, and V. M. Terpigoreva, *Mathematical Analysis (Advanced Topics), Volume 2. Applications of Some Methods of Mathematical and Functional Analysis*. Mir Publ., Moscow, 1985.

219. P. A. M. Dirac, Quantised singularities in the electromagnetic field, *Proc. R. Soc. London A* **133**, 60–72 (1931).

220. K. A. Nugent and D. Paganin, Matter-wave phase measurement: A non-interferometric approach. *Phys. Rev. A*. in press (2000).

221. F. Zernike, Phase contrast, a new method for the microscopic observation of transparent objects, *Physica* **9**, 686–693 (1942).

222. H. W. Fuller and M. E. Hale, *J. Appl. Phys.* **31**, 238–248 (1960).

223. P. N. T. Unwin, *Philos. Trans. R. Soc. London, B* **261**, 95 (1971).

224. A. Barty, K. A. Nugent, D. Paganin, and A. Roberts, Quantitative optical phase microscopy. *Opt. Lett.* **23**, 187–819 (1998).

225. G. Missiroli, G. Pozzi, and U. Valdre, Electron interferometry and interference electron microscopy. *J. Phys. E: Sci. Instrum.* **14**, 649–671 (1981).

226. D. Paganin, *Studies in Phase Retrieval*. Ph.D. Thesis, University of Melbourne, 1999.

227. J. B. Tiller, A. Barty, D. Paganin, and K. A. Nugent, The holographic twin image problem: A deterministic phase solution. submitted (2000).

228. S. Wischnitzer, *Introduction to Electron Microscopy*, 3rd ed. Maxwell McMillan, New York, 1989.

229. D. L. Misell, The phase problem in electron microscopy. In *Advances in Optical and Electron Microscopy* (V. E. Cosslett and R. Barer, eds.), pp. 185–279. Academic Press, (1978).

230. D. Van Dyck and W. Coene, A new procedure for wave function restoration in high resolution electron microscopy, *Optik* **7**, 125–128 (1987).

231. J. Kessler, *Polarized Electrons*, 2nd Ed. Springer-Verlag, Berlin, 1985.

232. R. Feder, *Polarized Electrons in Surface Physics*. World Scientific, Singapore, 1985.

233. R. J. Celotta and D. T. Pierce, *Science* **234**, 333–340 (1985).

234. J. Kirschner, *Polarized Electrons at Surfaces*. Springer-Verlag, Berlin, 1985.

235. K. Koike and K. Hayakawa, *Jpn. J. Appl. Phys.* **23**, L187–L188 (1984).

236. J. Unguris, G. Hembree, R. J. Celotta, and D. T. Pierce, *J. Microsc.* **139**, RP1–RP2 (1985).

237. H. P. Oepen and J. Kirschner, *Scanning Microsc.* **5**, 1–16 (1991).

238. J. Unguris, M. R. Scheinfein, R. J. Celotta, nd D. T. Pierce, in *Chemistry and Physics of Solid Surfaces VIII*, (R. Vanselow and R. Howe, eds.), pp. 239–262. Springer-Verlag, Berlin, 1990.

239. D. T. Pierce, J. Unguris, and R. J. Celotta, *MRS Bull.* **XIII-6**, 19–23 (1988).

240. M. R. Scheinfein, J. Unguris, R. J. Celotta, and D. T. Pierce, *Phys. Rev. Lett.* **63**, 668–671 (1990).

241. R. Allenspach, *Spin-Polarizing Scanning Electron Microscopy*, IBM Research Report rz3151.ps, 1999.

242. R. J. Celotta, J. Unguris, M. H. Kelley, and D. T. Pierce, *Methods in Materials Research: A Current Protocols Publication*. Wiley, New York, in press.

243. E. D. Dahlberg and R. Proksch, *J. Magn. Magn. Mater.* **200**, 720–728 (1999).

244. E. Kisker, W. Gudat, and K. Schröder, *Solid State Commun.* **44**, 591–595 (1982).

245. H. Hopster, R. Raue, E. Kisker, G. Guntherodt, and M. Campagna, *Phys. Rev. Lett.* **50**, 70–73 (1983)

246. D. R. Penn, S. P. Apell, and S. M. Girvin, *Phys. Rev. Lett* **55**, 518–521 (1985).

247. D. R. Penn, S. P. Apell, and S. M. Girvin, *Phys. Rev. B* **32**, 7753–7768 (1985).

248. J. Unguris, R. J. Celotta, and D. T. Pierce, *Phys. Rev. Lett.* **69**, 1125–1128 (1992).

249. J. Unguris, R. J. Celotta, and D. T. Pierce, *J. Magn. Magn. Mater.* **127**, 205–213 (1993).

250. H. Matsuyama and K. Koike, *J. Electron Microsc. Jpn.* **43**, 157–163 (1994).

251. M. R. Scheinfein, *Optik* **82**, 99–113 (1989).

252. T. Kohashi, H. Matsuyama, and K. Koike, *Rev. Sci. Instrum.* **66**, 5537–5543 (1995).

253. J. Unguris, D. T. Pierce, and R. J. Celotta, *Rev. Sci. Instrum.* **57**, 1314–1323 (1984).

254. D. T. Pierce, M. H. Kelley, R. J. Celotta, and J. Unguris, *Nucl. Instrum. Meth. A*, **266**, 550–559 (1988).

255. J. Barnes, L. Mei, B. M. Lairson, and F. B. Dunning, *Rev. Sci. Instrum.* **70**, 246–247 (1999).

256. M. R. Scheinfein, D. T. Pierce, J. Unguris, J. J. McClelland, and R. J. Celotta, *Rev. Sci. Instrum.* **60**, 1–11 (1989).

257. H. P. Oepen and J. Kirschner, *Phys. Rev. Lett.* **62**, 819–822 (1989).

258. M. R. Scheinfein, J. Unguris, M. H. Kelley, D. T. Pierce, and R. J. Celotta, *Rev. Sci. Instrum.* **61**, 2501–2526 (1990).

259. M. R. Scheinfein, J. Unguris, J. L. Blue, K. J. Coakley, D. T. Pierce, R. J. Celotta, and P. J. Ryan, *Phys. Rev. B* **43**, 3395–3422 (1991).

260. A. Gavrin and J. Unguris, *J. Magn. Magn. Mater.* **213**, 95–100 (2000).
261. K. Koike and K. Hayakawa, *J. Appl. Phys.* **57**, 4244–4248 (1985).
262. J. Unguris, M. R. Scheinfein, D. T. Pierce, and R. J. Celotta, *Appl. Phys. Lett.* **55**, 2553–2555 (1989).
263. W. J. Tseng, K. Koike, and J. C. M. Li, *J. Mater. Res.* **8**, 775–784 (1993).
264. M. R. Khan, S. Y Lee, J. L. Pressesky, D. Williams, S. L. Duan, R. D. Fisher, N. Heiman, M. R. Scheinfein, J. Unguris, D. T. Pierce, R. J. Celotta, and D. E. Speliotis, *IEEE Trans. Magn.* **26**, 2715–2717 (1990).
265. H. Matsuyama, K. Koike, F. Tomiyama, Y. Shiroshi, A. Ishikawa, and H. Aoi, *IEEE Trans. Magn.* **30**, 1327–1330 (1994).
266. M. Aeschlimann, M. R. Scheinfein, J. Unguris, F. J. A. M. Greidanus, and S. Klahn, *J. Appl. Phys.* **68**, 4710–4718 (1990).
267. T. Kohashi, H. Matsuyama, Y. Murakami, Y. Tanaka, and H. Awano, *Appl. Phys. Lett.* **72**, 124–126 (1998).
268. R. D. Gomez, Chapter 3 in this book.
269. P. Rice, S. E. Russek, J. Hoinville, and M. H. Kelley, *IEEE Trans. Magn.* **33**, 4065–4067 (1997).
270. W. J. M. de Jonge, P. J. H. Blomen, and F. J. A. den Broeder, in *Ultrathin Magnetic Structures I* (J. A. C. Bland and B. Heinrich, eds.), pp. 65–86. Springer-Verlag, Berlin, 1994.
271. H. P. Oepen, M. Speckman, Y. Millev, and J. Kirschner, *Phys. Rev. B* **55**, 2752–2755 (1997).
272. R. Allenspach and A. Bischof, *Phys. Rev. Lett.* **69**, 3385–3388 (1992).
273. D. P. Pappas, C. R. Brundle, and H. Hopster, *Phys. Rev. B* **45**, 8169–8172 (1992).
274. S. S. P. Parkin, in *Ultrathin Magnetic Structures II* (B. Heinrich and J. A. C. Bland, eds.), pp. 148–185. Springer-Verlag, Berlin, 1994.
275. G. A. Prinz, *J. Magn. Magn. Mater.* **200**, 57–68 (1999).
276. J. Unguris, R. J. Celotta, D. A. Tulchinsky, and D. T. Pierce, *J. Magn. Magn. Mater.* **198–199**, 396–401 (1999).
277. R. Allenspach and W. Weber, *IBM J. Res. Dev.* **42**, 7–23 (1998).
278. D. T. Pierce, J. Unguris, R. J. Celotta, and M. D. Stiles, *J. Magn. Magn. Mater.* **200**, 290–321 (1999).
279. A. S. Arrott, B. Heinrich, and S. T. Purcell, *Kinetics of Ordering and Growth at Surfaces* in (M. G. Lagally, ed.), pp. 321–341. Plenum, New York, 1990.
280. J. A. Stroscio and D. T. Pierce, *J. Vacuum Sci. Technol. B* **12**, 1783 (1994).
281. D. T. Pierce, J. Unguris, and R. J. Celotta, *Ultrathin Magnetic Structures II* in (B. Heinrich and J. A. C. Bland, eds.), pp. 117–148. Springer-Verlag, Berlin, 1994.
282. D. T. Pierce, J. A. Stroscio, J. Unguris, and R. J. Celotta, *Phys. Rev. B* **49**, 14564–14572 (1994).

283. M. D. Stiles, *J. Magn. Magn. Mater.* **200**, 322–337 (1999).
284. W. P. Pratt, Jr., S.-F. Lee, J. M. Slaughter, R. Loloee, P. A. Schroeder, and Bass, *J. Phys. Rev. Lett.* **66**, 3060–3063 (1991).
285. J. Unguris, D. A. Tulchinsky, M. H. Kelley, J. A. Borchers, J. A. Dura, C. F. Majkrzak, Y. Hsu, R. Loloee, W. P. Pratt, Jr., and J. Bass, *J. Appl. Phys.* **87**, 6639–6643 (2000).
286. J. A. Borchers, J. A. Dura, J. Unguris, D. A. Tulchinsky, M. H. Kelley, C. F. Majkrzak, S. Y. Hsu, R. Loloee, W. P. Pratt, Jr., and J. Bass, *Phys. Rev. Lett.* **82**, 2796–2799 (1999).
287. D. A. Tulchinsky, M. H. Kelley, J. J. McClelland, R. Gupta, and R. J. Celotta, *J. Vacuum Sci. Technol. A* **16**, 1817–1819 (1998).
288. C. Stamm, F. Marty, A. Vaterlaus, V Welch, S. Egger, U. Maier, U. Ramsperger, H. Fuhrmann, and D. Pescia, *Science* **282**, 449–451 (1998).
289. H. Dekkers and H. de Lang, *Philips Tech. Rev.* **37**, 1–9 (1977).
290. K. Tsuno, *Rev. Solid State Sci.* **2**, 623–658 (1988).
291. I. R. McFadyen and J. N. Chapman, *EMSA Bull.* **22**, 64–75 (1992).
292. Y. Takahashi and Y. Yajima, *Jpn. J. Appl. Phys.* **32**, 3308–3311 (1993).
293. J. N. Chapman, I. R. McFadyen, and S. McVitie, *IEEE Trans. Magn.* **26**, 1506–1511 (1990).
294. Y. Takahashia and Y. Yajima, *J. Appl. Phys.* **76**, 7677–7681 (1994).
295. Y. Yajima, Y. Takahashi, M. Takeshita, T. Kobayashi, M. Ichikawa, Y. Hosoe, Y. Shiroishi, and Y. Sugita, *J. Appl. Phys.* **73**, 5811–5815 (1993).
296. S. Middelhoek, *J. Appl. Phys.* **34**, 1054–1059 (1963).
297. H. Shinada, H. Suzuki, S. Sasaki, H. Todokoro, H. Takano, and K. Shiiki, *IEEE Trans. Magn.* **28**, 3117–3121 (1992).
298. Y. Yajima and Y. Takahashi, *Microsc. Microanal.* **5** (Suppl. 2), 38–39 (1999).
299. Y. Yajima, Y. Takahashi, and K. Kuroda, *Jpn. J. Appl. Phys.* **35**, 2851–2854 (1996).
300. T. Leuthner, H. Lichte, and K.-H. Herrmann, *Phy6s. Status Solidi* **A116**, 113–121 (1989).
301. Y. Takahashi, Y. Yajima, M. Ichikawa, and K. Kuroda, *IEEE Trans. Magn.* **31**, 3367–3369 (1995).
302. Y. Takahashi, Y. Yajima, M. Ichikawa, and K. Kuroda, *Jpn. J. Appl. Phys.* **33**, L1352–L1354 (1994).
303. J. F. Herbst, *Rev. Mod. Phys.* **63**, 819–898 (1991).
304. E. Burzo, *Rep. Pro. Phys.* **61**, 1099–1266 (1998).
305. C. D. Fuerst and E. G. Brewer, *J. Appl. Phys.* **73**, 5751–5756 (1993).
306. C. D. Fuerst, E. G. Brewer, R. K. Mishra, Y. Zhu, and D. O. Welch, *J. Appl. Phys.* **75**, 4208–4213 (1994).
307. S. Hirosawa and M. Sagawa, *J. Appl. Phys.* **64**, 5553–5555 (1988).
308. L. Jahn, S. Hirosawa, V. Christoph, and K. Elk, *Jpn. J. Appl. Phys. Pt. 1* **30**, 489–492 (1991).

309. G. C. Hadjipanayis and A. Kim, *IEEE Trans. Magn.* **23**, 2533–2340 (1987).
310. B. M. Ma, E. B. Boltich, S. G. Sankar, and W. E. Wallace, *Phys. Rev. B* **40**, 7332–7335 (1989).
311. J. J. Croat, *J. Less-Common Met.* **148**, 7–15 (1989).
312. R. K. Mishra, E. G. Brewer, and R. W. Lee, *J. Appl. Phys.* **63**, 3528–3530 (1988).
313. R. K. Mishra, *J. Appl. Phys.* **62**, 967–971 (1987).
314. M. Sagawa and S. Hirosawa, in *High Performance Permanent Magnet Materials* (S. G. Sankar, J. F. Herbst, and N. C. Koon, eds.), *Mater. Res. Soc. Symp. Proc.* **96**, 161–166 (1987).
315. R. Ramesh, G. Thomas, and B. M. Ma, *Acta Metall.* **37**, 1421–1431 (1989).
316. A. Yan, X. Song, C. Chen, and X. Wang, *J. Magn. Magn. Mater.* **185**, 369–376 (1998).
317. M Sagawa, S. Fujimura, N. Togawa, H. Yamamoto, and Y. Matsuura, *J. Appl. Phys.* **55**, 2083–2087 (1984).
318. J. Hu, Y. Liu, M. Yin, Y. Wang, Y. Hu, and Z. Wang, *J. Alloys Compounds* **288**, 226–228 (1999).
319. K. S. V. L. Narasimhan, *J. Appl. Phys.* **57**, 4081–4085 (1985).
320. M. Sagawa, S. Hirosawa, H. Yamamoto, S. Fujimura, and Y. Matsuura, *Jpn. J. Appl. Phys.* **26**, 785–800 (1987).
321. W. Lee, *Appl. Phys. Lett.* **46**, '790–791 (1985).
322. R. W. Lee, E. G. Brewer, and N. A. Schaffel, *IEEE Trans. Magn.* **21**, 1958–1963 (1985).
323. Magnequench catalogue. Published by Magnequench Inc., 1994.
324. L. H. Lewis, J. Gao, D. C. Jiles, and D. O. Welch, *J. Appl. Phys.* **79**, 6470–6472 (1996).
325. L. H. Lewis, Y. Zhu, and D. O. Welch, *J. Appl. Phys.* **76**, 6235–6237 (1994).
326. D. D. Mishin, *Magnetic Materials* [in Russian]. Technical report, High School, Moscow, 1991.
327. R. K. Mishra, *Mater. Sci. Eng. B* **7**, 297–306 (1991).
328. V. V. Volkov and Y. Zhu, *J. Appl. Phys.* **85**, 3254–3263 (1999).
329. Y. Zhu, J. Tafto, L. H. Lewis, and D. O. Welch, *Philos. Mag. Lett.* **71**, 297–305 (1995).
330. Y. Zhu, H. Zhang, M. Suenaga, and D. O. Welch, *Philos. Mag. A* **68**, 1079–1089, 1993.
331. F. E. Pinkerton and C. D. Fuerst, *J. Magn. Magn. Mater.* **89**, 139–142 (1990).
332. F. E. Pinkerton and C. D. Fuerst, *J. Appl. Phys.* **69**, 5817–5819 (1991).
333. J. Tafto, R. H. Jones, and S. M. Heald, *J. Appl. Phys.* **60**, 4316–4318 (1986).
334. L. A. Bursill, J. C. Barry, and P. R. W. Hudson, *Philos. Mag. A* **37**, 789–812 (1978).

335. F. M. Ross and W. M. Stobbs, *Philos. Mag. A* **63**, 1–36 (1991).
336. C. A. Fowler, Jr., and E. M. Fryer, *Phys. Rev.* **86**, 426–430 (1952).
337. S. Methfessel, S. Middelhoek, and H. Thomas, *IBM J. Res. Dev.* **4**, 96–100 (1960).
338. R. Gemperle, V. Kambersky, J. Simsova, L. Murtinova, L. Pust, P. Gornert, and W. Schuppel, *J. Magn. Magn. Mater.* **118**, 295–301 (1993).
339. R. Gemperle, L. Murtinova, and V. Kambersky, *Phys. Stat. Sol. A* **158**, 229–246 (1996).
340. W. Szmaja, *J. Magn. Magn. Mater.* **153**, 215–223 (1996).
341. R. F. Egerton, *Electron Energy-Loss Spectroscopy*. Plenum, New York, 1986.
342. R. H. Wade, *Proc. Phys. Soc.* **79**, 1237–1240 (1962).
343. T. Schrefl and J. Fidler, *J. Appl. Phys.* **79**, 6458–6463 (1996).
344. M. K. Griffiths, J. E. L. Bishop, J. W. Tucker, and H. A. Davies, *J. Magn. Magn. Mater.* **183**, 49–67 (1998).
345. Y. Zhu and M. R. McCartney, *J. Appl. Phys.* **84**, 3267–3272 (1998).
346. S. Hirosawa, Y. Matsuura, H. Yamamoto, S. Fujimura, M. Sagawa, and H. Yamauchi, *J. Appl. Phys.* **59**, 873–879 (1986).
347. R. Grossinger, R. Krewenka, X. K. Sun, R. Eibler, H. R. Kirchmayr, and K. H. J. Buschow, *J. Less-Common Met.* **124**, 165–170 (1986).
348. V. V. Volkov and Y. Zhu, *Proc. Microsc. Microanal.* **4** (Suppl. 2), 404–405 (1998).
349. L. H. Lewis, J.-Y. Wang, and P. Canfield, *J. Appl. Phys.* **83**, 6843–6845 (1998).
350. P. Politi, *Comments Cond. Mater. Phys.* **18**, 191–221 (1998).
351. Z. X. Cai and Y. Zhu, *Microstructures and Defects in High-Temperature Superconductors*. World Scientific Publ., New Jersey, 1998.
352. V. V. Volkov, D. C. Crew, Y. Zhu, and L. H. Lewis, *Proc. Microsc. Microanal.* **5**(Suppl. 2) 46–47 (1999).
353. V. V. Volkov and Y. Zhu, unpublished results (2000).
354. V. V. Volkov and Y. Zhu, *J. Magn. Magn. Mater.* **214**(3), 204–216 (2000).
355. Y. R. Wang, S. Guruswamy, and V. Panchanathan, *J. Appl. Phys.* **81**, 4450–4452 (1997).
356. H. Kronmuller, *NATO ASI Ser. B, Phys.* **259**, 657–675 (1991).

Index

A

Aberrations, negation in TEM, 161–164
Acquisition time, for SEMPA image, 176–178
Aharonov–Bohm effect, 40
Alignment
  antiferromagnetic, 190
  grain, and magnetic domains, 238–240
Amplitude variation, nonmagnetic, 212–213
Anisotropies
  balanced, at spin transition, 185–186
  effect on magnetization distribution, 254–255
  uniaxial, crystalline grains, 17
Aperture
  function, 47
  positioning in Foucault mode, 37
Astigmatism, Foucault images, 54
Atomic force microscope, as force gradient mapper, 73–74

B

Back focal plane, objective lens, 38
Bacterium, magnetotactic, 126–127
Bicrystal structure, thin film recording medium, 20–21, 25
Biprism, electrostatic, 113–115, 159
Bohr magnetons, 169

C

Chemical analyses, thin intergranular phase, 232–234
Classification, island magnetization patterns, 99–101
Cliff–Lorimer ratio technique, 233
Closure domains
  formed at zero field, 99–100
  patterns, 103–104
Cobalt
  Co/Cu multilayers, magnetization depth profiling, 189–191
  nanostructured islands, 106–109
  small magnetized squares
    electron holographic analysis, 158–160
    for phase imaging, 153–154
Coercivity field, Nd–Fe–B samples, 266–268
Color wheel
  mapping phase gradients into, 123
  representing in-plane magnetization, 129
Component-resolved imaging, and probe hysteretic effects, 93–96
Conducting strip, nonmagnetic, 87
Constant height mode, mapping of surface topography, 73–74
Contrast
  magnetic, spin polarized secondary electron, 168–170
  reversal, 95–96
  from stray fields, 249–251
Crystalline grains
  behavior as single-domain particle, 6
  local magnetic domain configurations, 236
  uniaxial anisotropy, 17
Crystals
  orientation, effect on magnetic imaging, 252–253
  thickness, domain width as function of, 240–242
Curl-free component, magnetization vector field, 11–12, 86

D

Damping envelope, 47
Data acquisition, effect of systematic errors, 149–152

Deep-gap field, MFM, 91–93
Defect layers, $RE_2Fe_{14}B$ magnets, 230–231
Deflection, *see also* Lorentz deflection angle
  Lorentz, 35, 59
    detection, 204–206
    small, 196–201, 217–218
    spatially varying, 206–208
  in MFM, 71–72
Defocus
  in Lorentz microscopy, 33–35, 53–54
  precise calibration of, 155–165
  small to vanishingly small, 63–64
Demagnetization, sample, 255
Depth profiling, of magnetization of Co/Cu multilayers, 189–191
Detector, performance in Lorentz STEM, 208–212
Die-upset, $RE_2Fe_{14}B$ magnets produced by, 228
  correlation with sintered magnets, 268–269
  grain boundaries, 231–236
  microstructure, 229–231
Differential phase-contrast
  comparison with MFM, 12–14
  magnetic induction mapping, 58–60
  mode of electron holography, 133–134
Diffraction mode, in Lorentz microscopy, 38–39
Divergence-free component, magnetization vector field, 11–12
Domains
  dipole, 261–262
  magnetic, and grain alignment, 238–240
  nucleation, near grain boundaries, 262–263
  reorientation under thermal cycles, 256–258
  striped, 121–122
  width, as function of crystal thickness, 240–242
Domain structure
  dependence on geometry, 104
  effect of film thickness, 192–193
  micromagnetic simulations, 251–252
  patterned magnetic thin film elements, 15–18
  quantifying local magnetization, 236–238
  simulated, 8–14
Domain wall
  Bloch, Nèel, and cross-tie, 101–102
  convergent and divergent, 33–34
  cross-tie, 122
  energy, 248–249
  magnetic, 56–58
  missing magnetization at, 175–176
  motion, 258
    small NiFe elements, 98–106
  slope changes, 42
  width
    estimation, 34–35
    measurement, 243–249
Double boundary pinning center, 260–261
DPC, *see* Differential phase-contrast

E

Easy axes
  crystalline, 109
  dispersion, 107–108
  magnetic crystallites, 20–22, 24–25
  magnetization, 103–104
  in SEMPA imaging, 174
Electron beam
  Lorentz deflection, 197
  probe, 199–200
Electron beam deposition
  in SEMPA, 170
  tips, 85
Electron holography, 55
  applications to
    DPC mode, 133–134
    fields in vacuum, 123–124
    hard magnets, 118–123
    layered thin films in cross section, 131–132
    layered thin films in plan view, 132–133
    nanostructured elements, 127–131
    small magnetic particles, 124–127
  development, 112–113
  measurement of phase excursions, 160–161
  off-axis technique, 113–118, 135–136, 158–160
Electron microscope
  image magnification, 150–151
  and specimen chamber: for SEMPA, 170–172
Electron phase microscopy
  phase retrieval in, 144–152
  TIE technique for imaging, 165

Electron spins
  contribution to magnetization, 168
  cooperative behavior, 1
Electrostatic force microscopy, 88–89
Electrostatic potential
  2-dimensional, 113
  at domain wall, 57–58
Energy density, in ferromagnetic system, 2–3
Energy flow, connection with phase, 142
Erasure process, thin film medium, 96–98
Exchange coupling, Fe/Cr/Fe, 186–189
Exchange energy, in continuous magnetization distribution, 3–4
Extended tip model, dipolar and monopolar interaction, 80–83
External field
  island interaction with, 108
  MFM imaging in presence of, 91–96

F

Field emission gun, 112–113
Field-free region
  Lorentz microscopy, 29
  TEM, 31
Field interaction model, 79–80
Field tensor, magnetostatic, 5
Flow lines, time-average, 141–142
Flux change, enclosed, 223–225
Flux lines
  delineation by Lorentz STEM, 218–220
  within ferrite particles, 125
  visualized with electron holography, 128
Flux quantum, 42
Force–distance curve, in MFM, 70–73
Force gradient
  detection, 74–76
  measurement, 73–74
  tip–surface, 76
Foucault mode
  image of magnetic domain structure, 239
  Lorentz microscopy, 36–38
  quantum aspects, 44–45
  zero-loss images, 60
Free space, magnetic field in: Lorentz STEM, 216–218
Fresnel imaging
  Lorentz image of domain wall width, 243–245
  low-angle dark-field, 234–236
Fresnel mode
  comparison with holographic methods, 247–248
  Lorentz microscopy, 32–36
  out-of-focus electron microscope in, 163–164
  quantum aspects, 44–45
Fringes
  cosinusoidal, 115–116
  holographic, 133–134
  interference
    at convergent wall images, 39
    Foucault mode, 37
    Fresnel mode, 35
  visibility maximization, 154
Fringing fields, between nanostructured elements, 129

G

GB domain, nucleated, 263
Giant magnetoresistance, 186, 189, 191
Grain boundary
  in die-upset magnets, 231–236
  domain nucleation near, 262–263
  effect on magnetization, 122
Grains
  alignment, and magnetic domains, 238–240
  crystalline
    behavior as single-domain particle, 6
    local magnetic domain configurations, 236
    uniaxial anisotropy, 17
  die-upset $RE_2Fe_{14}B$ magnets, 229–231
  interior, dipole domains growing in, 261–262
  magnetic moments, 265
Gyromagnetic motion, inclusion in micromagnetic modeling, 26

H

Hard materials, magnetic induction mapping, 215–216

High resolution imaging, domain walls, 101–102
Holography
  comparison with Fresnel method, 247–248
  in-line, twin-image problem, 164
  measuring domain wall width, 245–247
Hydrodynamic formulation, quantum mechanics, 139

I

Image formation theory, for TEM, 45–48
Images
  acquisition time, in SEMPA, 176–178
  Fresnel, underfocused, 120
  MFM, in presence of external field, 91–96
  normalization, 152
  perturbations on, 78
  reconstruction, 85–86
  recovered, qualitative evaluation, 158
  rotation, and phase distribution, 151–152
  shifts and magnification, in electron microscope, 150–151
  underfocus and overfocus, 63–64
Image simulation process
  domain configuration in Terfenol-D, 55–58
  uniformly magnetized sphere, 48–55
In-plane magnetic induction
  in Lorentz microscopy, 29–31, 43, 66
  magnitude, 132
Instrumental asymmetries, in SEMPA, 176
Instrumentation
  Lorentz STEM, 213
  SEMPA imaging system, 170–179
Interference microscopy, combination with Lorentz STEM, 220–225
Intergranular exchange coupling, elimination, 19–20
Intergranular phase
  grain boundaries in die-upset magnets, 231
  nanoscale chemical analyses, 232–234
Iron
  coatings, for polarization enhancement, 179–180
  enrichment, at grain boundaries, 234
  Fe/Cr/Fe exchange coupling, 186–189
  nanostructured islands, 106–109
  very fine wires, domain structure, 192
Island magnetization patterns, 99–101

L

Lens configurations
  Lorentz microscopy, 31–32
  probe-forming, 203
  TEM, 113
L-J potential, in MFM, 71, 73
Lorentz deflection angle, 29–31, 42–44, 210, 244
Lorentz microscopy
  classical approach, 28–39
    diffraction mode, 38–39
    experimental methods, 31–32
    Foucault mode, 36–38
    Fresnel mode, 32–36
  magnetic induction mapping methods, 58–61
  phase retrieval
    application to negation of aberrations, 161–164
    comparison with electron holography, 158–161
    evaluation of recovered images, 158
    experimental results, 154–158
    sample selection and preparation, 153–154
    theory, 138–144
  quantum mechanical formulation, 39–58
    Fresnel and Foucault modes, 44–45
    image formation, 45–48
    image simulations, 48–58
    strong phase objects, 40–44
  STEM features essential for, 201–213
  TIE equation, 61–66
Lorentz STEM
  combination with interference microscopy, 220–225
  instrument, 213
  magnetic induction mapping, 214–218
  nature of small Lorentz deflections, 196–201
  phase contour delineation, 218–220
  STEM features essential for Lorentz microscopy, 201–213

M

Magnetic anisotropy energy, in micromagnetic modeling, 3
Magnetic characteristics, small NiFe elements, 98–106
Magnetic charge equation, 90
Magnetic component, Foucault contrast, 54
Magnetic contrast, spin polarized secondary electron, 168–170
Magnetic elements, nanostructured, 127–131
Magnetic field
  applied to nanostructured elements, 128–129
  direction, effect on sample, 92–93
  in free space: Lorentz STEM, 216–218
  SEMPA, 178–179
Magnetic force microscopy
  comparison with SEMPA, 182–184
  erasure process of thin film medium, 96–98
  force gradient detection, 74–76
  images
    comparison with micromagnetic simulations, 8–12
    reconstruction, 85–86
  imaging in presence of external field, 91–96
  models for probe magnetization, 79–85
  nanostructured Co and Fe islands, 106–109
  probes for, 69
  quantification techniques, 86–91
  rare earth magnet tip, 124
  small NiFe elements, 98–106
  surface forces and force–distance curve, 70–73
  theory, 76–78
  topography and force gradient measurement, 73–74
Magnetic imaging, stray field effects, 249–255
Magnetic materials, electron holographic applications, 118–134
Magnetic pole density, simulated, 9–10
Magnetic structures, patterned, application of SEMPA, 191–193
Magnetic vector potential, 48
Magnetization
  Co/Cu multilayers, depth profiling, 189–191
  distribution, 97
  electron spin contribution to, 168
  in-plane and out-of-plane components, 172, 181–182
  island, patterns, 99–101
  missing, at domain walls, 175–176
  probe
    changing, 94–95
    models for, 79–85
  single-domain particles, 105–106
Magnetization processes
  experiments in domain wall motion, 102–106
  modeling, 5–8
  in thin film recording media, 18–25
Magnetization reversal
  in bicrystal films, 21–25
  domains, 259–260
  and domain wall motion, 102–106
  DPC images during, 12–14
  in electron holography, 126–127
  initial stages, 97
  loop, remanent states formed from, 130
Magnetization vector, 3-dimensional imaging, 180–182
Magnetization vector field, 9, 11
Magnetization vortex
  circular pattern, 34
  fluctuations, 19–20
  in middle of domain wall, 16
  during reversal process, 13–14
Magnetostatic energy
  for mesh cell, 4–5
  minimum, 102
Magnetostrictive strains, at room and very-low temperature, 27–28
Magnets
  hard, electron holographic applications, 118–123
  rare earth, 27–28
  $RE_2Fe_{14}B$, *see* $RE_2Fe_{14}B$ magnets (RE = Nd, Pr)
  sintered and die-upset, correlation, 268–269
Magnification
  electron holography, 114–115
  image, in electron microscope, 150–151
Mapping
  magnetic induction
    by Lorentz STEM, 214–218
    methods, 58–61
    within sample, 120–121
  phase gradients, into color wheel, 123

Melt spinning, rapid-solidification technique, 228
MFM, *see* Magnetic force microscopy
Micromagnetic modeling
  domain configurations, 15–18
  inclusion of gyromagnetic motion, 26
  magnetization processes in thin film recording media, 18–25
  in material microstructure engineering, 1
  simulated domain structures, 8–14
  theory and computation, 2–8
Micromagnetic simulations, domain structures, 251–252
Microstructure, $RE_2Fe_{14}B$ magnets
  produced by die-upset, 229–231
  sensitivity to, 227–228, 264–266

N

Nanostructured elements, electron holographic applications, 127–131
Nanostructured islands, cobalt and iron, 106–109
Negative pulse, at bright edges, 95
Noise, effects on recovered phase, 146–148
Nucleation
  domain, near grain boundaries, 262–263
  reverse domain, 23–24
  transverse domain, 25

P

Particles
  single-domain, magnetization, 105–106
  small magnetic, electron holographic applications, 124–127
  SW, 104, 108
Permalloy film
  domain configurations in, 15–17
  patterned elements, MFM images, 8–12
  small elements, wall motion, 98–106
Perpendicular medium, 107
Phase
  contour, delineation by Lorentz STEM, 218–220
  distribution, and image rotation, 151–152
  generalized definition of, 140–143
  recovered
    and defocus distance, 156–157
    stability and effects of noise, 146–148
Phase contrast, propagation-induced, 144, 146
Phase map
  dimpled appearance, 157
  qualitatively correct, 65–66
Phase ramp, generated across image, 150
Phase retrieval
  in electron phase microscopy, 144–152
  holographic techniques, 160–161
  theory, 138–144
Phase shift
  A–B, 40–42, 50
  of electron wave function, 116
  magnetic component, 43
  in vacuum, 125
Phase unwrapping, 117
Photoresist, applied to cobalt wafer, 154
Pinning centers, application of $Nd_2Fe_{14}B$ magnets, 258–261
Point charge, tip appearing as, 81
Point–dipole model, 79–80, 82, 90, 93–94
Polarization
  background Fe, 188–189
  enhancement, iron coatings for, 179–180
  spin, *see* Spin polarization
Powder metallurgy, production of $RE_2Fe_{14}B$ magnets, 228
Probability density
  flow of, 137
  focused on propagation, 140
  nonzero time-averaged, 142
Probe field, extent of, 84–85
Probe moment, determination with MFM, 89–91
Probes
  formation, for Lorentz STEM, 202–204
  hysteretic effects, 93–96
  magnetic moment of, 87
  magnetization, models for, 79–85
  for MFM, 69
  vibration along *z*-axis, 77–78
Probe–sample interaction, in MFM, 70–71, 75

Q

Quantification techniques, MFM, 86–91
Quantum mechanical formulation, Lorentz microscopy, 39–58

R

Reconstruction
  hologram, 117
  image, 85–86
Recording media
  SEMPA applied to, 182–184
  thin film, magnetization processes in, 18–25
$RE_2Fe_{14}B$ magnets (RE = Nd, Pr)
  coercivity field, 266–268
  die-upset
    correlation with sintered magnets, 268–269
    grain boundaries, 231–236
    microstructure, 229–231
  domain structure, 236–249
  in situ experiments, 255–263
  remanence, 264–266
  sensitive to microstructure, 227–228
  stray field effect on magnetic imaging, 249–255
Relative-thickness image, 121
Remanence, $RE_2Fe_{14}B$ permanent magnets, 264–266
Remanent states
  formation, 130
  ripple structure formed at, 17
  saturation, 22–25
Resolution
  for magnetic materials, 118
  relationship to image acquisition time, 177–178
  and sensitivity limits, probe magnetization, 83–84
Retraction, tip: retract curve, 72–73
Ripple structure
  formed at remanent states, 17
  in Fresnel image, 215–216
Rotation, image, and phase distribution, 151–152
Roughening, dark areas, 97–98

S

Samples
  demagnetization, 255
  for phase imaging, selection and preparation, 153–154
  with rough surfaces: SEMPA, 182
Saturation remanent state, 22–25
Scanning electron microscopy with polarization analysis, *see* SEMPA
Scanning transmission electron microscopy
  features essential for Lorentz microscopy, 201–213
  magnetic induction mapping, 58–60
SEMPA
  applied to patterned magnetic structures, 191–193
  depth profiling of magnetic structures, 189–191
  3-dimensional magnetization imaging, 180–182
  Fe/Cr/Fe exchange coupling, 186–189
  imaging rough surfaces, 182
  instrumentation, 170–179
  iron coatings for polarization enhancement, 179–180
  magnetic contrast in, 168–170
  recording media applications, 182–184
  spin reorientation transitions, 184–186
Sensitivity
  $RE_2Fe_{14}B$ magnets to microstructure, 227–228, 264–266
  spin, 173–174
Sensitivity limits, and resolution, probe magnetization, 83–84
Sidebands, complex conjugate, 116–117
Silicon nitride membrane, thickness, 159–160
Sintering, $RE_2Fe_{14}B$ magnets produced by, 228
  correlation with die-upset magnets, 268–269
Soft materials, magnetic induction mapping, 214–215

Spatial discretization mesh
cell magnetization, 2
hexagonal and square, 6–8
Spatial frequency
Lorentz deflection, signal dependence on, 210–212
low, recovered phase of noise dominated by, 148
Spatial frequency limit, high and low, 82–83
Specimen chamber, and electron microscope: for SEMPA, 170–172
Spin polarization
detectors, 172–174
secondary electron magnetic contrast, 168–170
Spin reorientation transitions, SEMPA applied to, 184–186
Stability, of phase recovery: and noise contamination, 146–148
STEM, *see* Scanning transmission electron microscopy
Stray fields, effects on magnetic imaging, 249–255
Strong phase objects, in Lorentz microscopy, 40–44
Surface forces, in MFM, 70–73
Switching field, Co and Fe islands, 108–109

T

TEM, *see* Transmission electron microscope
Temperature
MFM imaging in presence of external field, 91–92
room and very-low, magnetostrictive strains, 27–28
sample, for electron holography, 119
Theoretical considerations
MFM, 76–78
phase retrieval, 138–144
Thermal cycles, domain reorientation under, 256–258
Thin film
atom positions, 52–53
elements
magnetization vector, 12–14
patterned magnetic, 15–18
layered
in cross section, 131–132
in plan view, 132–133
magnetization configurations in, 55–58
medium, erasure process of, 96–98
polycrystalline soft magnetic, 6, 8
recording media, magnetization processes in, 18–25
SEMPA and MFM images of test patterns, 183
TIE, *see* Transport-of-intensity equation
Tip–sample spacing, in MFM, 70–71, 80–81, 84–85
Topography, and force gradient measurement, in MFM, 73–74
Transfer function
Lorentz, 51, 56
microscope, 46–47
phase, 47
specimen, 205, 212
Transient domain configurations, 9–10
Transmission electron microscope
field-free region, 31
image formation, 45–48
magnetic induction mapping, 60
negation of aberrations, 161–164
phase recovery, 153–160
stray field effects on magnetic imaging, 249–255
Transport-of-intensity equation, 61–66, 138–140, 143–144, 165
Transverse domain, expansion, 23–24
TV camera, signal, feeding to liquid crystal panel, 136

V

Vacuum, fields in, 123–124
Vector potential, and small Lorentz deflections, 196–197

W

Wave fields
in definition of phase, 140–141
discontinuous phase of, 143

Wave function
  aberrated, 163
  electron, 196
    phase shift of, 116
  exit, 50–51
  paraxial, 62–63
  in TEM work, 46
Whiskers, iron, 187–188
Width
  domain, as function of crystal thickness, 240–242
  domain wall
    estimation, 34–35
    measurement, 243–249

Z

Zeeman energy, 5